Biopharmaceutical Drug Design and Development

Biopharmaceutical Drug Design and Development

SECOND EDITION

Edited by

Susanna Wu-Pong, PhD

Director, Pharmaceutical Sciences Graduate Programs
Department of Pharmaceutics, School of Pharmacy
Virginia Commonwealth University, Richmond, VA

and

Yon Rojanasakul, PhD

Professor of Pharmaceutical and Pharmacological Sciences
Department of Basic Pharmaceutical Sciences
West Virginia University, Morgantown, WV

 Humana Press

Editors
Susanna Wu-Pong, PhD
Box 980581
Medical College of Virginia Campus
Virginia Commonwealth University
Richmond VA 23298
email: swupong@vcu.edu

Yon Rojanasakul, PhD
Department of Basic Pharmaceutical Sciences
West Virginia University
P.O. Box 9530
Morgantown, WV 26506-9530
yrojan@hsc.wvu.edu

ISBN: 978-1-58829-716-7   e-ISBN: 978-1-59745-532-9

Library of Congress Control Numberr: 2008924320

Cover illustration: High-content protein microarrays used to discover cancer biomarkers (*see* Fig. 1 in Chapter 5 and discussion on p. 75).

Printed on acid-free paper

9 8 7 6 5 4 3 2 1

springer.com

Preface

The first edition of *Biopharmaceutical Drug Development*, published in 1999, was intended to provide a comprehensive overview of biotechnology in pharmaceutical drug development in one concise volume. At the time, the field of biotechnology was expanding faster than many professionals even within the field could comprehend.

Since the publication of the first edition, the field of biotechnology has continued to grow in the manner of a continually inflating ball, with the leading edges of the ball widening every year. These edges represent seemingly endless new frontiers in all areas of biotechnology, including health care, agriculture, and the food and energy industries. Since 1999, dozens of new biopharmaceutical drugs have become available for therapeutic use. The main categories of these drugs include monoclonal antibodies for in vivo use, cytokines, growth factors, enzymes, immunomodulators, thrombolytics, and immunotherapies, including vaccines. As professionals in the health care industry, we must monitor, understand, and unravel the implications of these technological developments in drug development and patient care.

Though gene therapy and recombinant DNA technology have been introduced decades ago, society continues to grapple with the therapeutic opportunities and ethical ramifications of these and even newer technological developments. Many fundamental problems in certain disciplines have yet to be resolved. Indeed, despite years of development and clinical trials, a gene therapy product has yet to be approved by the US Food and Drug Administration (FDA), perhaps in part because of the death of a patient enrolled in a gene therapy trial in 1999. However, according to the *Journal of Gene Medicine*, 29 Phase III clinical trials are ongoing as of Spring 2007 compared to only 4 in 2001, with 1283 active protocols in Phases I, II, and III compared to 596 in 2001. In contrast, many new biopharmaceutical drugs have been approved for therapeutic use in the last five years, providing new treatment opportunities for cancer and many inherited diseases. One of the fastest growing biopharmaceuticals is the monoclonal antibody, which is now used to treat diseases such as multiple sclerosis and cancer (www.fda.gov).

An even more contentious issue than gene cloning is now stem cell research, a topic that was even debated at the national level in the 2004 presidential election. At issue is the use of therapeutic stem cells for the treatment of a variety of diseases and illnesses where healthy cell replacement would be therapeutically beneficial, Alzheimer's disease being the classic example. However, stem cells are usually derived from human embryos. The controversial nature of the source of stem cells has created heated and emotional debate in the United States over the appropriate use of embryos vs the potential therapeutic benefit of the technology.

Since 1999, the debate about human cloning has also heated up. Early in 2004, the Raelian religious sect claimed to have successfully clon a human, and that a woman gave birth to a 7 pound baby clone called "Eve." Though the results to this day remain unsubstantiated, the announcement reopened the debate about the wisdom and morality of cloning technology when the methods may be used inappropriately to create cloned babies.

The wisdom of such an endeavor is truly questionable, especially as scientists have discovered the premature aging of the original cloned mammal, "Dolly." Dolly's rapid senescence is believed to be the result of the shortening of DNA molecules that happens each time DNA replication occurs. Cloned humans would therefore be unlikely to have the life span of a naturally conceived baby given the limitations of today's technology.

Though privately funded efforts to create human clones appear limited, several groups are cloning human cells for medical research. Therapeutic cloning, though illegal in the United States, has been legal in the United Kingdom since 2001. Dr Ian Wilmut of the Roslin Institute (creator of "Dolly") has recently announced that his group will be cloning embryos to study disease development and expressing no intent to create cloned humans.

Another revolutionary new invention since 1999 has been microarray or "chip" technology. Previous sequence identification methods required probing a sample one sequence at a time. Microarray technology now allows simultaneous testing of tens of thousands of probes at once using microscopic arrays of DNA or other probes (including proteins, cells, antibodies). This incredible technology has already changed the way we think about the role of genes in heredity or disease, from the monogenic (one gene results in one disease) to a polygenic view of disease etiology. Microarray technology is profoundly affecting the manner in which scientists now view the genetic basis, and thus the treatment and detection of disease.

Also since 1999, the first draft of the sequence of the human genome was announced, having been completed ahead of schedule. The speed of completion and implications of this accomplishment are stunning. Though much more work lies ahead for fully understanding the results and implications of the Human Genome Project, access to the entire human genome sequence will accelerate our understanding of human genetics, its effect on disease, and open multitude of doors for the creation of drugs tailored to a patient's genotype (pharmacogenetics).

The advancements discussed here are only an overview of a few projects that are currently emerging from the biopharmaceutical sciences. The second edition of *Biopharmaceutical Drug Development* is a natural sequel to a discussion on a dynamic, exciting field of biotechnology.

Contents

Contributors

NEELAM AZAD • *Department of Pharmaceutical Sciences, School of Pharmacy, West Virginia University, Morgantown, WV*

ROBERT G. BELL, PhD • *Drug and Biotechnology Development, LLC, Clearwater, FL*

KAI I. CHEANG, PhD • *Department of Pharmacy, Virginia Commonwealth University, Richmond, VA*

WAGNER DOS SANTOS, PhD • *Department of Neurosurgery, Harold F. Young Neurosurgical Center, Virginia Commonwealth University Health System, Virginia Commonwealth University, Richmond, VA*

HELEN L. FILLMORE, PhD • *Department of Neurosurgery, Harold F. Young Neurosurgical Center, Virginia Commonwealth University Health System, Virginia Commonwealth University, Richmond, VA*

JINGJIAO GUAN, PhD • *Center for Affordable Nanoengineering of Polymer Biomedical Devices (CANPBD), The Ohio State University, Columbus, OH*

ROBERT L. HAINING, PhD • *Department of Basic Pharmaceutical Sciences, West Virginia University, Morgantown, WV*

ZHI HONG, PhD • *Infectious Diseases Center of Excellence for Drug Discovery, GlaxoSmithKline*

GLEN E. KELLOGG, PhD • *Department of Medical Chemistry, School of Pharmacy, Virginia Commonwealth University, Richmond, VA*

SUNG-KWON KIM, PhD • *Immunology Drug Discovery, Valeant Pharmaceuticals, Inc., Costa Mesa, CA*

ANAND KRISHNAN V. IYER • *Department of Pathology, University of Pittsburgh School of Medicine, Pittsburgh, PA*

L. JAMES LEE, PhD • *Center for Affordable Nanoengineering of Polymer Biomedical Devices (CANPBD); Department of Chemical and Biomolecular Engineering, The Ohio State University, Columbus, OH*

ROBERT J. LEE, PhD • *Division of Pharmaceutics, College of Pharmacy, The Ohio State University, Columbus, OH*

RICHARD E. LOWENTHAL, MSC, MSEL • *Pacific Link Consulting, San Diego, CA*

MIRIANA MORAN, PhD • *Immunology Drug Discovery, Valeant Pharmaceuticals, Inc., Costa Mesa, CA*

PHILIP D. MOSIER, PhD • *Department of Medical Chemistry, School of Pharmacy, Virginia Commonwealth University, Richmond, VA*

XIAOGANG PAN, MS • *Center for Affordable Nanoengineering of Polymer Biomedical Devices (CANPBD); Division of Pharmaceutics, College of Pharmacy, The Ohio State University, Columbus, OH*

YON ROJANASAKUL, PhD • *Department of Basic Pharmaceutical Sciences, West Virginia University School of Pharmacy, Morgantown, WV*

MICHAEL L. SAMUELS, PhD • *Senior Scientist, RainDance Technologies, Inc. Guilford, CT*

RITA SHIANG, PhD • *Associate Professor of Human Genetics, School of Medicine, Virginia Commonwealth University, Richmond, VA*

ROBERT C. TAM, PhD • *Immunology Drug Discovery, Valeant Pharmaceuticals, Inc., Costa Mesa, CA*

ANDREI VARNAVSKI, PhD • *Drug Discovery, Valeant Pharmaceuticals, Inc., Costa Mesa, CA*

SUSANNA WU-PONG • *Director, Pharmaceutical Sciences Graduate Program, Virginia Commonwealth University School of Pharmacy, Richmond, VA*

ALENA YIN ZHANG, PharmaD, PhD • *Postdoctoral Fellow, Abramson Research Center, Children's Hospital of Philadelphia, Philadelphia, PA*

1

Biotechnology: 2008 and Beyond

Susanna Wu-Pong

Abstract

The emergence of molecular medicine and genomics is transforming health care in ways unimaginable 50 years earlier. Traditional medical practice is often empiric and reactive; the future of medicine will involve a detailed understanding of the role each individual's genetics plays on their susceptibility to disease and their response to medications. Furthermore, new types of drugs that allow much more specific therapy for diseases and disorders with genetic components are in development. This chapter examines how the genomics era and biotechnology are changing biomedical science and health care.

Key Words: Biotechnology; microarray; human genome project; drug development; pharmacogenetics.

1. Introduction

Despite decades of technological advances and billions of dollars of investment, applying the term "practice" to modern medical care is still as literally appropriate now as it was hundreds of years ago because of the unfortunate amount of guesswork that is still inherent in diagnosis and treatment (Fig. 1A). This uncertainty contributes to the rising cost of health care. Add the aging of the baby-boom generation to this mixture and the result is fewer Americans who can afford insurance or pay their medical bills. Furthermore, the rising cost of medical care also limits the time a care provider can devote to education of the patient on their care or wellness. As a consequence, many consumers have had to become medically savvy, accessing Internet and print resources to monitor and understand their own, increasingly complicated, medical treatments. It is not just medical care that suffers from rising costs and insufficient time; health maintenance is largely still up to the patient. Therefore, modern patients are responsible for their own education and wellness plan; their primary sources

From: *Biopharmaceutical Drug Design and Development*
Edited by: S. Wu-Pong and Y. Rojanasakul © Humana Press Inc., Totowa, NJ

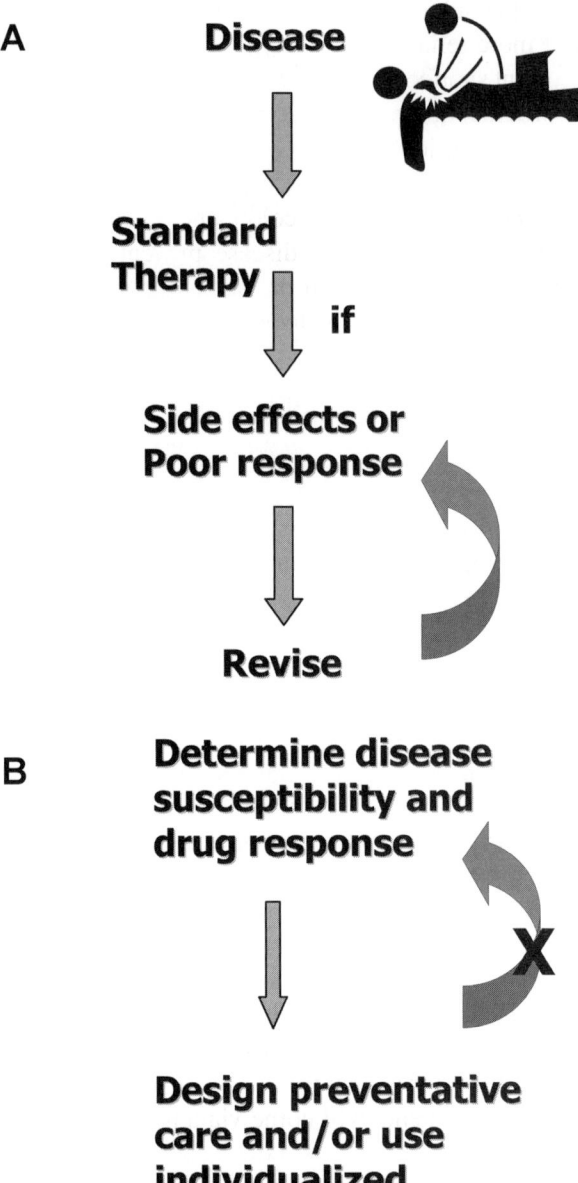

Fig. 1.

of information on the subjects are often obtained from the media, friends, or family instead of more reliable professional healthcare resources.

Despite the current shortcomings in preventative care, the medical community currently does provide limited routine medical screenings and prophylactic treatments as part of America's wellness plan. For most Americans, routine and widespread screening for disease prevention occurs aggressively in childhood and thereafter as annual breast, pelvic, colon, and prostate cancer screenings for adults. Pharmaceutical treatment for disease prevention is mostly limited to childhood vaccines unless a physical or medical exam shows early stages of disease progression, such as in hypertension, hyperlipidemia, or diabetes. But for the most part, medical intervention is primarily in reaction to diseases once symptoms have already begun to occur.

Once disease has been identified, the patient then usually receives the standard of care for their particular diagnosis. For example, according to the National Heart, Lung, and Blood Institute *(1)*, the standard of care for uncomplicated Stage I hypertension involves use of a thiazide diuretic with or without the addition of a second drug such as a beta blocker. The patient is then monitored for response to treatment or adverse reactions. Unsatisfactory therapeutic response results in a change in the dose or drug until a satisfactory clinical outcome is obtained.

Hypertension is usually a straightforward diagnosis. However, in some situations, the diagnosis may not be given with much confidence. In this situation, a working diagnosis is applied until care providers are otherwise proven wrong, usually via either failure of treatment or onset of new symptoms.

This trial-and-error approach often works well for many patients and many disease states. For others, the experience is frustrating, costly, or even fraught with medical misadventures that can sometimes be deadly. The Adverse Drug Effect Prevention Study Group estimates that 3500 adverse drug reactions or potential adverse drug reactions are likely to occur per hospital per year, and that approximately 1% of all adverse drug reactions are deadly *(2,3)*. Therefore, from a quality-of-life and financial perspective, the American medical system still has much room for improvement.

Some of these improvements will come via a new medical model involving molecular medicine, where molecular biology is applied to the field of health care to improve disease diagnosis, disease prevention, and creation and selection of treatments for individual patients. In this model, genetic analysis will be available to all patients such that disease predisposition can be identified and treated early using medical or lifestyle interventions before the onset of symptoms. For example, in the case of the patient predisposed to hypertension, low-salt and low-fat diets can be initiated well before organ or vascular damage occurs. Blood pressure monitoring can begin as a part of the patient's lifestyle in their youth, and medical intervention, if necessary, can also be initiated early and aggressively as the disease develops (Fig. 1B).

In addition to identification of disease predisposition, medication management will also be scientifically based on genetic information. Instead of drug treatment trial-and-error, which can result in thousands of unnecessary adverse drug reactions and millions of dollars in unnecessary expense, genetic analysis will predict which drugs will or will not be effective or safe for certain patients. The correct dose and the correct drug can be determined in advance, and the trial-and-error cycle will never be initiated. The genetic analysis itself, although costly initially, will pay for itself multifold over the patient's lifetime. Integration of this type of genetic analysis into the drug development process will also allow subject prescreening such that early clinical trials testing occurs only in patients who are likely to receive pharmacological benefit with minimal adverse reactions. This approach will therefore save the drug companies millions of dollars in drug development costs.

The shift from trial-and-error health care to a more proactive, preventative model will require a commitment from both government and private agencies to invest the necessary resources into biotechnology research and technology development. In addition, both the public and healthcare professionals will also be required to become facile with an alphabet soup of new terminology. This second volume of *Biopharmaceutical Drug Design and Development* was written to provide some insight into this dynamic and important field of biotechnology.

2. The Present: The Status of Biotechnology Today

During my recent review of the first volume of *Biopharmaceutical Drug Design and Development*, I found it somewhat difficult to recall the state of the biopharmaceutical industry back in 1998–1999, when the book was in press, when so much has transpired and changed us during the intervening period. At that time, the industry epitomized the word "potential", with the possible options seeming to expand exponentially with every year and every new discovery.

2.1. The Drugs

In the last decade, the focus in the industry was on the development of new recombinant protein drugs and novel classes of DNA-based drugs, especially for the treatment of many diseases that have previously had a limited array of pharmaceutical treatment options. Unlike empirical treatments, these newer technologies allowed one to design a biopharmaceutical based on a natural product and the genetics of a disease, thereby reducing most of the side effects that accompany traditional small molecule drugs. For example, recombinant drugs such as insulin and erythropoietin have exponentially improved replacement therapy with much safer and specific products compared to, say, bovine insulin. As a result, biotechnology-derived therapeutics has been one of the fasting growing classes of new drugs. A total of 226 biologics have been approved during the years 1982–2004 *(4)*. Most of these drugs have been recombinant proteins or antibodies. A small number of DNA-based drugs have also been approved such as oligonucleotide drugs Vitravene and Macugen, though to date the much-hyped gene therapy has not yet achieved the much-coveted FDA approval.

2.2. The Finances

Surprisingly, despite the hundreds of new products on the market since the early 1980s and hundreds more in advanced clinical trials, biotechnology as an industry has yet to make a profit. In 2004, over $6 billion was lost overall in the worldwide industry according to Ernst & Young. However, Ernst also predicts that the industry will enjoy a net profit in 2008 *(5)*. This is not to say, of course, that individual companies have not seen profits in their own portfolios. Overall, the US biotechnology industry earned almost $30 billion in revenues over 1466 companies in 2002, approximately double the revenues of $14.6 billion earned in 1996. The majority of these revenues have come from recombinant DNA (rDNA) protein drugs *(4)*.

The initially spectacular growth of the biotechnology industry in the 1990s appears to have reached a tentative equilibrium since the market decline post-2000, with a market capitalization stabilizing over $200–300 billion, total financing approximately $15–16 billion, patents per year approximately 7000–8000, and the number of US companies stabilizing at 1300–1400 including both private and public entities *(4)*. Successful biotech companies often collaborate with large pharmaceutical companies to bring a product to market, with approximately half of the new drugs approved in 2004 resulting from such collaborations *(6)*.

Given these successes and disappointments, where is the biopharmaceutical industry now? Biotechnology will once again revolutionize the healthcare industry, but now in a way completely different from that envisioned a decade ago. Instead of the emphasis on new drugs to treat old, previously empirically treated diseases, genetic analysis will be used to create a new model for healthcare, where genetic predispositions to disease, drug metabolism, drug-induced adverse reactions, and drug interactions will be tested prior to the first signs of disease and the first drug dose given. New drugs will be designed specifically for groups of people with similar genetic profiles. In short, disease prevention and treatment will be individualized for patients as part of wellness and personalized disease and drug management.

3. The Future: The Potential of Biotechnology

3.1. Pharmacogenetics

The future of medicine lies with the use of genetic information to optimize health care for individuals or groups of patients, i.e., pharmacogenomics. The term *pharmacogenomics* and a related subject, pharmacogenetics, has as many definitions as the number of individuals involved in the discussion. Therefore, despite the risk of sounding like a freshman English essay, Webster's dictionary was consulted (Dictionary.com). Pharmacogenomics was defined as a "biotechnological science that combines the techniques of medicine, pharmacology, and genomics and is concerned with developing drug therapies to compensate for genetic differences in patients which cause varied responses to a single therapeutic regimen." Pharmacogenetics was defined as "study of genetic factors that influence an organism's reaction to a drug." To illustrate the confusion in the terminology in this field, Stedman's online dictionary defined the two words synonymously: "the study of genetically determined variations in responses to drugs in humans or in laboratory organisms."

Regardless of the nomenclature, the concept of pharmacogenomics is applicable to the future of medicine and how genetics will be used to transform medical care in the 21st century. The challenges to implement this vision are daunting and will inevitably involve multidisciplinary cooperation. Scientists in both the public and private sectors are identifying linkages between genetics and patient care. For example, the Human Genome Project (HGP; *see* Section 3.2.) is an overwhelmingly successful collaboration between government and private business. The HGP is an international effort led by the National Human Genome Research Institute (NHGRI) and includes 20 different universities and research centers across the USA, Europe, and Asia. A parallel effort by a private company, Celera Genomics, has been equally successful in human genome sequencing. NHGRI envisions the future of molecular medicine in a document entitled "Vision for the Future of Genomics Research," which is available on their website (http://www.genome.gov). The vision outlines an ambitious role of genomics in health care and society. Six "grand challenges" to health care for translating genome-based knowledge into health benefits includes:

(1) Develop robust strategies for identifying the genetic contributions to disease and drug response;
(2) Develop strategies to identify gene variants that contribute to good health and resistance to disease;
(3) Develop genome-based approaches to prediction of disease susceptibility and drug response, early detection of illness, and molecular taxonomy of disease states;

(4) Use new understanding of genes and pathways to develop powerful new therapeutic approaches to disease;

(5) Investigate how genetic risk information is conveyed in clinical settings, how that information influences health strategies and behaviors, and how these affect health outcomes and costs; and

(6) Develop genome-based tools that improve the health of all.

Almost $500 million has been proposed for the fiscal year 2006 to continue the NHGRI mission of using genetic technologies to study disease (http://www.genome.gov).

Collaborations such as the HGP by NHGRI and others have already begun to pay off in the application of genetic information to individualized drug therapy to patients. However, pharmacogenomic concepts have been in practice well before the HGP was initiated. An early example of pharmacogenetics is epitomized by the classic acetylation polymorphism, where different acetylation phenotypes (fast vs slow drug metabolism via acetylation) affect a large fraction of the population (e.g., 50–60% Caucasians are slow acetylators). Despite widespread knowledge of the importance of this polymorphism (normal genetic variation), routine screening for the polymorphism is not available for patients initiating therapy on the affected drugs (e.g., hydralazine, isoniazid, sulfonamides, and procainamide).

Another classic example of polymorphic variations in drug metabolism is of course the cytochrome P450 enzymes or CYP450. CYP450 enzymes are found predominantly in the liver, but are also expressed to a lesser degree throughout the body. CYP450 enzymes are largely responsible for drug metabolism, though not all CYP450 enzymes are involved in drug metabolism. Because CYP450s are also expressed in the gut, polymorphisms in enzyme expression or activity may also have a profound effect on bioavailability of orally administered drugs. Because of the importance and prevalence of genetic polymorphisms on drug metabolism, a huge commercial market is potentially available for products that can quickly and inexpensively genotype the relevant CYP450 polymorphisms that can affect drug metabolism and response (*see* Section 3.4).

The CYP450 enzymes have extensive arrays of polymorphisms that are now being sequenced and identified demographically. Relevant examples include CYP1A2, CYP3A4, and CYP2D6. Because these drug metabolizing enzymes have polymorphic variability, a patient's rate of drug metabolism can depend on their genetic profile of the respective enzyme. In addition, coadministration of a drug with a competing substrate, inhibitor, or inducer of that enzyme can have profound effects on the plasma concentrations, and thus pharmacologic or toxicologic outcomes, of that drug. The genetic determination of the

CYP450 enzyme activity in individual patients is an important contemporary example of the role of genetics in drug therapy. Summaries of these known CYP450 drug interactions are available from some excellent Internet resources, such as http://medicine.iupui.edu/flockhart/.

In addition to metabolic polymorphisms, reports of new pharmacologic polymorphisms are also becoming more frequently reported. For example, scientists have reported that subgroups of patients with non-small-cell lung cancer that have an epidermal growth factor receptor mutation may be more responsive to treatment with Iressa (gefitinib), a small molecule drug *(7)*. Similarly, subsets of patients with specific polymorphisms in O-6-methylguanine-DNA methyltransferase have also been shown to respond to temozolomide therapy in glioblastoma *(8)*. Other examples of efficacy pharmacogenetics have been nicely described in available review papers *(9,10)*.

Clearly, enhancing our understanding of individual genetic differences in drug metabolism (CYP enzymes) and drug effects (such as Iressa) provides new opportunities for improving drug safety and efficacy. New drugs can be designed specifically for genetic subgroups (as in the Webster definition of pharmacogenomics) and dosed based on the genetic profile of their metabolic enzymes (pharmacogenetics). The result will be safer, more effective use of drugs.

3.2. Human Genome Project

Using genetic information to improve health care requires a better understanding of the effect of genetic differences or polymorphisms on disease and wellness. The HGP is an ambitious effort to sequence the entire human genome including all of its variations or polymorphisms. In this new era of molecular medicine, the information gleaned from the project will allow an unprecedented opportunity to understand the molecular mechanisms of biological processes in the human body, and therefore, disease etiology and treatment. Begun in 1990, Phase I of the project was intended to obtain a first draft of the human genome and was completed ahead of schedule in April 2003.

Phase II of the project is certain to be the most difficult phase, focusing on mapping variations in the 0.1% of the genome that comprises the genetic differences that make each of us unique. This uniqueness is in part characterized by our outward appearance: hair color, skin color, height. More importantly, however, these differences can also account for why some people are predisposed to certain diseases or conditions. These differences also determine pharmacologic, adverse, or lethal response to drugs. For example, Makita et al. *(11)* report that subsets of patients with congenital long-QT syndrome have mutations that predispose them to drug-induced arrhythmias. Similarly, some patients with certain serotonin receptor subtypes are susceptible to antipsychotic-induced dyskinesias and weight gain *(12)*. The demand for these genetic identifiers has spawned a new venture to assist scientists to identify the genetic polymorphisms that distinguish groups of patients who are predisposed to disease or who will respond appropriately to drug therapy (e.g., Perlegen Sciences; http://www.perlegen.com).

Completion of the HGP will really just provide the template for the interpretation of the 10^9 nucleotides that make up the human genome. Just as when Watson and Crick solved the structure of DNA, deciphering the genetic code was only the beginning of our understanding of the way DNA makes proteins and how those proteins provide the basic function and structure of the cell. Similarly, the HGP will accelerate the determination of which polymorphisms are clinically relevant in terms of disease predisposition and development of new drug treatments. Relevant polymorphisms can be further studied to determine the role of that genetic variant on disease development, progression, or drug effects. Toward this end, the International HapMap Project (http://www.hapmap.org) is an international organization devoted to describing common patterns of genetic variation in patients for the purposes of improving the treatment of disease. International collaborations such as the International HapMap Project will be necessary to efficiently translate the vast array of genetic information into useful medical diagnostics and therapeutics.

As stated earlier, Phase I was completed ahead of schedule because of the innovation of Celera and NHGRI sequencing methods. These efforts have focused on exons, the parts of the gene that code for mRNA and protein transcripts. However, sequencing the 0.1% of the genome containing polymorphic regions in Phase II may be even more difficult than originally believed, because the intronic (noncoding) regions that are normally considered to be noncoding and "filler" are demonstrating unexpectedly high conservation (e.g., the sequences are preserved). Because genetic sequences that are found repeatedly in nature are believed to have biological importance, highly conserved noncoding regions imply that useful genetic information may be present and may therefore require further examination. As a result, NHGRI will be studying, among other things, DNA sequences that are highly conserved in noncoding regions in an effort to determine their function in genome.

3.3. Bioinformatics and Database Management

Because of the ambitious nature of the HGP, sequencing the human genome would appear to be an end point in and of itself. However, from a drug development perspective, the work will have only just begin once the sequence is obtained. First, consider that the human genome is comprised of a trillion nucleotides. Each person has a unique sequence. How will the sequence of those trillion nucleotides in individual or groups of patients be used to improve human health?

One approach would be to use new sequence information to produce novel insights about a gene of ongoing interest. New genetic information might illuminate the role of a specific gene in disease development or the role of a gene in drug disposition. However, one could also explore this new set of data without necessarily starting with a known gene. The HGP will yield the sequences of genes we know little or nothing about. Computer programs are being used to search for new genes in DNA databases by searching for elements that are common to all genes, such as promoters, enhancers, and polyadenylation signals. Newly discovered elements that mark the presence of a gene are being discovered and used to improve the rate of discovery of new genes. Once a new human gene candidate has been identified, homology (matching sequences) to other organisms can be examined to try to understand the function of this new gene. In addition, the structure and thus the function of the gene's encoded protein can be predicted (though with not a great deal of accuracy yet). Gene-browsing tools and actual gene sequences for a variety of species are available free on the Internet (e.g., http://www.ensembl.org). Such tools will continue to be used to uncover genetic treasures that are still buried in known DNA sequences.

3.4. DNA Chip

The HGP goal of deciphering the therapeutic implications of all the polymorphisms of the human genome is a daunting task. In the recent past, scientists only had the tools to detect or measure one gene sequence at a time and therefore allowed researchers to only scratch the surface of the understanding of this incredibly complex system. Today, newer tools enable an unprecedented rate of progress in the identification of the most important polymorphisms and their impact on human health. A new field called "theranostics" is now evolving which integrates genetically based diagnostics and therapeutics.

Perhaps the technology most central to this transformation is arguably the microarray. Just as Watson and Crick revolutionized genetics with the discovery of the genetic code, microarray technology is allowing genetic analysis of thousands of gene sequences simultaneously in a matter of minutes. Previously, DNA or RNA sequences were detected and analyzed one at a time via Southern or Northern blotting or, more recently, polymerase chain reaction. The old paradigm of one disease resulting from one genetic mutation was supported for the most part using the standard one gene analysis. The invention of the DNA microarray ("DNA chip") provided the analytical power necessary to detect the expression of thousands of genes simultaneously. Microarrays therefore allow one to compare expression levels of thousands of genes between normal and diseased tissues for one patient, or between individual patients. As a result, microarrays have enabled the identification of dozens or even hundreds of genes that may have differential expression in a single disease, compared to the

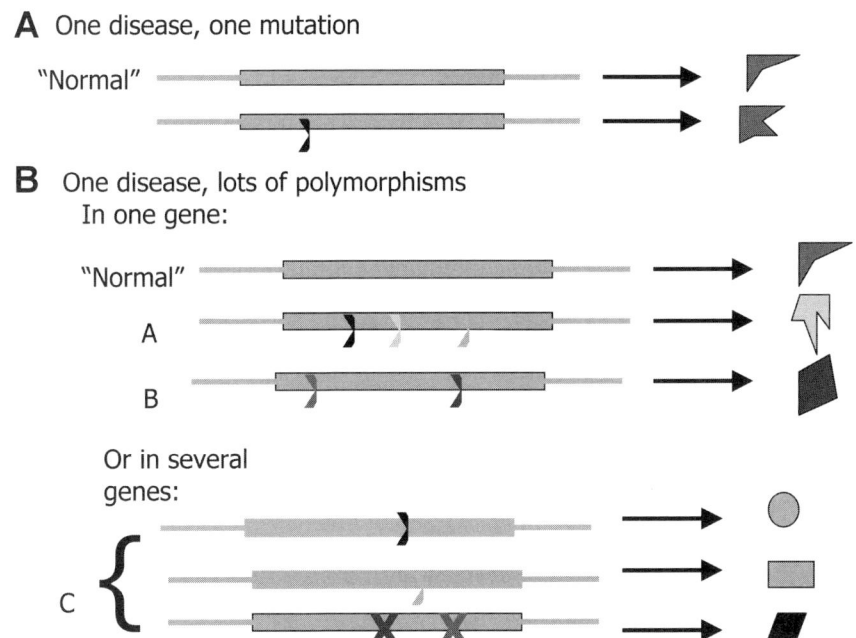

Fig. 2. Genetic basis of disease paradigms. (**A**) One genetic mutation - one disease; (**B**) multigenic model of disease.

one-gene methods and theories used previously. For example, Staudt et al. report that differences in treatment response for non-Hodgkin's lymphoma could be traced to about 1000 genes *(13)*. Scientists still do not know the clinical implications of differential expression for each of these genes, but such studies illustrate the complexity of disease that was previously unfathomable using traditional molecular assays in previous years.

3.5. Pharmacogenomic Drugs

The results of the early studies using microarray technology have been nothing short of astounding. The one gene → one protein → one disease paradigm, although true for a limited number of disorders, has been shown to be hopelessly simplistic for most common diseases (Fig. 2A). Rather, differential expression has been observed for dozens of genes in diseases such as cancer, hypertension, epilepsy, and Parkinson's disease (Fig. 2B). Each of these genes represents a new opportunity to learn about the genetic etiology of a disease, and possibly a new pharmacologic drug target.

Probably the earliest example of the successful application of pharmacogenetics and pharmacogenomics has been to the breast cancer subtype HER-2 (human epidermal growth factor receptor) overexpression. Patients who are HER-2 positive, and thereby overexpress HER-2, are candidates for treatment with Herceptin, a monoclonal antibody that targets cells expressing HER-2. Patients who are negative for HER-2 would derive no benefit from Herceptin treatment. Other breast cancer subtypes have been identified, such as BRCA (breast cancer susceptibility), and are starting to lead to improved prevention, diagnosis, or treatment therapies based on these genetic distinctions.

Herceptin is an example of a whole new generation of drugs that will be designed for patients with a specific genetic profile within a certain disease. Knowing *a priori* which patients will benefit from which drugs will have the long-term consequence of more efficacious, more cost-effective drug therapy.

About the same time Herceptin was approved by the FDA for the treatment of breast cancer with HER-2 overexpression, the first oligonucleotide drug Vitravene (fomivirsen) was also approved for the treatment of cytomegalovirus infections of the retina. Oligonucleotide drugs such as Vitravene are designed to bind to and downregulate the expression of genes containing unique sequences. A second oligonucleotide drug Macugen (pegaptanib), targeting vascular endothelial growth factor, has been approved in early 2005 for the treatment of neovascular macular degeneration. Because of the gene sequence-specific nature of these therapies, oligonucleotides continue to have increasing potential to participate in pharmacogenomic, gene-specific drug therapy.

Gene therapy is another form of gene-specific drug therapy but, unlike oligonucleotides, has still not received FDA approval. As of June 2007, 1283 studies are in clinical trials, 29 of which are in Phase III. The industry has made some significant accomplishments and setbacks in the last decade, the most notable (public) failure being the unfortunate death of a patient in February 2000 believed to be the result of an inflammatory response to the gene therapy itself (University of Pennsylvania). The trial was halted by the FDA once the problem was discovered. In addition, two patients from the early adenosine deaminase trials developed leukemia as a result of transgene (the administered gene) insertion into the patient's DNA. Though this problem has always remained a theoretical possibility, this was the first demonstration of gene transformation in the clinic.

3.6. Cell-Based Therapies

Biotechnology-derived therapies are not limited solely to protein or nucleic acid drugs. Modified cells are a source of great therapeutic potential and controversy. The best example of modern cell-based therapies is stem cells. Stem cells have such enormous potential because these cells can be induced to differentiate into any cell type, and also divide in culture for long periods. Embryonic stem cells can divide and grow in culture for up to a year; adult stem cells do not. Thus, the controversy surrounding stem cell development is related to objections surrounding the use of fetuses to obtain embryonic stem cells.

Stem cells are important for both scientific reasons and their therapeutic potential. Because the cells can be induced to differentiate, they serve as a useful model to study cell differentiation and dedifferentiation, a process very important to cancer cell research. Stem cells can also provide a wide range of cell types available for laboratory research.

Cell replacement is the obvious therapeutic application of selectively differentiated stem cells, especially for neuronal or brain tissue that are slow to repair or regenerate, or tissues that are otherwise defective. For example, stem cells induced to differentiate into normal pancreatic tissue might one day be used to treat diabetes.

The value of the application of stem cell technology is beyond debate. However, stem cells have become the lightning rod for political and social controversy that gene therapy and cloning were a decade ago. Like cloning and gene-based therapies, technological advancements in the use of stem cells outpace our ability to comprehend or solve the ethical dilemmas associated with the creation and use of these products.

4. Conclusion

In the last decade, the changes in the landscape of biotechnology have been remarkable. These changes are the result of rapid advances in technology and knowledge in the industry, including HGP, microarray technology, and a larger than ever arsenal of biologic drugs. The result is a brand new paradigm for medical care based on molecular medicine: pharmacogenetics and pharmacogenomics. If the pace of change continues at this rate for the next 20–50 years, it would indeed be difficult to even imagine the potential that will reside within our healthcare model in the future.

Despite the pace of change, some things remain the same since a decade ago, such as poor scientific literacy, uncertainties about allocation of resources, product safety, and ethics of gene or cell manipulation. These unanswered questions will undoubtedly continue to provoke much needed discussion and debate for the foreseeable future.

References

1. "Seventh Report of the Joint National Committee on Prevention, Detection, Evaluation, and Treatment of High Blood Pressure (JNC 7) Express" National Heart, Lung, and Blood Institute (December 2003), http://www.nhlbi.nih.gov/guidelines/hypertension/express.pdf.
2. Bates, D. W., Cullen, D. J., Laird, N., et al. (1995) Incidence of adverse drug events and potential adverse drug events: implications for prevention. *JAMA* **274**(1), 29–34.

3. Leape, L. L., Bates, D. W., Cullen, D. J., et al. (1995) Systems analysis of adverse drug events. *JAMA* **274**(1), 35–43.
4. Bio Editors' and Reporters' Guide, Biotechnology Industry Association, 2004–2005, www.bio.org.
5. "Ernst and Young Global Biotechnology Reports Track Dramatic Industry Rebound"; Ernst and Young Press Release, May 12, 2004.
6. Agres, T. (2004) US biotech may leap into the black. *Drug Discov. Dev.* www.dddmag.com.
7. Bell, D. W., Lynch, T. J., Haserlat, S. M., et al. (2005) Epidermal growth factor receptor mutations and gene amplification in non-small-cell lung cancer: molecular analysis of the IDEAL/INTACT gefitinib trials. *J. Clin. Oncol.* **23**(31), 8081–8092.
8. Sordella, R., Bell, D. W., Haber, D. A., and Settleman, J. (2004) Gefitinib-Sensitizing EGFR mutations in lung cancer activate anti-apoptotic pathways. *Science* **305**, 1163–1167.
9. Schmith, V. D., Campbell, D. A., and Sehgal, S. (2003) Pharmacogenetics and disease genetics of complex diseases. *Cell. Mol. Life Sci.* **60**, 1636–1646.
10. Meisel, C., Gerloff, T., and Kirchheiner, J. (2003) Implications of pharmacogenetics for individualizing drug treatment and for study design. *J. Mol. Med.* **81**, 154–167.
11. Makita, N., Horie, N., and Nakamura, T. (2002) Drug-induced long-QT syndrome associated with a subclinical SCN5A mutation. *Circulation* **106**, 1269–1274.
12. Reynolds, G. P., Templeman, L. A., and Zhang, Z. J. (2005) The role of 5-HT2C receptor polymorphisms in the pharmacogenetics of antipsychotic drug treatment. *Prog. Neuropsychopharmacol. Biol. Psychiatry* **29**(6), 1021–1028.
13. "Cracking the Cancer Code" (2002) U.S. News and World Report, June 24, 49–58.

2

The Human Genome Project and Drug Development

Susanna Wu-Pong

Abstract

The Human Genome Project (HGP) was a 13-yr international effort whose primary goal was to sequence the 3 billion nucleotides of the human genome. Other goals included sequencing other genomes, developing new related technology, making the technology widely accessible, and examining the ethical, legal, and social implications of the project. The implications of the HGP on the current methods used in biomedical research and its impact on future healthcare are vast and far-reaching. This chapter reviews these topics and also provides a glimpse into the post-HGP era.

Key Words: Human genome project; genomics; annotation; ethics; sequencing.

1. Introduction

The 21st century is already being called the "Biology Century" because of the implications of the Human Genome Project (HGP) and the field of genomics. The Biology Century sprung from the five decades separating the discovery of Watson and Crick's structure of the DNA molecule (published in 1953) and the completion of HGP in 2003, the effort to sequence the 3 billion nucleotides comprising the human genome.

How did the HGP contribute to the emergence of the Biology Century, and what is the implication of this effort? This 13-yr effort to sequence the human genome was initially envisioned in 1985 by Robert Sinsheimer but at that time considered "crazy" and "premature", but soon endorsed by the National Research Council in 1986, then coordinated by the US Department of Energy and National Institutes of Health starting in 1988. This multinational effort that included the UK, EU, China, and Japan began in the early to mid-1990s, and by 2003, approximately 99% of the human genome's gene-containing regions was sequenced to 99.99% accuracy. As a result, new sequencing technology, new methods to identify, annotate, and analyze genetic information, and insights

From: *Biopharmaceutical Drug Design and Development*
Edited by: S. Wu-Pong and Y. Rojanasakul © Humana Press Inc., Totowa, NJ

into gene variation and protein function have emerged, as well as a seemingly endless stream of data that defines the key to life and illness. In addition, other nonhuman organism genomes were also sequenced, such as *Escherichia coli*, mouse, roundworm, and fruit fly. Readers are encouraged to visit http://www.ornl.gov/sci/techresources/Human_Genome/home.shtml for more information on these related topics.

The emergence of new information and technology is only the tip of the iceberg. The HGP will have a continuing impact on medical science because of its commitment to transfer the technology to the private sector to aid in the development of new medical applications. Deciphering the mystery of the genetics of human health and illness will also have social, legal, and ethical implications, a topic also addressed by the HGP.

2. The Race: Public Vs Private

The Human Genome Project's remarkable success and early completion (2 years ahead of schedule) was the result of rapidly advancing technologies and a race between public and private efforts. In June 2000, the rough draft (one-third) of the human genome was completed a year ahead of schedule. In 1993, the first 5-yr plan was revised to account for the unexpected progress. The Final Plan (1998) was the third effort and included the following goals:

- Identify all the approximately 25,000–30,000 genes in human DNA;
- Determine the sequences of the 3 billion chemical basepairs that make up human DNA;
- Store this information in databases;
- Improve tools for data analysis;
- Transfer related technologies to the private sector; and
- Address the ethical, legal, and social issues (ELSI) linked to the project.

One rarely encounters projects that finish in advance of projected timelines and under budget especially for projects of this size, which makes the achievements of the HGP even more astounding. The working draft of 90% of the genome was published in 2001 *(1,2)*, and the complete, high-quality genomic sequence was published in 2003 *(3,4)* (Tables 1 and 2). The consortium agreed that sequences emerging from the HGP must also be highly accurate and largely continuous (1 error per 10,000 bases). The other goals were also accomplished in this time frame and continue to be developed and refined, including identifying common genetic variants, creating a single nucleotide polymorphism (SNP) map of at least 100,000 markers, developing tools and methodologies, and training and developing scientists in these areas.

The surprising and remarkable pace of progress could be at least partially attributed to the competition and conflict that initially threatened to undermine

Table 1
Human Genome Project Goals and Completion Dates

Area	HGP goal	Standard achieved	Date achieved
Genetic map	2- to 5-cM resolution map (600–1500 markers)	1-cM resolution map (3,000 markers)	September 1994
Functional analysis	Develop genomic-scale technologies	High-throughput oligonucleotide synthesis	1994
		DNA microarrays	1996
Physical map	30,000 sequence tagged sites (STS)	52,000 STS	October 1998
Capacity and cost	Sequence 500 Mb per year at <$0.25 per base	Sequence >1,400 Mb per year at <$0.09 per base	November 2002
Human sequence variation	100,000 SNPs	3.7 million SNPs	February 2003
Human gene identification	Full-length cDNAs	15,000 full-length cDNAs	March 2003
Human DNA sequence	95% of gene-containing genome finished to 99.99% accuracy	99% of gene-containing genome finished to 99.99% accuracy	April 2003
Model organisms	Complete genome sequences of *E. coli*, yeast, Drosophila, and others	Finished genome sequences of *E. coli*, yeast, Drosophila, and others	April 2003

Source: Ref. *(5)* and http://www.ornl.gov/sci/techresources/Human_Genome/home.shtml.

the effort. A major dispute arose between the private company Celera Genomics, who embarked on independent genome sequencing, and the public effort of the HGP consortium. Among the issues was the use of different strategies (improved Sanger dideoxy and capillary sequencing of mapped bacterial artificial chromosome (BAC) clones vs shotgun sequencing) to sequence the genome. Moreover, the question of sequence ownership and access when private companies were involved became a major issue. The new sequencing technologies combined with the race to be the first to complete milestones spurred the race forward despite the conflict. The animosity eventually ended in 2000 with a truce between Celera and HGP resulting in joint announcement of the working draft and simultaneous publication of the final sequence.

The initial friction between Celera and HGP by no means reflected the overall nature of HGP–private sector collaborations. For example, capillary electrophoretic methods for sequencing, creation of EST and SNP public domains, and sequencing

Table 2
Timeline of Chromosome Sequencing

	Chromosome sequenced
December 1999	22
May 2000	21
March 2000	Drosophila
December 2001	20
December 2002	Mouse
January 2003	14
June 2003	Y
July 2003	7
October 2003	6
March 2004	19
	13
May 2004	10
September 2004	5
December 2004	16
March 2005	X
April 2005	4
	2

of other genomes were the result of successful relationships with the private sector. The DNA microarray which analyzes thousands of sequences concurrently has similarly emerged from the HGP, which then laid the foundation for further development by Affymetrix (http://www.affymetrix.com).

3. Sequencing the Genome

Prior to the HGP, sequencing was done one small single strand of DNA at a time using traditional methods like the Sanger dideoxy method combined with gel electrophoresis. This method, though accurate, was very slow and completion of the genome sequence using this method would likely have taken several decades. Several innovations in sequencing technology emerged from the HGP that enabled the remarkable pace of sequencing. Both public and private efforts used similar automation and sequencing technology. However, the groups differed in the approach to sequencing. In the "hierarchical shotgun," individual large DNA fragments of known position are shotgun sequenced with the use of BAC vectors that are fingerprinted to mark the chromosomal location of the DNA to be sequenced. In contrast, in "whole genome shotgun" the entire genome is digested into small fragments that are sequenced. In both cases, the sequence is reassembled or aligned based on sequence overlaps to produce a long, contiguous sequence. Sequence alignment is facilitated by use of landmarks contained in the physical map produced by the HGP. This complex process is complicated by the great abundance of repetitive sequences that could be present in multiple locations throughout the genome. Therefore, to reduce error, each section of the genome must be read 6–12 times to compensate for the error caused by repetitive sequences. The multiple reads on the sequence combined with reassembly of overlapping sequences provide enough redundancy and information to produce the draft of the genome.

Both the hierarchical and whole genome methods have advantages and disadvantages. The hierarchical method allows the chromosomal location of each individual sequence to be known with certainty, but mapping BAC clones prior to sequencing results in a relatively slower and more expensive method. The whole shotgun method does not require insertion and cloning of DNA using large insert BAC vectors, but instead sequences much smaller clone libraries, which produces results with higher error when sequences are reassembled.

4. Storing the Genome

Identifying the human gene sequence is only one objective of the HGP. The second objective is to store the information that emerges from the project. To illustrate the size of the 3 billion basepairs that comprise the human genome, one can consider the amount of time or memory that would be required to read or store the sequence. A sequence consisting of 3 billion basepairs requires 3 gigabytes of computer storage space, equivalent to 200 copies of the Manhattan phone book, and 95 years to read aloud, 1 base at a time. If one chooses to focus on functional units in the genome instead of basepairs, the number of units reduces to a "mere" 30,000 genes translated in 100,000 or more proteins. Fortunately sequences, genes, and proteins are no longer transmitted verbally and rarely on hard copy. All data along with further biological relevant context are now available online for query and analysis at the European Bioinformatics Institute (EBI; http://www.ebi.ac.uk) and National Center for Biotechnology Information (NCBI; http://www.ncbi.nih.gov) websites.

The primary function of databases such as NCBI is to store data, but additionally, databases are also used to process data and allow for data visualization. Databases should also use accepted scientific standards for the type of data in storage. For example, specific databases exist just for microarray or SAGE (serial analysis of gene expression) data. Microarrays, discussed in Chapters 4 and 5, allow for simultaneous probing of thousands of sequences, resulting in a large amount of information emerging from a single assay. The application of standards for database content allows for efficient and effective database use, qualities most important when dealing with large amounts of information as is the case with microarrays or shotgun sequencing. In the case of microarray databases, standards formalized by the Microarray Gene Expression Data Consortium allow data import from different sources and addition of annotation and complex queries using standard terminology. The ability to group data according to structure or function, for example, allows an investigator to obtain a lateral view of the function of a particular molecule and gain a fresh perspective on its biological function.

Storing by sequence is only one way to maintain large amounts of genetic information. Databases that group data by specific criteria are useful for objective-driven excursions into sequence data and hypothesis testing of genome data. Table 3 lists some databases found on the NCBI website and elsewhere that are pertinent to human health.

5. Annotating the Genome

Sequencing and storing genetic information is an academic exercise unless those data can be translated into useful information. Therefore, at the minimum the alphabet soup of the genome must therefore be identified of its genes, how those genes and gene products collaborate to create biological processes, the variability inherent in those genes (0.1% of the sequence is normal variability), and what those variations mean in terms of human health or disease. Clearly, these previously unimaginable and hugely ambitious goals are now within reach because of the foundation that the HGP provides. Though many of these worthy objectives are still somewhat in their infancy, the degree to which they are in practice now will be briefly reviewed here and in Chapters 3, 6, and 12 of this book.

The method used for identifying regions of DNA that is translated or expressed into protein is different depending on the species from which the DNA originates. For example, a prokaryotic gene can be defined as the longest open reading frame for a given region. Similarly, simple eukaryotes have small and few intronic regions, making gene identification relatively easy. However, multicellular eukaryotes generally have complex gene organization; as a result, identification is also complex. Furthermore, gene identification should include both exons and introns in order to enable reconstruction of the resulting mRNA. Gene identification in multicellular eukaryotes is complicated by the presence of large intronic regions, existence of splice variants and polymorphisms, and the presence of pseudogenes and sequencing errors. Splice site consensus sequences can be used to identify intron/exon boundaries, although atypical splice sites may be present and result in inefficient and inaccurate splice site determination.

The difficulties with gene recognition have resulted in the development of software that typically relies on one, or a combination, of two methods. First is gene prediction ab initio which relies on statistical parameters such as DNA sequence or gene structure analysis for gene identification (Table 4). In contrast,

Table 3
Databases Relevant to Human Health

Area	Database names and description	
Genes and health	HapMap: catalog of haplotypes (shared genetic variance)	OMIM: a guide to human genes and inherited disorders
	GeneCards: a database of human genes, their products and their involvement in diseases	dbSNP: a database of SNPs
Genome sequence	BLAST: sequence comparison to other sequences and their products	RefSeq: reference sequences of human chromosomes, genomic contiguous sequences, mRNAs, and proteins
	Clone Registry: a centralized registry of genomic clones	
cDNA	GEO: gene expression omnibus, a public repository for gene expression and hybridization data	UniGene: organizes transcribed sequences into gene-based clusters
		SAGEmap: SAGE (serial analysis of gene expression) tags mapped to mRNA sequences
Comparative genomics	HomoloGene: evolutionarily related genes on large-scale comparative sequence analysis	Homology Map: conserved gene arrangements between mouse and human
Cancer	PEDB: the prostate expression database	CancerGene: a catalog of cellular genes involved in different cancers

homology-based methods such as DNA:protein or genome:genome alignment use algorithms such as BLAST to identify homologous sequences in gene databases. Examples of homology-based programs available include INFO, AAT, or Procrustes. This method is especially conducive for prokaryote genomes, but is less useful when used alone for the more complex eukaryote systems.

In addition to gene identification, the gene locations and other landmarks on chromosomes were also mapped by the HGP (*see* Section 3). Landmarks on chromosome maps include genes, transcripts, NCBI contigs (the "Contig" map comprised of overlapping sequences), the BAC tiling path (the "Component" map),

Table 4
Software for ab initio Gene Recognition in Prokaryotes and Eukaryotes

	Human	Mouse	Rat	Drosophila	Yeast	Bacteria
EBI	X	X	X	X	X	X
Ensembl	X	X	X	X	X	
GDB	X					
GeneMark	X	X	X	X	X	X
Grail	X	X				X
GenScan	X					
Genie	X			X		
GeneFinder Fgenes, Fgenesh	X			X	X	
GeneID	X			X		
GeneFinder, MZEF	X	X				
HMMgene	X					
NCBI	X	X	X	X	X	X

Adapted from *Sequence-Evolution-Function. Computational Approaches in Comparative Genomics*, Koonin, E. V. and Galperin, M. Y., Kluwer Academic Publishers.

(STSs), FISH-mapped clones, ESTs, SNPs, and transcripts from several different organisms. The annotated genome information is available in many forms. A database for the many types of genomic components is available at the genome database (GDB) (http://www.gdb.org; Fig. 1). To date of publication, the chromosomal location of over 20,000 genes have been identified and stored in GDB, including pseudogenes (inactive genes) and putative genes (include EST transcript clusters and syntenic regions). For visualization of genomic components, the genome can be viewed at two sites, the Genome Browser (http://genome.ucsc.edu/) and the European site, Ensembl (http://www.ensembl.org/index.html). Both sites allow the user to insert and view their own data on the genome. Additional information for genome components can be found at EBI and NCBI. Information from these sites is intertwined and linked to each other.

Since the completion of HGP, further refinements have been made on the sequence. When first published, the sequence was interrupted by approximately 150,000 gaps. Recently, the consortium has reported a refined sequence, now determined to be comprised of 2,851,330,913 basepairs and 20,000–25,000 protein-coding genes. The sequence is 99% complete, interrupted by only 341 gaps, and is accurate to roughly 1 event per 100,000 bases *(6)*. The consortium defined a finished sequence as having an error rate of no more than 1 event per 10,000 bases for at least 95% of the euchromatic genome, with the only regions refractory to sequencing (using all available techniques) remaining.

Fig. 1. Chromosome map of the human genome.

6. Improve Tools for Data Analysis

As the human genome sequence increases in accuracy, the focus has shifted from data acquisition toward a more complete understanding of genetic and cellular function. The elucidation of complex genetic and protein interrelationships that are required for biological processes is increasingly within reach as the wide variety of state-of-the-art information management tools continue to evolve. However, for this to occur, the development of new tools for analyzing large amounts of data became necessary for gene identification (*see* Section 5), determining gene function or genetic variations in biological processes. These tools have been enhanced by the growth of the Internet which has allowed rapid global access to computational tools and databases. Bioinformatics is a new field that has emerged from this need to use and manage these databases. This topic, though inexorably linked to the results of the HGP, is discussed in Chapter 3.

7. Transfer of Technologies to the Private Sector/Implications

As mentioned earlier, an important feature of the HGP is the translation of sequence information into discoveries that could benefit society or healthcare. Therefore, the HGP was dedicated to the transfer of technology to the private sector. By licensing technologies and awarding grants for innovation, HGP catalyzed the biotechnology industry and fostered biomedical research for the new century.

It is difficult to find a sector of biomedical research that has not been profoundly influenced by HGP and like efforts. Technology such as the microarray has profoundly altered our conception of disease predisposition and progression;

thus, the way new treatments are developed will also reflect this changing paradigm. Pharmacogenomics is a new field (also described in detail in Chapters 6 and 12) devoted to the rational use of genetic information to design specific drugs. As thousands of SNPs and haplotypes (linked genetic variation) are catalogued and, even more importantly, selectively identified as clinically significant, the era of individualized medicine is becoming more of a reality. Currently, the NCBI lists over 10 million unique SNPs in the SNP human database, and the number is expected to grow approximately 90 SNPs per month (http://www.ncbi.nlm.nih.gov). These polymorphisms, or normal variations in DNA sequences, somehow contribute to our individual uniqueness, not only in outward appearance, but more importantly of whether we are predisposed to certain diseases or drug-related adverse reactions. Determining the clinical relevance of 10 million or so of these single nucleotide variations will indeed be a Herculean task.

One way to reduce the sheer number of SNPs that must be evaluated for their clinical relevance is to determine which SNPs tend to vary as a linked group. These haplotypes are being identified and catalogued by the HapMap Project (http://www.hapmap.org), estimated to be only 300,000–600,000 compared to the millions of SNPs identified.

Therefore, as new information such as the clinical relevance of a SNP or haplotype emerge, not only will we alter the way we think about the design of new drugs and how they should be effectively used, but we will also re-examine the use of pre-existing therapies. Normally medications are dosed based on population studies; each patient is assumed to behave like the "average" patient unless proven otherwise. In truth, we are as unique in drug disposition and metabolism, and therefore drug response, as we are in our appearance. In other words, the success or failure of a drug and its dosing regimen is conceivably predetermined by our genetic makeup. The question remains as to which genes and polymorphisms are relevant to clinical situation.

The consortium authors of the most recent paper elegantly state regarding the HGP: "It allows systematic searches for the causes of disease—for example, to find all key heritable factors predisposing to diabetes or somatic mutations underlying breast cancer—with confidence that little can escape detection. It facilitates experimental tools to recognize cellular components—for example, detectors for mRNAs based on specific oligonucleotide probes or mass-spectrometric identification of proteins based on specific peptide sequences—with confidence that these features provide a unique signature. It allows sophisticated computational analyses—for example, to study genome structure and evolution—with confidence that subtle results will not be swamped or swayed by noisy data. At a practical level, it eliminates tedious confirmatory work by researchers, who can now rely on highly accurate information. At a conceptual level, the near-complete picture makes it reasonable for the first time to contemplate system approaches to cellular circuitry, without fear that major components are missing" *(6)*.

The genomics revolution has enabled many pharmaceutical companies to develop innovative methods of identification of new drug targets and diagnostic biomarkers. Because many of the traditional pharmaceutical titans have historically been based on small molecule drug development, they have lacked the expertise to exploit new molecular techniques for target identification. As a result, many companies have jumped on the bandwagon in the rush to buy smaller companies that own and develop these novel technologies.

8. Ethical, Legal, and Social Issues

As with any new technology, the consequences, both intended and unintended, of the work are not always readily apparent. The social implications of a comprehensive project such as HGP are profound and wide-ranging especially when compared to a relatively "simple" project like a human gene therapy trial. Gene therapy raises the questions about the ethical issues of genetic manipulation in the laboratory and human patients, the wisdom of engineering the virulence of viruses, the question of cost and scarcity of resources, the risk/benefit ratio, and so forth. The complexity and scope of the HGP exponentially amplifies the social implications of the project. A few of the major highlights will be discussed here.

The HGP realized in advance that the outcome of the research will have vast ethical, social, and legal implications. Therefore 3–5% of the HGP budget was allocated to research in this area resulting in the largest bioethics program in the world.

One theme that arose from ELSI is fairness in the use of genetic information. Sequencing of the human genome and creation of tools that allow rapid genotyping increase the likelihood of widespread genetic testing as part of routine medical care. The question of who should have access to an individual's genetic information arises. First, should the patient automatically have the right to their genetic information? What are the implications of having access to the crystal ball of one's future health? What kind of counseling and education should be provided to these individuals who choose to gaze into their genetic program? What are the ethical ramifications of informing an individual they may possibly, but not definitely, develop a debilitating disease? What if this disease is heritable; what are the implications for fetal genetic testing or reproduction? Should family members, employers, or insurers be allowed access to this information?

Interestingly, to date there is no federal legislation that protects the rights of privacy regarding genetic information. Bills have been introduced, but not passed, to protect individuals from discrimination based on genetic information from insurers or employers. However, in 2000, President Clinton signed an executive order prohibiting the use of genetic testing or other genetic information in employment decisions of federal employees but also providing strong privacy protection to genetic information gathered for the purposes of medical treatment or research.

The first lawsuit involving genetic discrimination in the workplace was in 2001 when the Equal Employment Opportunity Commission (EEOC) filed suit against the Burlington Northern Santa Fe Railroad (BNSF) for secretly testing their employees for a rare genetic condition that could allegedly predispose employees to workplace injuries and screening for several other more common conditions such as diabetes and alcoholism. An employee who refused to be tested was threatened with termination. The EEOC used the Americans with Disabilities Act to argue that the tests were unlawful because they were not job related, and any medical conditions discovered by these evaluations would result in illegal discrimination based on disability. BNSF agreed to EEOC demands and the lawsuit was settled.

A case like EEOC vs BNSF shows that the Americans with Disabilities Act can to some degree protect individuals against workplace discrimination based on genetic information. In addition, the Health Insurance Portability and Accountability Act (HIPAA) directly addresses the issue of genetic discrimination as it pertains to insurance coverage. However, HIPPAA applies to employer-based and commercially issued group health insurance only. Clearly, at the minimum, legislation should be created to protect individuals from insurance discrimination based on genetic information. Similarly, in 1998 the Clinton administration recommended future legislations that ensured HGP information should not be used to discriminate against workers or their families.

Allowing an individual to access their genetic information can have some unintended consequences as discussed earlier. In addition to altering their perception of their future, health, and well-being, this information can also impact how they view their role in society or how others may view the individual. For example, even in our "enlightened" age, significant social stigmas still exist for mental illness, especially for the more serious disorders such as schizophrenia. A diagnosis of a predisposition to this debilitating disorder can have profound psychological ramifications, not only for the patient but for the patient's offspring when this information is weighed into reproductive decisions.

In addition, defined genetic traits may also predominate in certain ethnic or minority groups. Ethnicity and race can still be charged topics in this enlightened age, and the discovery of genetic information with racial identifiers may not necessarily be in a person's self-interest, especially if some of those identifiers are attached to stigma or stereotypes. On the other hand, race-related genetic information may also potentially be used to understand the genetic dimension of disease in minority groups and therefore improve diagnosis, treatment, and prevention. An excellent discussion on this topic is available in *Nature Genetics Supplement*, Nov 2004, **36**(11).

As discussed throughout this book, genetic information will also be used to make clinical decisions. Under ideal circumstances, this information would minimize or prevent drug misadventures and enable optimal preventative care. Indeed, genetic tests are commercially available for over a thousand diseases (http://www.genetests.org). However, a lag time will exist between when genetic sequence information is readily available and when the clinical implications of those sequences are full elucidated. During that time period, diagnostic information may be of questionable accuracy or predictive value. The medical community must also be properly educated to properly interpret the information. Furthermore, the cost of the test may far exceed the benefit of the treatment, if it is even available. Clearly, even after a firm understanding of the role of certain genetic variations in disease management is established, the proper use of this information must continue to be evaluated.

Another ELSI topic that emerged was data ownership. This issue became hotly contested during genome sequencing as described earlier and was distilled to the question of whether genetic sequences can be intellectual property. The HGP's intent was to make the human genome publicly available to both public and private parties, for the purposes of developing innovative technologies and products. However, private companies who invest substantial sums of money also have a right to a return on their investment vis-à-vis patent protection, even if a genetic sequence is involved.

Where does the federal government currently stand on the issue of sequence patenting? Prior to 1980, the government considered life forms not patentable because they were part of nature. In 1980, in the case Chakrabarty vs Diamond the Supreme Court ruled that genetically engineered organisms were patentable because they did not occur naturally. Since then, the patent office has issued patents for whole gene sequences and has published guidelines about the submission of partial sequences. Sequencing of partial sequences, especially if the function of the sequence is unknown, is likely to generate legal challenges if more complete sequence information later becomes available and commercially viable. The patent office has since issued guidelines in 2001 that sequence submissions must now state specifically how the product functions in nature.

The question of the patentability of SNPs is being proactively addressed by TSC (The SNP Consortium), a consortium composed of 10 large pharmaceutical companies and the UK Wellcome Trust philanthropy, founded in 1999. The TSC has published a publicly available map that includes several million SNPs and intends to patent the SNPs to prevent others from patenting and owning the information.

9. The Future and HGP

Most would agree that the HGP has fulfilled all of its goals above and beyond expectation. The human genome and several others have been sequenced faster and with much greater accuracy than the project's own predictions. This information has spawned multiple databases which are treasure troves of information awaiting harvest for biomedical research. The project affirms the notion of community and collaborative efforts working synergistically to accomplish much more than the sum of the individual efforts.

The completion of the HGP is not the end; it is the beginning of a new era of molecular medicine and genomics. Though the genome has now been sequenced to 99% with thousands of haplotypes and millions of SNPs catalogued, the difficulty of piecing together the biological puzzle for each disease or treatment effect remains. It is estimated that 0.1% of the genome accounts for interindividual variation. Yet that small fraction of the genome still represents 3 million bases whose variability somehow codes for our individual uniqueness. Determining which of those polymorphisms are relevant for which disease or treatment will undoubtedly, unlike the HGP, take decades or longer to complete.

In parallel with the identification of polymorphisms that contribute to human health and disease will be the elucidation of the genetics and mechanisms of biological processes. Whereas absolute completion of both of these lofty goals is unlikely, progress in these areas will substantially contribute to medical science's ability to enhance the quality of human life and yield insights into the beauty and complexity of nature.

10. Conclusion

The HGP and contemporary genomics advances have changed our perception of basic biological processes and our vision for the future of individualized medicine. With the emergence of the accompanying new technology, the HGP has also altered the way biomedical research is conducted as well as the research in ancillary fields. In sum, these insights have already changed both the pipeline of new drugs in development and the new products on the market both in medicine and other areas such as agriculture, food industry, textiles, and other chemicals in the field of biotechnology. The prediction of phrenicea.com is already in evidence: "If the 20th century could be labeled the 'Century of the Computer,' then the 21st century will become 'The Century of DNA'" (http://www.phrenicea.com).

Acknowledgment

Special thanks to Dr. Rita Shiang, Department of Human Genetics, VCU, for her valuable input and feedback on this chapter.

References

1. International Human Genome Sequencing Consortium (2001) Initial sequencing and analysis of the human genome. *Nature* **409,** 860–922.
2. Nadeau, J. H., Balling, R., Barsh, G., et al. (2001) Functional annotation of mouse genome sequences. (Human Genome Project) *Science* **291**(5507)**,** 1251–1254.
3. *Science* Apr 11 2003 issue.
4. Collins, F. S., Green, E. D., Guttmacher, A. E., and Guyer, M. S. (2003) A vision for the future of genomics research. A blueprint for the genomic era. *Nature* **422** (6934)**,** 835–849.
5. Collins, F., Morgan, M., and Patrinos, A. (2003) The human genome project: lessons from large scale biology. *Science* **300**(5617)**,** 286–289.
6. International Human Genome Sequencing Consortium (2004) Finishing the euchromatic sequence of the human genome. *Nature* **431,** 931–945.

3

The Use of Bioinformatics and Chemogenomics in Drug Discovery

Susanna Wu-Pong and Rita Shiang

Abstract

The Human Genome Project and the emergence of accompanying technology are beginning to transform the drug discovery process. Sequence and structure databases are offering new avenues for the identification of novel genes and drug targets. This communication reviews how bioinformatics and chemogenomics are used for new gene identification, selection of genes or proteins that may be potential drug targets, and the screening of libraries, both in vitro and *in silico*, for designation of new lead drug compounds.

Key Words: Structure; annotation; drug design; bioinformatics; chemogenomics; drug discovery; high-throughput screening.

1. Introduction

As discussed in Chapter 2, a major aim of the Human Genome Project (HGP) was to sequence the human genome and allow public access to the resulting data for scientific research and development of new technology. It became clear to the HGP Consortium that the vast amount of data that emerged from the effort would require improvements in the management and analysis of the 3 billion nucleotides that comprise the human genome. Therefore, goals directed specifically at bioinformatics were proposed: (1) improve content and utility of databases; (2) develop better tools for data generation, capture, and annotation; (3) develop and improve tools and databases for comprehensive functional studies; (4) develop and improve tools for representing and analyzing sequence similarity and variation; and (5) create mechanisms to support effective approaches for producing robust, exportable software that can be widely shared.

This list represents the HGP's general goals in the area of bioinformatics which may appear academic to some in the absence of a practical application. From the perspective of the topic of this book, two global outcomes from the

From: *Biopharmaceutical Drug Design and Development*
Edited by: S. Wu-Pong and Y. Rojanasakul © Humana Press Inc., Totowa, NJ

HGP are of paramount interest. First, because the HGP provides the ingredients for all human biologic processes, the identification of these genes (now estimated to be 20,000–25,000) and gene products (over 100,000) holds the key to the future of medicine. These genes and gene products, and all of the variation therein, represent all the "druggable" targets available, with the exception of infectious diseases.

Though this statement "all the 'druggable' targets available" may seem unusually bold, the putative targets represent an almost unimaginable array of sequences (proteins and nucleic acids) and structures (all translated amino acid sequences with their subsequent posttranslational modifications), especially given the inclusion of genetic variability of normal and diseased conditions. Making sense of this information and channeling it into new drug discovery using bioinformatics is the subject of this chapter. In other words, how can the biological database information be used to improve human health?

2. Target Development

The traditional method for drug discovery usually begins with the identification of a potential drug target, typically a human, viral or bacterial protein that has a property we wish to modify for therapeutic purposes. Drug candidates are usually identified by chemical library screening, where thousands of chemicals are systematically applied to an assay that allows rapid identification of compounds that produced the desired effect. This approach of evaluating thousands of compounds quickly is referred to as high-throughput screening. Alternatively, rational drug design can be used to engineer a drug molecule whose physico-chemical properties will fit the active site of the target molecule, typically producing either an inhibitory or stimulatory effect.

Biomedical scientists continue to seek new and improved drug targets and to develop new methods to identify novel drug targets. The completion of the HGP provides a seemingly unlimited resource to explore for a new Achilles' heel for old diseases. One perspective on novel target identification involves attempting to discover either novel members of known gene families or entirely new gene families.

3. Gene Family Research

Sequencing the entire human genome has provided scientists a genome-wide perspective for the field of genomics and is enabling the identification and classification of all members of every gene family. A gene family is a set of genes defined by presumed homology, in other words, genes that have evolved from a common ancestral gene and generally share some biochemical activity, sequence motifs, and/or structure. Homology can be ascertained by inspecting and comparing gene or protein sequences or protein structures. Sequence examination alone is insufficient when assigning homology to two genes. For example, in addition to sharing similar gene sequences, the positions of introns within the coding sequence can be used to infer common ancestry and therefore homology. Similarly, protein sequences can also yield information about ancestry or homology even if gene sequences differ. Knowledge of the protein's secondary structure also gives further information about ancestry, because the organization of secondary structural elements presumably would be conserved even if the amino acid sequence changes considerably.

3.1. Comparative Genomics for Gene Family Research

Comparative genomics examines a gene's biological context by studying it across different species (orthologs) and thus facilitates identification of new genes and gene families. Obtaining orthologs has been considerably facilitated by the creation of databases (*see* Table 1 for a list of databases and tools used in genomics-based drug discovery) that are widely available commercially or on the Internet. Orthologous sequences or structures can be sorted using different clustering methods, then the properties of the clusters could be examined for insight about the molecules in terms of their evolutionary origin, their function, or their taxonomical source.

New members of known gene families are discovered using homolog identification. Homologous nucleotide or amino acid sequences are searched using pairwise sequence-search methods such as Basic Local Alignment Search Tool (BLAST; http://www.ncbi.nih.gov/) or FASTA (Table 1). Sequences with homology greater than 30% are relatively easy to find using these tools. However, remote homolog identification requires use of algorithms such as PSI-BLAST (Position-Specific Iterated BLAST) or sequence alignment modeler (SAM) because these programs are more iterative and therefore more likely to detect distant homologies. PSI-BLAST takes all statistically significant protein alignments found by BLAST and combines them into a multiple alignment, from which a position-specific score matrix (PSSM) is constructed. This matrix is used to search the database for additional significant alignments, and the process may be iterated until no new alignments are found. SAM uses a linear hidden Markov model to characterize an entire family of sequences which can then be used to determine if a new sequence belongs to a particular family. Frequently, more than one search algorithm is used and all results from these programs are evaluated to try to obtain as many leads as possible. The list may be narrowed by focusing on hits that demonstrate complexity and redundancy, both indications of essential biological function.

In addition to sequence analysis, studies directly on the proteome can be used to identify new genes. However, because of the vast number of proteins in the proteome (approximately 100,000 in humans), the number of proteins to be

Table 1
Databases and Software Used for Drug Discovery

Type	Name		URL
Chemical database	Available chemicals directory	>484,000 compounds	http://cds.dl.ac.uk/cds/datasets/ orgchem/isis/acd.html
	ZINC	>3.3 million compounds	http://blaster.docking.org/zinc/
Drug database	Comprehensive medicinal chemistry database	7000 drugs	Commercially available at mdl.com
	MACCS-II drug data report	>132,000 drugs	Commercially available at mdl.com
	World drug index	80,000 drugs	Commercially available at http://www.daylight.com
Ligand binding	Binding database		Bindingdb.org
	Relibase		http://www.relibase.ccdc.cam.ac.uk
Small molecule alignment	FlexS		Commercially available at tripos.com
	GASP		http://bioinformatics.rcsi.ie/ ~redwards/gasp/gasp_input.htm
Sequence alignment	BLAST		www.ncbi.nih.gov/
	FASTA		http://fasta.bioch.virginia.edu/
	PSI-BLAST		http://www.ncbi.nlm.nih.gov
	SAM		http://www.cse.ucsc.edu/research/ compbio/sam.html
Sequence databases	FlyBase	Drosophila	flybase.bio.indiana.edu/
	SGD	Yeast	http://www.yeastgenome.org/
	MGD	Mouse	http://www.ncbi.nlm.nih.gov
Structure database	PDB	>35,000 structures	rcsb.org
	MMDB		http://NCBI.nlm.nih.gov
Structure alignment	MAMMOTH		http://fulcrum.physbio.mssm.edu: 8083/mammoth/
	SCOP		http://scop.mrc-lmb.cam.ac.uk/scop/ index.html

(Continued)

Table 1 (*Continued*)

Type	Name	URL
Structure prediction	DOCK	http://www.cmpharm.ucsf.edu/ kuntz/dock.html.
	AUTODOCK	http://www.scripps.edu/pub/ olson-web/doc/autodock/

screened must be reduced by some rational means, usually either subcellular fractionation or an affinity method. Differential protein expression can then be used to compare diseased and normal tissues to determine which genes are regulated differently in diseased states. 2D gel electrophoresis or differential in-gel electrophoresis (DIGE) followed by mass spectrometry is used to identify such proteins. Protein arrays are also used in the development of proteomics research (Chapter 5).

An alternative and more rapid method of measuring differential gene expression involves the use of DNA microarrays (Chapter 4). A limitation of sequencing technology in general is that genetic analysis occurs one gene at a time; thus microarray technology has revolutionized bioinformatics and genomics. Ideally, one would prefer to test for the presence of thousands of genes simultaneously in a miniaturized system. Consequently, the DNA microarray, which detects the presence of thousands of sequences concurrently, has emerged from the HGP, which then laid the foundation for further development by Affymetrix (http://www.affymetrix.com). Because this technology monitors thousands of genes concurrently, it can encompass the entire genome of simple organisms such as yeast on a single chip (Agilent) so that researchers can simultaneously examine the expression of thousands of genes. In fact, Nimblegen has recently introduced a whole human genome microarray that probes 37,000 genes at a time (http://www.nimblegen.com). This approach, however, fails to differentiate between genes which cause the disease vs genes that are affected by the disease. Furthermore, mRNA expression analysis may not necessarily correlate with gene product translation. The amount of data that emerges from this and other high-capacity technologies should be used to continuously update the human genome databases.

A useful application of microarrays in comparative genomics is comparative genomic hybridization (CGH). CGH allows direct comparison of gene expression in normal vs tumor cells. Normal and tumor cell DNA are labeled with two different fluorescent probes and are then applied together to the microarray. The relative fluorescence from normal and tumor cells is measured using quantitative image analysis. Novel genes may be discovered in this manner if they are differentially amplified or deleted in tumor cells and the corresponding sequences are present on the microarray. This approach has been useful in cancer classification and possibly even determination of tumor sensitivity. New methods such as subtractive hybridization techniques further develop the use of microarrays for gene discovery (for review *see [1]*).

3.2. Structure Prediction

In small molecule drug discovery, knowledge of the structure of the target protein facilitates drug design. Protein structure is determined once the protein is isolated, identified, and crystallized, using X-ray crystallography or NMR techniques. In contrast, when using a bioinformatics approach, one is starting with one or more gene or amino acid sequences rather than a purified protein, thus, accurate prediction of the resulting protein structure would be very useful when ascertaining the potential value as a drug target.

For a given test sequence, determination of the protein's primary structure is trivial; prediction of the 3D structure of the protein is not. Yet knowledge of protein structure is critical in understanding protein function. Protein structure prediction based on primary structure is still fraught with error, and the majority of proteins with known sequences still have undefined structures. Molecular modeling database (MMDB) and protein data bank (PDB) are databases containing 3D structures of macromolecules (Table 1), yet for the majority of proteins in the database the folded structure remains unknown. As of April 2006, PDB listed almost 36,000 structures in its database.

Three primary approaches are used for structure prediction. First, ab initio or *de novo* prediction can be used to predict structure based only on the laws of physics and chemistry. This method is used for protein sequences that lack comparable structures. The second method involves homology modeling, which assumes that homologous sequences will produce similar structures. A limitation of this approach is that homologous sequences can also produce structures that differ substantially *(2)*.

The third method involves the use of fold recognition or "threading". This method is especially useful when proteins have similar 3D structure but differ in their primary sequence. Fold recognition therefore determines if the unknown can be reasonably aligned structurally to a known protein structure. The root mean square distance between corresponding amino acids is calculated to determine how well the two molecules align structurally. Programs such as MAMMOTH and SCOP are used for structural alignment (Table 1).

The most accurate approach, however, to protein structure prediction is a combination of methods. For example, if a protein is suspected to be homologous to another protein with known structure, the sequence of the unknown can be aligned to the structure of the known protein. If one assumes that the structure is conserved to a greater degree than the primary sequence, there will be a similarity between the two proteins if they are homologous. For example, both human and legume hemoglobin (leghemoglobin) transport oxygen in the respective organism. Though the proteins have vastly different primary sequences, their protein structures are virtually identical. Energy minimization and molecular dynamic simulations may be used to refine and test the structure prediction.

3.3. Gene Annotation

The new technologies such as microarrays (Chapters 4 and 5) and shotgun sequencing (Chapter 2) have dramatically elevated the rate at which genomic and genetic information is emerging. As a result, the classification and organization of this information have been critical in scientists' ability to use the information effectively. In addition, because genetic information has evolved over time and is highly conserved across the animal kingdom (for example, human genome is 95% homologous to the baboon), the understanding and interpretation of biological processes require perspective from across a wide range of different species.

The range of information available also necessitates the use of numerous databases that support the various needs of the scientific community. The annotation of the database is critical to its utility as a research tool. Annotation of a gene by sequence or name only has limited value because gene names or other terminology may vary widely and may be very discipline specific. Therefore, novel ways to describe or sort the data are required to allow efficient searching by scientists in different fields. For example, other ways to annotate a gene could include function, location, structural components, publication number, markers, phenotypes, role in a biological process, and so on.

The gene ontology (GO) project (http://www.geneontology.org) is a collaborative effort to provide consistent descriptions of gene products across different databases. GO began in 1998 as a collaboration between FlyBase (Drosophila), the Saccharomyces Genome Database (SGD) and the mouse genome database (MGD). Since then, the GO Consortium has grown to include 14 (as of Spring 2006) repositories for plant, animal, and microbial genomes. The project collaborators are developing uniform vocabularies (ontologies) that describe gene products in terms of how they behave in a cellular context (either molecular function, biological process, or cellular component) that is independent of the species of origin. Collaborating databases then use GO terms to annotate their genes or gene products, thus providing the users with a uniform structure and vocabulary. Other efforts are also in place to standardize classification terminology such as HUPO-PSI (http://psidev.sourceforge.net).

Databases such as GO are useful tools for evaluating the context of a gene's role within a biological process or across species. Multiple databases can also be merged to query information from areas that may be missing from the use of a single database.

3.4. Functional Genomics—Determining Gene Function

The ultimate utility of gene annotation lies in its ability to predict gene function. The human genome still has thousands of genes whose function is not known, and even genes that have been functionally annotated have only been generally defined and therefore require refinement.

In computational biology, a major approach to functional genomics is comparison of a gene across different species (orthology). As organisms have evolved, their genes have also evolved in a manner that is suitable for that organism. However, the ancestral genes will retain some level of homology, thus enabling scientists to trace genes across species and the gene's evolution. These orthologs also retain functional similarities. Therefore, a newly identified gene with homology across several species is also likely to have similar biologic function to its orthologs. The scientist must be able to differentiate between homologous, which suggests an evolutionary relationship, and similar sequences.

An attractive approach to comparative genomics is the theory that orthologous genes also have predictable roles in a biochemical process, or phyletic pattern. In other words, genes that evolve retain similar relationships to each other in biological processes or "pathways" like signaling or synthetic pathways. If true, then one may examine not only the homologous sequences across species, but also the pathway information to determine gene function. In practice, however, pathways can differ significantly across species, even among critical biochemical schemes like glycolysis. In conclusion, phyletic pattern analysis has limited use unless examined with other data.

In some cases, proteins that have a functional interaction may, in some species, present as a single molecule that retains the function(s) of the original molecules. The discovery of such fused domains also offers new perspectives and fresh insights in the determination of gene function.

The pharmaceutical industry and biomedical researchers are obviously most interested in genes that play significant roles in the development of disease. As mentioned earlier, differential expression is a valuable approach in identifying genes that behave differently in disease, yet the actual role of that gene in the disease is critical in determining its importance as a potential drug target. A gene's cellular function and role in disease is determined using an experimental approach. Specific downregulation of a gene, first in cell lines, then in wild-type animal and transgenic models, is necessary to determine the protein's function and thus, appropriateness for development as a drug target.

Ideally, one would determine the role of a gene in a disease by either inducing expression in a null animal or inhibiting expression in a system where the phenotype is present. Gene induction is best accomplished by delivering an exogenous gene expressing the protein of interest to either somatic cells (gene therapy, Chapters 8 and 14) or germline or blastocyst cells (transgenics, Chapter 7). Antisense and siRNA are typically used in cultured cells or in animal models to specifically inhibit target gene expression (Chapter 10). The next step would involve the use of gene therapy or inhibitory sequences in rats or mice, though delivery challenges are significant and may complicate data interpretation (Chapter 14). The investigator can examine the outcome of gene inhibition or induction on the cells' properties. These proof-of-principle studies are important in the establishment of the role of a gene in disease prior to proceeding to the more difficult and labor-intensive knockout models, where germline cells are modified to create an organism that has reduced expression of the gene of interest. These time-consuming and intensive studies are required before the gene or protein can be validated as a druggable target. Drug targets that usually emerge from such studies are most commonly G-protein-coupled receptors, enzymes, or hormones.

4. High-Throughput and Virtual Drug Screening

If a drug target that has gone through the phases of identification, structure and function determination, in vitro and in vivo testing still remains viable, then a variety of approaches may be used to develop drug candidates for that target. Table 2 lists drugs that were developed using computational approaches to drug discovery.

4.1. High-Throughput Screening

Traditional methods to identify potential drug candidates involve the use of high-throughput screening, where libraries containing thousands of chemicals are systematically tested for activity in vitro. According to Hann and Oprea *(3)*, up to 100,000 molecules per day can be screened for activity using this method. In excess of 1 million chemicals might be screened in a high-throughput assay. This process is automated so that the in vitro assay that provides easy and rapid measurement is ideally suited for maximum screening efficiency. The assay could be cell free, cell based or involve the use of organisms.

After the initial screening, the appropriate "hits" undergo a second round where their biological activity, structure, and mechanism of action are determined to ascertain that the observed effects are pharmacologically favorable. Compounds that survive this round are identified as "leads" and a new library of compounds may be generated using combinatorial chemistry based on the structure of the lead molecule.

Table 2
Chemogenetic Drugs in Development

Genomics company	Marketed	Clinical trials (therapeutic indication)	Preclinical
Vertex	Lexiva (HIV) — w/ GlaxoSmithKline	VX 950 (HCV) VX 883 (bacterial)	VX 770 (Cystic fibrosis) VX 409 (pain) – w/ GlaxoSmithKline
		VX 702 (rheumatoid arthritis) VX 680 (cancer) — w/Merck Brecanavir (HIV) — w/GlaxoSmithKline	
Ambit Bioscience			AC220 (cancer) AB087 (stroke)
Acadia		ACP103 (Schizophrenia, Parkinson's)	AC262271 (glaucoma) w/Allergan
Avalon		AVN944 (cancer)	
Astex		AT7519 (cancer) w/ Novartis AT9283 (cancer) AT9311 (cancer) w/ Novartis	
MorphoChem		Oxaquin (antiinfective)	
Infinity		IPI — 504 (cancer) IPI — 609 (cancer)	
Amphora			AKT (cancer) TTK (cancer) P38-a (inflammation) GSK3-b (pain)
Avalon		VX 994 (cancer)	

4.2. Virtual Screening

This very lengthy and expensive process of high-throughput screening has fueled the emergence of a new field called chemogenomics that uses in silico methods for virtual drug screening. The goal of chemogenomics is to discover potential drug candidates from different and often disparate databases containing sequence or structure information. Chemogenomics relies on the use of gene families to construct predictive 3D models for protein families; the more complete and accurate the data set of gene family members, the more powerful the chemogenomics approach. Essentially, structural information about the target protein is used in combination with combinatorial and medicinal chemistry to create new classes of chemicals. Therefore, good estimates for protein structure prediction (discussed earlier) are critical for optimal virtual screening.

4.2.1. Virtual Screening by Docking

Once a target has been identified and its structure determined, the binding pockets on the molecule must also be characterized for drug screening. Modeling programs, such as DOCK and AUTODOCK (Table 1), can be used to visualize the landscape of the protein surface and predict which molecules will bind to which pockets. Such programs can also be used to screen databases (e.g., ZINC or Available Chemicals Database) for potential ligands. ZINC contains chemicals that are available for purchase and is accompanied by physicochemical property information on the molecules to aid in screening. The process of using the 3D structure of the target to screen ligands is called high-throughput docking.

The qualities of side chains in the binding pocket should also be considered when designing or selecting a ligand. Side chain configurations that are highly conserved tend to have biological relevance and are more likely to be found in other proteins' binding pockets. Drugs targeted to these configurations will therefore tend to be nonspecific; drugs that target nonconserved regions are predicted to have greater specificity. The active site may also differ between members in a gene family, so a diverse library of compounds should be screened for the desired activity.

4.2.2. Virtual Screening by Similarity

High-throughput docking or "virtual screening by docking" is one example of how chemogenomics can be used to mine sequence data for the purpose of drug discovery. Another approach which does not rely on the availability of the structure of the target protein is called "similarity-based virtual screening". For example, for a given biological target one may determine a specific arrangement of molecular properties that are critical for biological activity, or pharmacophore (Fig. 1). The pharmacophore is generated by examination of the protein's ligands, which may often differ structurally, for commonality in terms of structure or properties. Use of pharmacophores to virtually screen compounds enables reduction in size of the library of compounds used for the first in vitro screening round.

In addition to pharmacophore searching, a known ligand can be used to screen for other potential binding substrates using small molecule alignment (GASP, FlexS). The difficulty with this approach is that both the test and reference molecules tend to be flexible, resulting in many possible alignment

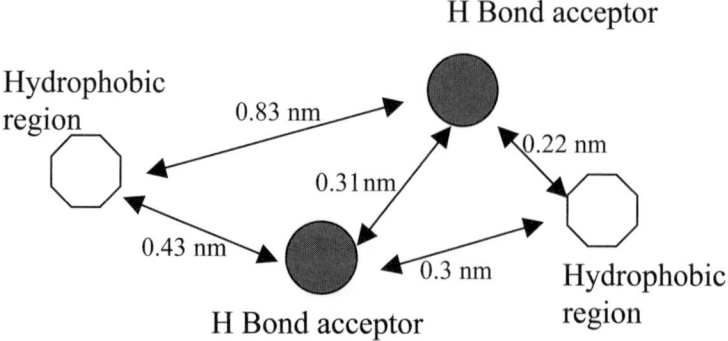

Fig. 1. Example of a simple pharmacophore.

options. Alignments are scored based on internal strain and overlap of molecular groups. "Chemistry space" can also be used, which assigns values to certain molecular properties and places them as points in a 3D space for each drug. However, as stated earlier, molecules are flexible and multiple conformations might be needed to adequately describe the molecule.

Instead of using just the pharmacophore or docking, the structure of the ligand binding site can also be used to identify new potential drug targets. This approach uses a combination of protein structure and ligand binding affinity data (e.g., Binding Database). Proteins with similar functions may also have similar binding pockets. A comprehensive examination of similar pockets and how they interact with ligands can help formulate the necessary elements for ligand binding. Tools such as Relibase (Table 1) can be used for ligand binding analysis and can account for factors such as pocket conformations, water molecule interactions, and ligand specificity.

All of the virtual screening methods could potentially benefit from specially designed screening libraries. For instance, scientists are beginning to screen and design compounds based on their pharmacokinetic properties such as blood–brain barrier permeability. Such library design preselects compounds that show potentially good in vivo properties even before one begins searching for pharmacophores or ligand binding. The library may be further refined by establishing criteria for which drug molecules seem to universally share: "drug likeness". Drug databases are available for developing these criteria (Comprehensive Medicinal Chemistry Database, MACCS-II Drug Data Report), which have included properties such as hydrogen binding properties, log P, the presence of certain functional groups such as an amine, and molecular weight. Library design has the obvious limitation of being either too restrictive or not restrictive enough, resulting either in the possible exclusion of good leads or not reducing the size of the library effectively.

4.3. Combining High Throughput with Bioinformatics

So far we have reviewed how protein target or ligand identification combined with virtual screening can aid in selecting new leads for drug candidates. A hybrid approach combines traditional high-throughput screening with bioinformatics. Traditional library screening can be used in tandem with gene expression information to gain more elaborate and detailed information from the screen. For example, when screening a library in an in vitro assay, microarray analysis can be used simultaneously to measure the drugs' effect on the expression of thousands of relevant genes instead of simply measuring a single phenomenon like receptor binding. Leads can then be selected based on the desired effect on individual genes.

One of the earliest efforts in this area emerged from the National Cancer Institute which used microarrays to measure expression of approximately 8,000 genes in 60 human cancer cell lines. The microarray gene expression data were evaluated using cluster analysis and visualization techniques. The investigators found that under drug-free conditions, the cells' gene expression clustered into groups based on the tissue of origin. High-throughput methods were then used to apply over 70,000 compounds to the cells, and gene expression was again measured and compared to untreated controls. The investigators found that gene expression now clustered based mostly on the drugs' presumed mechanism of action, rather than organ origin *(4)*. Similarities in cell physiologic properties such as doubling time also corresponded to specific gene expression patterns. In addition, cultured breast and leukemia cells were compared to tumor biopsy samples, which both demonstrated similar gene expression profiles, suggesting similarity in properties despite transfer to an in vitro environment *(5)*.

An example of the use of this approach was published recently by Dai et al. *(6)* who used NCI60 to study the effect of antitumor drugs on gene expression in the cancer cell collection. The authors used 119 anticancer drugs in the NCI60 cells and determined that 343 genes correlated with drug cytotoxicity and included members of the growth factors and receptors, metalloproteinases, and ras-like GTPase families. The genes were further culled to identify the 13 genes whose expression profiles can be used to predict drug potency.

Commercial products such as DrugMatrix (Iconix) are also available for the evaluation of high-throughput screening output consisting of gene expression analysis or other cellular responses. Iconix compiles gene expression or cell response profiles that result from drug treatment into a profile called Drug Signatures™, which can then be used to screen for other potential drugs that have a similar impact on gene expression. One can conceivably use such an approach for a variety of purposes, including identification of new genes involved in disease pathophysiology, finding potential new drug candidates, or determining the mechanism of action or toxicity of drugs.

4.4. Drug Design for Selected Targets

Target selection is a lengthy and complex process that is now expedited and improved by bioinformatics. As discussed earlier, drug design to the target can start with the structure and specific physicochemical properties of the protein. If one were to then proceed using a traditional rational drug design approach, sites available for molecular interaction (e.g., proton donors or acceptors, hydrophobic regions, etc.) would be mapped, and potential drug molecules designed for optimum interaction with those sites. Computational technology or computer-aided drug design can be used to create new molecules and classes of drugs that are designed to fit the properties of the binding pocket. This approach will yield numerous candidates which must then be prioritized by predicting binding free energy. Additional considerations may be included, such as predicted in vivo properties such as bioavailability. Assuming the computationally derived structures can be synthesized in the laboratory, they may be screened for activity, presumably now with a higher hit rate than a random library of molecules. After a lead is identified, combinatorial chemistry is used to synthesize a new library of compounds derived from the lead, and the library is then screened to select the candidate with the best binding affinity.

Similarly, when using a ligand-based approach, several candidate molecules will emerge from modeling, and the candidates must be scored to assign priority. This will be even more true if a broader approach, such as the use of pharmacophores, is used for the selection or design of candidate molecules. Several examples of *de novo* design exist in the literature and are described in reviews on this topic *(7,8)*.

A more direct chemogenomic approach is based on designing drugs to target the actual sequences of the target genes. For example, siRNA (small interfering RNA) and antisense oligonucleotides are sequence-specific inhibitors of gene expression (Chapter 10). Antisense oligonucleotides are simply designed to bind by Watson–Crick basepairing to complementary mRNA sequences and usually include backbone, base, or end-modifications for improved in vivo properties. Thus, compared to the virtual screening and de novo methods of drug candidate design, siRNA and oligonucleotide designs seem trivial, though as discussed elsewhere in this publication, delivery, stability, and cost issues can be major hurdles.

Antisense and siRNA are useful for inhibiting expression of unwanted genes, typically from a viral or cancer source. Overexpression of "normal" genes can also produce unfavorable therapeutic endpoints; thus antisense or siRNA strategies can also be useful in these situations, such as when anticoagulation is desirable. In contrast, sometimes genes fail to produce sufficient amounts of functional protein. Thus, desirable gene function can be supplemented using gene therapy (Chapter 8). Gene therapy presents a new range of technical issues including sufficient gene expression and safety.

5. Conclusion

In this chapter, we have reviewed the use of computational methods and databases in bioinformatics for the identification of drug targets, potential lead compounds, and design of drug candidates. The rapid emergence of bioinformatics tools and technology has only become recently available in tandem with sequencing efforts such as the HGP and sequencing tools like microarrays. This field is still in its infancy but the pace at which chemogenomics transforms the drug discovery process will only continue to grow.

References

1. Dougherty, J. D. and Geschwind, D. H. (2002) Subtraction-coupled custom microarray analysis for gene discovery and gene expression studies in the CNS. *Chem. Senses* **27**(3), 293–298.
2. Friedberg, I. and Godzik, A. (2005) Connecting the protein structure universe by using sparse recurring fragments. *Structure* **13**(8), 1213–1224.
3. Hann, M. M. and Oprea, T. I. (2004) Pursuing the leadlikeness concept in pharmaceutical research. *Curr. Opin. Chem. Biol.* **8**, 255–263.
4. Scherf, U., Ross, D. T., Waltham, M., et al. (2000) A gene expression database for the molecular pharmacology of cancer. *Nature Genet.* **24**(3), 236–244.
5. Ross, D. T., Scherf, U., Eisen, M. B., et al. (2000) Systematic variation in gene expression patterns in human cancer cell lines. *Nat. Genet.* **24**(3), 227–235.
6. Dai, Z., Barbacioru, C., Huang, Y., and Sadee, W. (2006) Prediction of anticancer drug potency from expression of genes involved in growth factor signaling. *Pharm. Res.* **23**(2), 336–349.
7. Schneider, G. and Fechner, U. (2005) Computer-based de novo design of drug-like molecules. *Nat. Rev. Drug Discov.* **4**(8), 649–663.
8. Ortiz, A. R., Gomez-Puertas, P., Leo-Macias, A., et al. (2006) Computational approaches to model ligand selectivity in drug design. *Curr. Top Med. Chem.* **6**(1), 41–55.

4

DNA Microarrays in Drug Discovery and Development

Neelam Azad, Anand Krishnan V. Iyer, and Yon Rojanasakul

Abstract

Ever since the completion of the human genome project, there has been great interest in the research community toward addressing the role played by multiple genes to orchestrate complex cellular functions. This requires techniques that allow high-throughput analysis of such target genes. Low manufacturing and application costs, flexibility, and speed of analyses in a high-throughput fashion make DNA microarrays one of the most invaluable tools in this endeavor. DNA microarrays have revolutionized genomic and pharmacologic investigations by allowing simultaneous monitoring of all the genes in different genomes, thus linking the entire genome expression with the function of the whole organism. Microarrays are widely used to address a plethora of scientific questions in the pharmaceutical industry, particularly in drug discovery and development. The technique has immense potential and promises to play a key role in furthering research in a number of fields, as discussed in this chapter.

Key Words: DNA; microarrays; genomics; drug discovery; pharmacogenomics.

1. Introduction

The completion and success of the human genome project has increased our understanding of intricate biological processes and related biomedical sciences. Although thousands of genes control the development and functioning of living beings, molecular biologists have traditionally been limited to single-gene studies that divulge the role of an individual gene in a particular physiological response which may truly be a cumulative effect of several gene interactions. Single-gene studies are extremely time-consuming and face major challenges such as their incapacity in explaining complex gene interactions that may be critical in organism function (1). The early gene expression methods focused either on measuring mRNA expression levels for individual genes or on determining the transcriptional profiles of several active genes simultaneously (2–4). Therefore,

From: *Biopharmaceutical Drug Design and Development*
Edited by: S. Wu-Pong and Y. Rojanasakul © Humana Press Inc., Totowa, NJ

these methods are ineffectual in meeting the basic requirements of effective pharmaceutical and biomedical applications as they cannot be used for conducting large-scale screening and developing expression profiles for cells or tissues *(5)*. The rapid identification of about 30,000–40,000 genes by the Human Genome Project, nearly all of which are possible drug targets, has brought us to the next important phase involving the identification of gene functions and the related biological pathways *(6)*.

The emergence of high-throughput screening methods such as microarray technology has contributed greatly toward this end, as DNA microarrays can measure the expression levels of thousands of genes simultaneously. DNA microarrays have revolutionized genomic and pharmacologic investigations by allowing simultaneous monitoring of all the genes in different genomes in a single experiment. The microarray itself is a small chip typically comprised of thousands of immobilized DNA sequences onto its surface. DNA microarrays are based on the principle of hybridization or basepairing of the unknown DNA sequences in the sample with complementary immobilized DNA probes with known sequences. This principle of hybridization of a DNA sample and a known labeled probe is based on the paper by E.M. Southern that first demonstrated the use of solid supports to examine nucleic acids *(7)*. This principle is also employed to detect single DNA species by Southern blotting or RNA by Northern blotting. However, completion of the various genome projects required techniques such as DNA microarrays that could detect and analyze multiple copies of DNA *(8)*. Initially, microarrays were limited to basic scientific applications such as studying fibroblasts response to serum or determining the genes induced during the yeast cell cycle *(9,10)*. However, considering the vast genomic application of DNA microarrays, it was clear from the very beginning that these technologies have potential application in research relevant to human diseases and clinical drugs.

DNA microarrays are mainly employed to study genetic variations in a sample or to determine the expression levels of genes *(11)*. Because the expression pattern of a gene is linked to its biological role, microarray studies can provide important information on the biochemical pathways involved, sites of gene expression, and most importantly the function of the gene in a particular organ as well as the whole organism *(5)*. By enabling the study of the expression of numerous genes under a range of experimental conditions, DNA microarrays provide researchers with a revolutionary new tool to ultimately link the entire genome expression with the function of the whole organism.

Additional applications of the DNA microarray technique include description of the genes involved in physiological and pathological processes, identification of signature genes indicative of a disease process, identification of disease-related genes that may become targets for therapeutic intervention, and monitoring

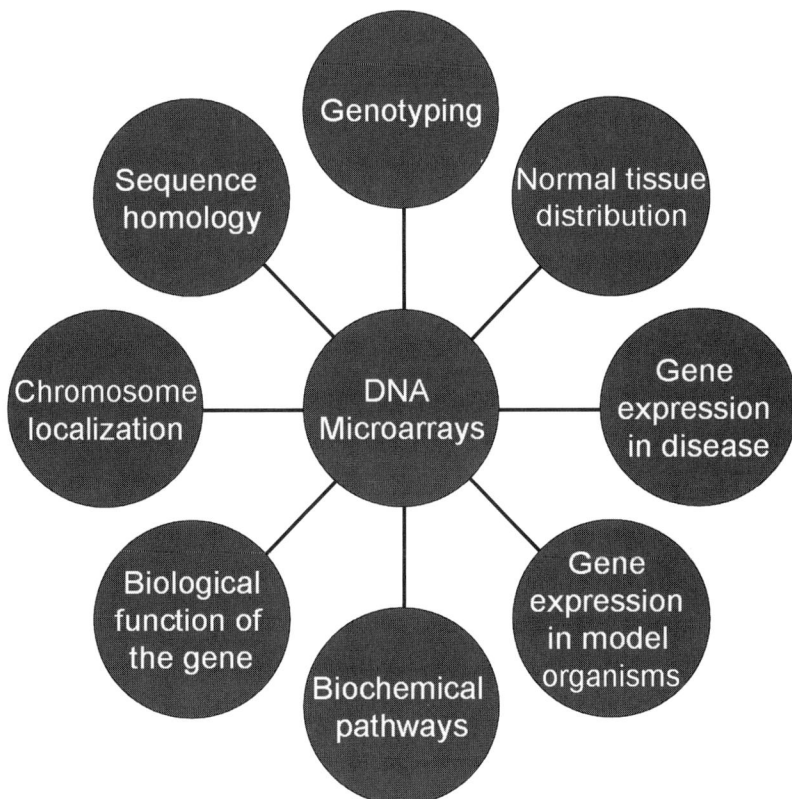

Fig. 1. Schematic representation of the various applications of DNA microarrays.

alterations in gene expression in response to drug treatments *(12–15)*. DNA microarrays may also assist in the identification of genes that demonstrate abnormal expression in a given cell or tissue in drug metabolism or when exposed to a drug or toxin *(11)*. In the past few years, DNA microarrays have been developed and applied in novel gene discovery, identifying side effects of a given drug, categorizing specific genes that are involved in drug reaction, and determining the effect of the drug on nontarget tissues *(5,16)*. Some of the applications of DNA microarrays are listed in Fig. 1.

Although the development of DNA microarrays has the potential to revolutionize therapeutics, a number of drawbacks have to be overcome before its successful applications. The first major shortcoming of DNA microarrays is that they can only probe genes that already have complementary sequences identified *(5)*. Furthermore, DNA microarray technology is extremely expensive. Even though self-made chips are significantly cheaper than the ready-made ones, the cost of system setup and other technical concerns hinder immediate implementation *(8)*. Besides cost, data management and analysis is another serious concern. Microarray experiments require specialized data management tools to store, analyze, and standardize the vast amounts of information obtained from all experiments. However, methods are being developed to overcome these drawbacks. For instance, standard expression database management systems are being developed that will lead to better experimental designs and data interpretation *(17)*. Despite the shortcomings, microarrays have numerous applications in several domains related to human health. For instance, microarrays facilitate rapid diagnosis of genetic diseases, aid in selection of potential targets for therapeutics, assist in drug discovery, and predict the activity of drugs and toxins *(11)*. As DNA microarray technology improves, costs will drop radically, enabling these tools to become available in most research laboratories. Over the last 5 yr, the utilization of DNA microarrays in academic laboratories, industrial service laboratories, and pharmaceutical companies has risen exponentially and is becoming a standard method for determining the molecular mechanisms involved in various biological processes. DNA microarrays have become an integral part of drug discovery, therapeutic optimization, and clinical validation. In this chapter, we will focus on the applications of DNA microarray technology in drug discovery, drug development, pharmacogenomics, and toxicogenomics.

2. DNA Microarrays

Gene expression analysis began with simple laboratory techniques developed to examine the expression of known individual genes. DNA microarrays revolutionized gene expression studies by enabling large-scale analysis of thousands of genes in model species where genomes are well characterized. Pioneered by Affymetrix Inc., this technology was first implemented for a comprehensive study of the gene expression profile of the plant *Arabidopsis (18)*. Since its advent, the use of this technology has increased exponentially, driven mainly by significant advancements in fabrication techniques, ease of use, and flexibility in application. DNA microarrays have truly come of age, with several thousand scientific articles published using this technology in various settings *(19)*.

As the name suggests, DNA microarrays consist of arrays of specific and well-characterized short DNA sequences, either single or double stranded, which are systematically arranged in huge numbers on silicon, glass, or plastic surfaces. These DNA sequences are used as probes to identify the presence or absence of thousands of complementary DNA sequences in a sample in a high-throughput fashion. The DNA probes on the chip surface may be between a few (as in the case of oligonucleotide microarrays) to several hundreds of basepairs (as in the case of complementary DNA (cDNA) microarrays) in length. Oligonucleotide microarray chips are typically used to generate a general expression profile of the sample using multiple sample-specific genome-wide probes for each gene, typically 18–25mers, derived from a library of known sequences available through databases such as GenBank and UniGene *(20)*. On the other hand, cDNA microarrays use specific sequences (more than 100 bases in length) that are derived from either custom cDNA libraries or are specified by the user. These sequences are complementary to specific gene sequences that are being assayed for in the sample—typically a single probe is used for assaying its complementary gene of interest with multiple spots on the chip surface *(19)*. However, all other aspects regarding fabrication of the chip, methodologies involving sample preparation, and analysis of data are similar for both types of DNA microarray chips.

DNA microarrays are commercially available from a number of vendors and may be tailor-made for user-defined applications in a cost-effective manner *(21)*. This has become possible because of advances in fabrication techniques, allowing for the production of DNA microarrays in a high-throughput manner. Light-directed synthesis (photolithographic and digital mirror based), inkjet printer-based synthesis, and electrode-directed synthesis are the main techniques employed by manufacturers *(21)*. The main areas of microarray application have been in measuring gene expression in different situations, including analysis of diseased vs normal tissues, mutation detection, genotyping, gene regulation studies, tumor profiling, and a number of other applications *(1,5,22,23)*. However, we shall focus on the applications of microarrays to the fields of drug discovery and development and related pharmaceutical areas.

3. DNA Microarray Methodology

In essence, DNA microarrays may be thought of as massive parallel versions of Northern blots, capable of analyzing the expression levels of thousands of genes in several different conditions in one single run. The fundamental basis of DNA microarrays is the process of hybridization, where two DNA strands hybridize if they are complementary to each other *(24,25)*. Though the nomenclature differs in literature dealing with DNA microarrays, we will refer to the fluorescently tagged unknown DNA sequence in the sample as the "target" and the known complementary sequence on the surface of the DNA microarray surface as the "probe". The stable binding of complementary DNA sequences allows for the quantification of the unknown target DNA sequence by means of the levels of fluorescent chemical label on the target, which can be detected by means of a light scanner. It is possible to place thousands of probes on a surface area of 1 cm^2, where each probe sequence matches a particular mRNA resulting from the expression of its corresponding gene. An expression profile of a sample may be obtained by simultaneously observing all the spots on the microarray.

As mentioned earlier, oligonucleotide arrays are composed of probes for all genes in the sample, giving a snapshot of the global expression profile in the sample. Multiple probes for different regions of the same target gene (called probe sets) are synthesized and presented on the surface. This redundancy increases the accuracy and reliability of the data generated. Thus, in order to compare two sets of samples, two separate oligonucleotide chips have to be used, one for each sample, and then analyzed separately to identify differences in gene expression *(26,27)*. Spotted microarrays, on the other hand, are customized to present specific probes of interest, generally cDNA probes, allowing for evaluation of relative responses between two samples on the same chip. This can be done by labeling the two samples with two separate, distinct dyes and applying a mixture onto the same chip. Differentially expressed genes may be identified by quantification of relative expression of the dyes *(28)*.

The methodology for performing DNA microarray experiments has evolved over the years and has been described in excellent detail by Bowtell et al. *(24,25)* in a couple of review articles as part of "The Chipping Forecast" series, published in *Nature Genetics* in 1999, 2002, and 2005. The general methodology, condensed from the articles published in the series, is described below and involves four major steps.

3.1. Target Preparation

Using several commercially available kits, messenger RNA is first isolated and purified from the sample. The principle behind this purification is that only fully transcribed and mature mRNA, which account for only 3% of total RNA content in a cell, have a poly-adenine (poly-A) tail and can be captured using a chromatographic column presenting complementary oligodeoxythymidine (oligo-dT) beads. Because mRNA are very sensitive and can be easily destroyed, they are reverse transcribed into more stable cDNA. In order to distinguish between the target cDNA sequences of the two samples, fluorescently labeled nucleotides with distinct fluorophores (typically Cy3 or Cy5) are used during the reverse transcription of mRNA from each sample so that they are automatically incorporated into the target cDNA sequence. Thus, for example, a particular experiment may have control sample targets tagged with Cy3 to emit green and the test sample targets tagged with Cy5 to emit red.

3.2. Hybridization

In this step, the cDNA samples from the control and test arms are mixed in equal volumes and applied directly onto the microarray slides, which present the DNA sequences of interest. The array may hold thousands of DNA probes, each with different DNA sequences. A cDNA target sequence from the sample mix will bind tightly to the probe sequence on the chip surface only if complementarities exist. After extensive washing to remove unbound targets, stable bonding between the target and its probe is achieved by exposing the microarray surface to ultraviolet light, which will crosslink the two complementary DNA sequences.

3.3. Data Acquisition

In order to determine the amount of target cDNA from the sample mix bound to each spot, the fluorescently tagged target sequences are stimulated to excitation by a confocal laser, causing the emission of light at particular wavelengths, specific to the fluorophore being used. The emitted light from spots on the microarray can be captured using a charge-coupled device (CCD) or a confocal microscope, and the intensity of light from each spot will be in direct relation with the amount of tagged target bound to the probe, and hence the amount of mRNA transcript present in the sample. It is important that the two fluorophores have different excitation and emission wavelengths, so that crosstalk between the two fluorescent channels is avoided. Simultaneous excitation over both channels will result in an image representing differential gene expression on a global level. Thus, mRNA sequences that are expressed equally in the control and test arms will bind equally to their complementary probes, emitting yellow. On the other hand, mRNA overexpressed in the test arm as compared to control will hybridize strongly with their complementary cDNA probes, emitting red. Such an example of a spotted microarray is shown in Fig. 2.

3.4. Image Analysis

Once a scanned image is obtained, a number of data analysis software dealing with image quantitation, filtering, and normalization may be applied to extract relevant information pertaining to gene expression. Gene clustering software may then be used to identify classes of genes that are differentially regulated between the samples. Following thresholding and gridding of the spots, a number of normalization procedures including LOWESS normalization, mean centering, total intensity normalization, ratio statistics, and standard deviation regularization may be applied *(29,30)*. Clustering programs such as Cluster and Tree View may be utilized to identify each spot and cluster them into separate classes based on function within the cell in particular settings *(31–34)*. Figure 3 is a general schematic representation summarizing the steps in a typical microarray experiments. DNA microarrays are easy to operate and eliminate the possibility of human errors as these techniques are completely automated *(21)*. Over time, adaptation of this methodology integrated with other principles will further expand the utility of DNA microarray experiments whereas still enhancing the biological discovery process.

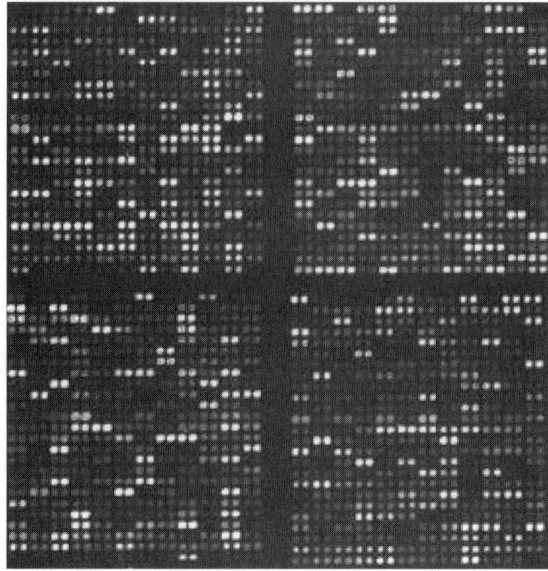

Fig. 2. A typical cDNA microarray, with different colors indicating the relative response of different genes in the test samples vs the control samples. This image was obtained from the NASA website (http://science.nasa.gov/headlines/y2004/images/radmicrobe/microarray.jpg) and is free from copyright protection.

4. Drug Discovery

Drug discovery was traditionally based on biochemical pathways that are implicated in physiological and pathological processes. A typical experimental procedure included characterization, purification, and screening of appropriate targets against a group of structurally dissimilar molecules that were relevant to the desired therapeutic activity, such as inhibiting a specific enzyme or receptor. This was followed by the optimization of the lead compounds for desirable properties such as bioavailability and target specificity (*12*). Although the conventional approach has led to the discovery of several effective drugs for a variety of diseases, this method is extremely lengthy, uncertain, and expensive. The advent of molecular biology during the past decades completely changed the process of drug discovery. It enabled the use of human targets which are more critical than animal samples as even a difference of single amino acid can render a compound ineffective (*35*). These techniques also made possible the employment of site-directed mutagenesis in the study of drug target interactions. Despite their advantages, standard molecular biology techniques were restricted

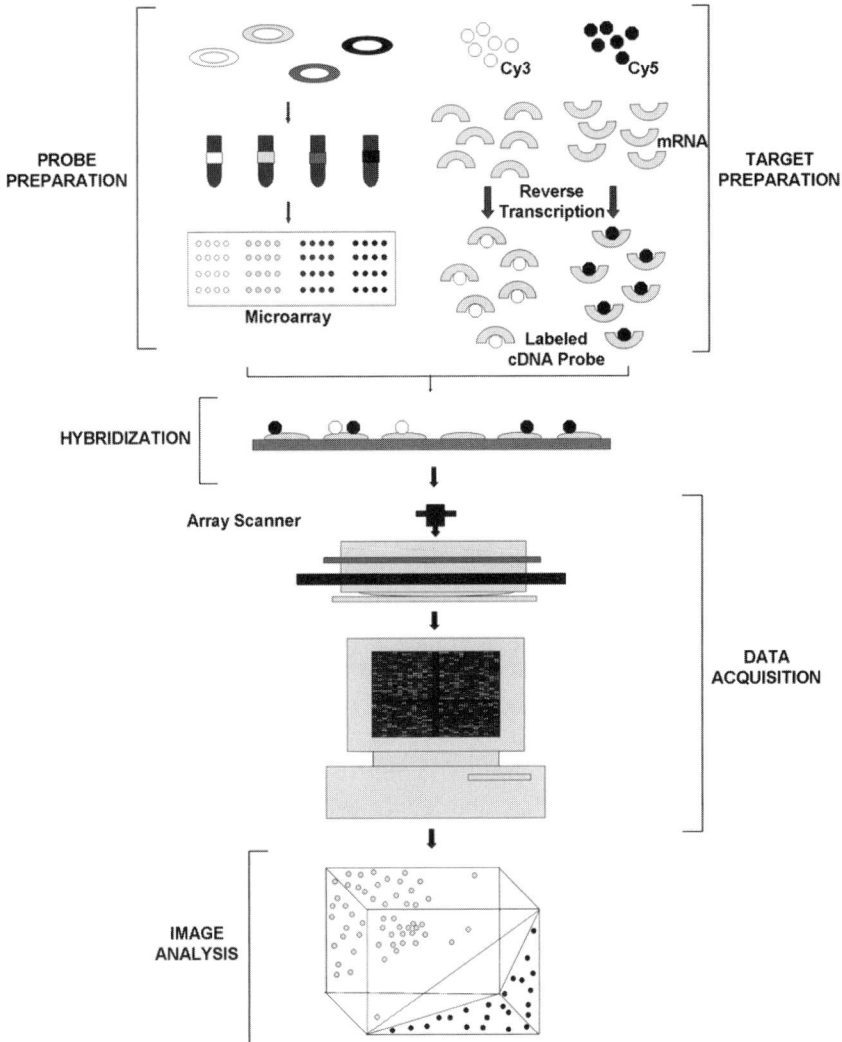

Fig. 3. The first step in a typical DNA microarray experiment is preparation of the labeled cDNA target, which is reverse transcribed from mRNA isolated from the test and control samples, which are both labeled with different fluorochromes (Cy3 and Cy5, respectively) so that they may be distinguished. The samples are mixed and hybridized to either custom cDNA or predesigned oligonucleotide microarrays. After stable binding of the complementary DNA strands, the chip surface is scanned and analyzed using various software available both commercially and through academic institutions.

by the limited number of potential targets and poor target validation as it required a detailed understanding of pathophysiological processes *(12)*.

Advances in medicinal chemistry and high-throughput gene expression profiling methods have been instrumental in surmounting these drawbacks. In the 1990s, the process for discovering novel therapeutic targets was based chiefly on large-scale genomics involving methods such as sequencing of expressed sequence tags (ESTs), differential display, and homology cloning *(6,36)*. However, in addition to these extremely vital bioinformatics methods, experimental approaches were also required to prioritize the potential therapeutic targets *(6)*.

Microarrays present an ideal technique for discovering novel therapeutic targets as they facilitate the identification of potential therapeutic targets from thousands of genes in a single experiment. This approach of searching differentially expressed genes is widely accepted as a valuable method in drug discovery. As mentioned earlier, DNA microarrays can determine the expression levels of genes, especially their function, thereby assisting in understanding the molecular basis of the disease, identifying new drug targets as well as examining the efficacy and toxicity of new drug candidates. DNA microarrays can scan every gene in a microbe to uncover the overall expression pattern of the pathogen, thus providing detailed information about the pathogens involved in the disease process *(5)*. DNA microarrays have already been employed in determining the specific genes that are abnormally regulated in pathological conditions. For instance, microarray study of approximately 100 genes demonstrated that the upregulation of genes encoding interleukin-6 and several matrix metalloproteinases plays a role in rheumatoid tissue *(37)*. A more important application of DNA microarrays would be to study and discover the susceptible gene in genetically complex diseases such as schizophrenia *(38)*. DNA microarrays are employed in drug discovery by various pharmaceutical companies to screen hundreds and thousands of compounds simultaneously to study the interactions of the compounds with their molecular targets. However, for proprietary reasons the exact utilization of DNA microarrays in the context of drug discovery is not clearly known and is hidden from the public with no published literature available *(6)*.

Nevertheless, it is generally believed that in drug discovery, investigators use arrays in primary screening to prioritize a few genes as potential therapeutic targets on the basis of various criteria including expression levels, disease specificity, and tissue or cell-type selectivity *(6)*. Generally, a combination of microarrays, bioinformatics, and simple validation experiments is used to prioritize a few potential candidate genes that are relevant to the desired therapeutic outcome from a series of congeners on the array. To further scrutinize the role of specific genes in the disease, secondary or tertiary screens are conducted to select compounds that exhibit a narrow range of pharmacological activity, i.e., those that exhibit

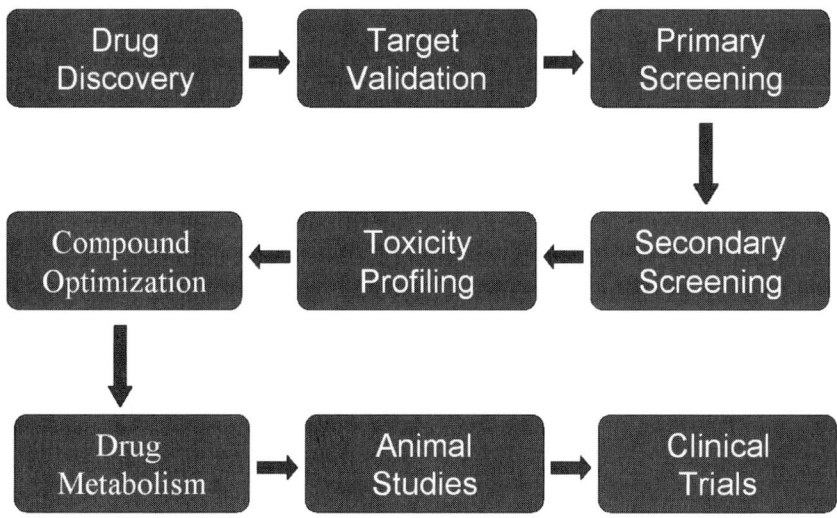

Fig. 4. Applications of DNA microarray technology in various stages of drug discovery and development.

100-fold selectivity for the specific receptor as compared to other related receptors *(39)*. These may include examining time and dose-dependent curves, diverse biological samples, animal models, or more diverse disease states *(6)*. Good experimental designs and thorough statistical analysis minimize the error caused by technical and biological variations including reproducibility and sensitivity problems *(40)*. Therefore, using sophisticated statistics and bioinformatics to validate statistically significant observations and to identify the complex patterns of biological pathways is an integral part of drug discovery using DNA microarrays. Such an integrated approach to drug discovery will undoubtedly produce significant information regarding the pharmacologic and toxicologic responses of the drug and will provide a comprehensive understanding of the biologic response to the therapeutic effects of novel target molecules.

5. Drug Development

Apart from drug discovery, another major use of gene expression information is to help with the development of newly discovered drugs. Analysis of gene expression levels by DNA microarrays has become a chief factor in many stages of the drug development (Fig. 4). Combinatorial chemistry and the expression profiles of genes obtained from DNA microarrays can together provide important information about the novel target compound including its solubility, absorption profile, interaction with drug-metabolizing enzymes, and its toxicity profile, thus aiding in new drug development *(16)*. DNA microarrays provide the most thorough and efficient method of understanding the molecular basis of physiological processes and thus are vital in characterizing and verifying the function of novel genes *(1)*. Once potential therapeutic agents have been recognized, several approaches including gene knockout and knockin strategies in cells and animal models can be undertaken to develop and validate these candidate therapeutic targets *(6)*. These knockout and knockin techniques play an important role in defining the potential action of drugs, identifying the pleiotropic effects as well as adverse effects of the target drug *(13)*. DNA microarrays can be used in this context to evaluate major consequences of the genetic intervention, thereby comparing compounds and identifying potential pleiotropic actions. For the microarray analysis of cell-based gene knockouts, gene deletion experiments in yeast and antisense technology have been implemented in the mammalian cell systems *(41,42)*. A classic example of the application of antisense technology is the suppression of protein kinase A RIα that led to the inhibition of proliferation-transformation genes and induction of differentiation-reverse transformation phenotype in cancer cells and tumors *(43)*. RNA interference (RNAi) is another strategy that has been employed for silencing genes and is considered less cumbersome and more efficient for gene inactivation than knockout or knockin techniques as it can potentially facilitate routine suppression of gene expression in living cells *(43,44)*. RNAi in combination with DNA microarrays is considered to provide an innovative strategy for high-throughput target validation *(45)*.

In mammalian cell cultures, the expression pattern of a therapeutic target gene is an important tool for defining its biological significance. DNA microarrays can be employed in combination with individual gene transfections using various vectors for characterizing the biological functions of potential therapeutic target genes. For example, the signaling pathways of platelet-derived growth factor-β (PDGF-β) receptor and fibroblast growth factor (FGF) receptor 1 were analyzed using a combination of gene transfection and microarrays *(46)*. In addition, this approach has been used in numerous studies for the identification of downstream consequences of altered expression of disease-related genes *(47,48)*. On the whole, transgenic and knockout studies using animal models have been very helpful in drug development. DNA microarray analysis may additionally substantiate the new drug target. Therefore, studies involving DNA microarrays and knockout experiments can validate the target specificity of potential drug candidates *(6)*.

DNA microarrays may also be employed in the identification of pathways involved in the metabolism of a specific drug. The transcriptional induction of drug-metabolizing genes can be identified by profiling either the livers of drug-treated rats or cultured primary human hepatocytes with DNA microarrays.

This can be very helpful in selecting potential candidates; for example, a drug candidate that is only metabolized by a polymorphic gene can be discarded as it can accumulate to toxic concentrations in individuals carrying specific polymorphisms *(6)*. Moreover, microarray experiments may also be useful in classifying similar acting drugs by comparison of the biological properties of the new compounds with the existing compounds by comparing their effects on gene expression profiles *(20)*. As opposed to the widespread perception that DNA microarray technologies are not delivering on their promise, these techniques are in fact significantly changing the ways in which drugs are discovered and developed.

6. Toxicogenomics

An important aspect of drug development is its toxicological effects on the patient. In fact, a study aimed at determining the common causes of failure of drugs during the developmental phase showed that adverse toxicological effects may account for nearly 20% of compound attrition *(49)*. In addition, adverse reactions triggered by drug interactions in patients taking multiple drugs have contributed significantly to the withdrawal of those drugs from the market because of unpredictable consequences. Thus, for a pharmaceutical company spending billions of dollars in drug discovery and development, drug toxicity is a serious concern.

In all cases, without exception, exposure to any drug compound or xenobiotic results in an alteration of gene expression, either as a direct or indirect effect of the compound *(50)*. The use of microarray technology to measure the transcriptional reprogramming as a consequence of drug or xenobiotic exposure is termed as "toxicogenomics" *(51)*. A potential advantage of toxicogenomics over conventional in vivo and in vitro techniques is that toxicogenomics may lead to early and reliable prediction of toxic liabilities of compounds, preventing the need for substantial investments into preclinical and clinical trials with potentially toxic compounds *(52)*. Thus, toxicogenomic analysis, when used in conjunction with established techniques, may provide an insight into the mechanism of organ toxicities and lead to faster drug development and production of cheaper and safer drugs. A number of organ tissues such as the liver, kidney, uterus, and the nervous system have been investigated for toxic effects of certain drugs *(53–58)*. For example, a causal relationship among the upregulation of CYP2B2, accumulation of fat in the liver, and inhibition in cholesterol synthesis was demonstrated using toxicogenomics *(59)*. Another study brought to the fore the mechanisms underlying cisplatin-induced nephrotoxicity *(60)*. Because DNA microarrays are used in profiling the expression levels of genes at the global level, toxicogenomics may prove useful for deciphering the synergistic effects on toxicity of multiple drugs administered simultaneously *(53)*.

An important aspect of toxicological analysis is the identification of specific patterns of gene expression related to distinct cellular processes such as immune system response, growth, development, and oxidative stress that are elicited because of the toxic effects of a drug under a set of experimental conditions. Custom microarrays such as the ToxBlot, ToxChip v 1.0, and ArrayPlate have been developed to investigate and characterize the potential mechanism of toxicity and results have been very promising so far *(23,51)*. In order to capitalize on this emerging field, the National Institute of Environmental Health Science (NIEHS) has now established the National Center for Toxicogenomics (http://www.niehs.nih.gov/nct/home.html), which aims to provide leadership for the development of a unified strategy for toxicogenomic study. Toxicogenomics may prove to be fundamentally informative in toxicology testing offering an ideal platform for a genome-wide analysis of drug toxicity.

7. Pharmacogenomics

Pharmacogenomics can be defined as the science of understanding the correlation between an individual's genotype and their response to drug treatment. It is a systematic approach that utilizes a variety of genomic technologies to discover the drug response determinants and also helps in understanding the genetic and molecular basis of the variable drug response observed among patients *(61)*. Certain drugs generate the desired therapeutic effect in a subset of patients whereas they may be ineffective in other populations. Therefore, studying the genetic basis of patient response to candidate drugs allows drug developers to effectively design better therapeutic molecules.

Although research on the human genome project has greatly advanced medical science, the detailed function and interaction of majority of the genes are still unclear. Drugs are often involved in complex metabolic pathways and sensitivity to drugs is determined by multiple genes, but most of the regulatory mechanisms of their expressions are not well understood *(61)*. Presently, the trend is to list as many genetic variations found within the human genome as possible. As suggested by the National Center of Biotechnology Information (NCBI), these variations or single-nucleotide polymorphisms (SNPs) can be used as a diagnostic tool to predict an individual's response to a certain drug. For this purpose, a patient's DNA must initially be sequenced for the presence of specific SNPs.

The major obstacle that has impeded the widespread use of SNPs as a diagnostic tool is that traditional gene sequencing technology is slow and expensive. DNA microarray analysis can be used to screen thousands of SNPs in few hours making it possible to identify specific SNPs found in a patient's genome, quickly and affordably. Because DNA microarrays can simultaneously provide expression patterns of thousands of genes that are characteristic of their function, it enables the documentation of detailed responses of cells and tissues to both disease and drug treatments *(62–64)*.

DNA microarrays are being increasingly used in pharmacogenomic studies to determine variations in gene expression that occur in cells on exposure to specific drugs. It makes possible the analysis of the response of practically the entire human genome to cellular drug exposure and also uncovers a wide variety of genes related with drug resistance *(20)*. Recent reports in this area range from small studies that characterize the genetic response of a specific cell line to a single therapeutic agent to larger investigations that examine the effect of several agents on a diverse set of cell lines. In addition, some researchers have already employed DNA microarrays to correlate changes in gene expression profiles with clinical outcomes in order to identify a collection of genes that may be predictive of clinical response to a specific drug regimen.

Pharmacogenomics of the response to doxorubicin is the best studied of almost all the drugs till date *(20)*. In one of the several studies, Kudoh et al. *(65)* used nylon-based arrays to identify genes in cells that were either responsive to doxorubicin or whose expression levels changed during selection for resistance to doxorubicin in MCF-7 breast cancer. In another study, DNA microarray analysis was used to investigate the protective effect of hepatocyte growth factor/scattering factor (HGF/SF) on MDA-MB-453 cells during exposure to doxorubicin *(66)*. DNA microarrays have also been employed in the study of other drugs. For example, this approach was used to determine if changes in gene expression could be used to elucidate the sensitivity of eight hepatoma cell lines to a panel of eight chemotherapeutic drugs *(67,68)*. Kumar et al. *(68)* used nylon-based arrays to analyze the changes in gene expression that occur as ovarian carcinoma cell lines become resistant to paclitaxel.

DNA microarrays also contribute to the understanding of the mechanism of drug action which is a major part of understanding the physiological responses to drug treatments. For example, DNA microarray analysis of yeast cells demonstrated that the immunosuppressive drug FK506 might have more than one target. It was reported that FK506 had the same effect on gene expression patterns even when the gene that FK506 suppresses was eliminated. Furthermore, in the absence of this gene, FK506 affected the expression levels in other ways *(69)*. The application of DNA microarrays to pharmacogenomics is likely to increase our understanding of the cellular response and mechanism of drug action in response to drug exposure. However, for successful application of this technique, researchers must understand its advantages, address its limitations, and finally successfully integrate it into drug-related areas.

8. Summary

DNA microarray technology is still in its infancy, but has the potential of rapidly identifying and validating novel therapeutic targets. These systems are redefining drug discovery and development by illuminating the complex workings of biological systems. DNA microarrays increase the output and speed of biological research considerably; however, it also identifies hundreds of genes that may or may not be relevant in vivo, especially when large, genetically diverse populations are studied. In the future, the challenge is to design experimental systems that minimize variability in the data, increase reproducibility, allow statistical analysis, and provide a clear link between drug response and the expression of specific genes *(20)*. DNA microarray technology will advance in several aspects including improvements in fabrication of microarrays, which will reduce the manufacturing costs and increase the throughput *(70)*. Additionally, there will be databases with expression profiles for hundreds and thousands of genes that may complement the human genome sequence databases *(71)*. DNA microarray analysis is a promising new technology that provides a comprehensive overview of the phenotypic variations in physiological as well as pathological conditions that may radically accelerate discovery in several realms of human health and medicine.

References

1. Wiseman, S. B. and Singer, T. D. (2002) Applications of DNA and protein microarrays in comparative physiology. *Biotechnol. Adv.* **20**, 379–389.
2. Greenberg, M. E. and Ziff, E. B. (1984) Stimulation of 3T3 cells induces transcription of the c-fos proto-oncogene. *Nature* **311**, 433–438.
3. Manley, J. L. and Gefter, M. L. (1981) Transcription of mammalian genes in vitro. *Gene Amplif. Anal.* **2**, 369–382.
4. Marzluff, W. F. Jr. (1978) Transcription of RNA in isolated nuclei. *Methods Cell Biol.* **19**, 317–332.
5. Lennon, G. G. (2000) High-throughput gene expression analysis for drug discovery. *Drug Discov. Today* **5**, 59–66.
6. Gerhold, D. L., Jensen, R. V., and Gullans, S. R. (2002) Better therapeutics through microarrays. *Nat. Genet.* **32**(Suppl), 547–551.
7. Southern, E. M. (1975) Detection of specific sequences among DNA fragments separated by gel electrophoresis. *J. Mol. Biol.* **98**, 503–517.
8. Cojocaru, G. S., Rechavi, G., and Kaminski, N. (2001) The use of microarrays in medicine. *Isr. Med. Assoc. J.* **3**, 292–296.
9. Iyer, V. R., Eisen, M. B., Ross, D. T., et al. (1999) The transcriptional program in the response of human fibroblasts to serum. *Science* **283**, 83–87.
10. Spellman, P. T., Sherlock, G., Zhang, M. Q., et al. (1998) Comprehensive identification of cell cycle-regulated genes of the yeast *Saccharomyces cerevisiae* by microarray hybridization. *Mol. Biol. Cell* **9**, 3273–3297.

11. Street, M. (2002) DNA microarrays: manufacture and applications. *Austr. Biotechnol.* **12,** 38–39.
12. Debouck, C. and Goodfellow, P. N. (1999) DNA microarrays in drug discovery and development. *Nat. Genet.* **21,** 48–50.
13. Meloni, R., Khalfallah, O., and Biguet, N. F. (2004) DNA microarrays and pharmacogenomics. *Pharmacol. Res.* **49,** 303–308.
14. Lock, C., Hermans, G., Pedotti, R., et al. (2002) Gene-microarray analysis of multiple sclerosis lesions yields new targets validated in autoimmune encephalomyelitis. *Nat. Med.* **8,** 500–508.
15. van 't Veer, L. J., Dai, H., van de Vijver, M. J., et al. (2002) Gene expression profiling predicts clinical outcome of breast cancer. *Nature* **415,** 530–536.
16. Chin, K. V. and Kong, A. N. (2002) Application of DNA microarrays in pharmacogenomics and toxicogenomics. *Pharm. Res.* **19,** 1773–1778.
17. Brazma, A. and Vilo, J. (2000) Gene expression data analysis. *FEBS Lett.* **480,** 17–24.
18. Schena, M., Shalon, D., Davis, R. W., and Brown, P. O. (1995) Quantitative monitoring of gene expression patterns with a complementary DNA microarray. *Science* **270,** 467–470.
19. Stoughton, R. B. (2005) Applications of DNA microarrays in biology. *Annu. Rev. Biochem.* **74,** 53–82.
20. Villeneuve, D. J. and Parissenti, A. M. (2004) The use of DNA microarrays to investigate the pharmacogenomics of drug response in living systems. *Curr. Top Med. Chem.* **4,** 1329–1345.
21. Heller, M. J. (2002) DNA microarray technology: devices, systems, and applications. *Annu. Rev. Biomed. Eng.* **4,** 129–153.
22. Advances in Clinical Research with Affymetrix GeneChip® RNA Expression and DNA Analysis Microarrays on World Wide Web.
23. Afshari, C. A., Nuwaysir, E. F., and Barrett, J. C. (1999) Application of complementary DNA microarray technology to carcinogen identification, toxicology, and drug safety evaluation. *Cancer Res.* **59,** 4759–4760.
24. Bowtell, D. D. (1999) Options available—from start to finish—for obtaining expression data by microarray. *Nat. Genet.* **21,** 25–32.
25. Holloway, A. J., van Laar, R. K., Tothill, R. W., and Bowtell, D. D. (2002) Options available- from start to finish- for obtaining data from DNA microarrays II. *Nat. Genet.* **32**(Suppl), 481–489.
26. Relogio, A., Schwager, C., Richter, A., Ansorge, W., and Valcarcel, J. (2002) Optimization of oligonucleotide-based DNA microarrays. *Nucleic Acids Res.* **30,** e51.
27. Rouillard, J. M., Herbert, C. J., and Zuker, M. (2002) OligoArray: genome-scale oligonucleotide design for microarrays. *Bioinformatics* **18,** 486–487.
28. Hegde, P., Qi, R., Abernathy, K., et al. (2000) A concise guide to cDNA microarray analysis. *Biotechniques* **29,** 548–550, 552–544, 556 passim.
29. Bilban, M., Buehler, L. K., Head, S., Desoye, G., and Quaranta, V. (2002) Normalizing DNA microarray data. *Curr. Issues Mol. Biol.* **4,** 57–64.
30. Quackenbush, J. (2002) Microarray data normalization and transformation. *Nat. Genet.* **32**(Suppl), 496–501.

31. Stoeckert, C. J. Jr., Causton, H. C., and Ball, C. A. (2002) Microarray databases: standards and ontologies. *Nat. Genet.* **32**(Suppl), 469–473.

32. Eisen, M. B., Spellman, P. T., Brown, P. O., and Botstein, D. (1998) Cluster analysis and display of genome-wide expression patterns. *Proc. Natl Acad. Sci. U. S. A.* **95,** 14,863–14,868.

33. Khan, J., Wei, J. S., Ringner, M., et al. (2001) Classification and diagnostic prediction of cancers using gene expression profiling and artificial neural networks. *Nat. Med.* **7,** 673–679.

34. Saal, L. H., Troein, C., Vallon-Christersson, J., Gruvberger, S., Borg, A., and Peterson, C. (2002) BioArray Software Environment (BASE): a platform for comprehensive management and analysis of microarray data. *Genome Biol.* **3,** SOFTWARE0003.

35. Oksenberg, D., Marsters, S. A., O'Dowd, B. F., et al. (1992) A single amino-acid difference confers major pharmacological variation between human and rodent 5-HT1B receptors. *Nature* **360,** 161–163.

36. Fryer, R. M., Randall, J., Yoshida, T., et al. (2002) Global analysis of gene expression: methods, interpretation, and pitfalls. *Exp. Nephrol.* **10,** 64–74.

37. Heller, R. A., Schena, M., Chai, A., et al. (1997) Discovery and analysis of inflammatory disease-related genes using cDNA microarrays. *Proc. Natl Acad. Sci. U. S. A.* **94,** 2150–2155.

38. Mirnics, K., Middleton, F. A., Lewis, D. A., and Levitt, P. (2001) Analysis of complex brain disorders with gene expression microarrays: schizophrenia as a disease of the synapse. *Trends Neurosci.* **24,** 479–486.

39. Bugelski, P. J. (2002) Gene expression profiling for pharmaceutical toxicology screening. *Curr. Opin. Drug Discov. Dev.* **5,** 79–89.

40. Yang, Y. H. and Speed ,T. (2002) Design issues for cDNA microarray experiments. *Nat. Rev. Genet.* **3,** 579–588.

41. Cho, Y. S., Kim, M. K., Cheadle, C., Neary, C., Becker, K. G., and Cho-Chung, Y. S. (2001) Antisense DNAs as multisite genomic modulators identified by DNA microarray. *Proc. Natl Acad. Sci. U. S. A.* **98,** 9819–9823.

42. Hughes, T. R., Marton, M. J., Jones, A. R., et al. (2000) Functional discovery via a compendium of expression profiles. *Cell* **102,** 109–126.

43. Cho-Chung, Y. S., Nesterova, M., Becker, K. G., et al. (2002) Dissecting the circuitry of protein kinase A and cAMP signaling in cancer genesis: antisense, microarray, gene overexpression, and transcription factor decoy. *Ann. N Y Acad. Sci.* **968,** 22–36.

44. Hannon, G. J. (2002) RNA interference. *Nature* **418,** 244–251.

45. O'Neil, N. J., Martin, R. L., Tomlinson, M. L., Jones, M. R., Coulson, A., and Kuwabara, P. E. (2001) RNA-mediated interference as a tool for identifying drug targets. *Am. J. Pharmacogenomics* **1,** 45–53.

46. Fambrough, D., McClure, K., Kazlauskas, A., and Lander, E. S. (1999) Diverse signaling pathways activated by growth factor receptors induce broadly overlapping, rather than independent, sets of genes. *Cell* **97,** 727–741.

47. Lee, S. B., Huang, K., Palmer, R., et al. (1999) The Wilms tumor suppressor WT1 encodes a transcriptional activator of amphiregulin. *Cell* **98,** 663–673.

48. Welcsh, P. L., Lee, M. K., Gonzalez-Hernandez. R. M., et al. (2002) BRCA1 transcriptionally regulates genes involved in breast tumorigenesis. *Proc. Natl Acad. Sci. U. S. A.* **99,** 7560–7565.

49. Luhe, A., Hildebrand, H., Bach, U., Dingermann, T., and Ahr, H. J. (2003) A new approach to studying ochratoxin A (OTA)-induced nephrotoxicity: expression profiling in vivo and in vitro employing cDNA microarrays. *Toxicol. Sci.* **73,** 315–328.

50. Nuwaysir, E. F., Bittner, M., Trent, J., Barrett, J. C., and Afshari, C. A. (1999) Microarrays and toxicology: the advent of toxicogenomics. *Mol. Carcinog.* **24,** 153–159.

51. Pennie, W. D. (2002) Custom cDNA microarrays; technologies and applications. *Toxicology* **181–182,** 551–554.

52. Suter, L., Babiss, L. E., and Wheeldon, E. B. (2004) Toxicogenomics in predictive toxicology in drug development. *Chem. Biol.* **11,** 161–171.

53. Luhe, A., Suter, L., Ruepp, S., Singer, T., Weiser, T., and Albertini, S. (2005) Toxicogenomics in the pharmaceutical industry: hollow promises or real benefit? *Mutat. Res.* **575,** 102–115.

54. Suter, L., Haiker, M., De Vera, M. C., and Albertini, S. (2003) Effect of two 5-HT6 receptor antagonists on the rat liver: a molecular approach. *Pharmacogenomics J.* **3,** 320–334.

55. Moggs, J. G., Tinwell, H., Spurway, T., et al. (2005) Phenotypic anchoring of gene expression changes during estrogen-induced uterine growth. *Environ. Health Perspect.* **112,** 1589–1606.

56. Jung, J. W., Park, J. S., Hwang, J. W., et al. (2004) Gene expression analysis of peroxisome proliferators- and phenytoin-induced hepatotoxicity using cDNA microarray. *J. Vet. Med. Sci.* **66,** 1329–1333.

57. Hamadeh, H. K., Jayadev, S., Gaillard, E. T., et al. (2004) Integration of clinical and gene expression endpoints to explore furan-mediated hepatotoxicity. *Mutat. Res.* **549,** 169–183.

58. Amin, R. P., Vickers, A. E., Sistare, F., et al. (2004) Identification of putative gene based markers of renal toxicity. *Environ. Health Perspect.* **112,** 465–479.

59. Ourlin, J. C., Handschin, C., Kaufmann, M., and Meyer, U. A. (2002) A Link between cholesterol levels and phenobarbital induction of cytochromes P450. *Biochem. Biophys. Res. Commun.* **291,** 378–384.

60. Huang, Q., Dunn, R. T. 2nd, Jayadev, S., et al. (2001) Assessment of cisplatin-induced nephrotoxicity by microarray technology. *Toxicol. Sci.* **63,** 196–207.

61. Nishiyama, M. (2005) Polygenetic pharmacogenomic strategies to identify drug sensitivity biomarkers. *Gan. Kagaku Ryoho* **32,** 1902–1907.

62. Golub, T. R., Slonim, D. K., Tamayo, P., et al. (1999) Molecular classification of cancer: class discovery and class prediction by gene expression monitoring. *Science* **286,** 531–537.

63. Gordon, G. J., Jensen, R. V., Hsiao, L. L., et al. (2002) Translation of microarray data into clinically relevant cancer diagnostic tests using gene expression ratios in lung cancer and mesothelioma. *Cancer Res.* **62,** 4963–4967.

64. Pomeroy, S. L., Tamayo, P., Gaasenbeek, M., et al. (2002) Prediction of central nervous system embryonal tumour outcome based on gene expression. *Nature* **415,** 436–442.

65. Kudoh, K., Ramanna, M., Ravatn, R., et al. (2000) Monitoring the expression profiles of doxorubicin-induced and doxorubicin-resistant cancer cells by cDNA microarray. *Cancer Res.* **60,** 4161–4166.

66. Yuan, R., Fan, S., Achary, M., Stewart, D. M., Goldberg, I. D., and Rosen, E. M. (2001) Altered gene expression pattern in cultured human breast cancer cells treated with hepatocyte growth factor/scatter factor in the setting of DNA damage. *Cancer Res.* **61,** 8022–8031.

67. Moriyama, M., Hoshida, Y., Otsuka, M., et al. (2003) Relevance network between chemosensitivity and transcriptome in human hepatoma cells. *Mol. Cancer Ther.* **2,** 199–205.

68. Kumar, A., Soprano, D. R., and Parekh, H. K. (2001) Cross-resistance to the synthetic retinoid CD437 in a paclitaxel-resistant human ovarian carcinoma cell line is independent of the overexpression of retinoic acid receptor-gamma. *Cancer Res.* **61,** 7552–7555.

69. Marton, M. J., DeRisi, J. L., Bennett, H. A., et al. (1998) Drug target validation and identification of secondary drug target effects using DNA microarrays. *Nat. Med.* **4,** 1293–1301.

70. Lemieux, B., Aharoni, A., and Schena, M. (1998) Overview of DNA chip technology. *Mol. Breeding* **4,** 277–289.

71. Hamadeh HaA, C. (2000) Gene chips and functional genomics. *Am. Scientist* **88,** 508–515.

5

Microarray Technology Using Proteins, Cells, and Tissues

Michael Samuels

Abstract

Microarray formatted assays have become well established for the global analysis of nucleic acids, but are only beginning to be adopted for proteomic analysis. This chapter reviews the current status of protein microarray assays and highlights efforts using cells and tissues on microarrays. These 'other' microarrays have the potential to yield highly parallel miniaturized assays that will accelerate drug discovery efforts and facilitate understanding of biological complexity.

Key Words: Protein microarray; proteomics; immobilized; kinase assay; binding assay; microarray; reverse phase microarray; antibody array; cell microarray; tissue microarray; reverse transfection.

1. Introduction

Analysis of biological material is critical for gaining a basic understanding of human health and disease, for clinical diagnosis, and for therapeutic drug development. Cataloging of human DNA and RNA species has accelerated with the recent sequencing of the human genome and ongoing compilations of the human transcriptome (1,2). In contrast to the high-content tools and bioinformatic strategies developed for analysis of the genome, however, it is only very recently that attempts have been made to analyze the proteome in a comprehensive manner. The reasons for this technological gap are manifold, but are highlighted by the complexities of 3D folding and the myriad posttranslational modifications (PTM) required for protein function. The need for a global analysis of the proteome is emphasized by studies showing that mRNA abundance is often not highly predictive of protein levels in the cell and by many observations that significant portions of cellular behavior are determined by transient protein PTM (3,4).

The microarray format has been very successfully utilized to collect comprehensive information about the transcriptome, and the more mature nucleic

From: *Biopharmaceutical Drug Design and Development*
Edited by: S. Wu-Pong and Y. Rojanasakul © Humana Press Inc., Totowa, NJ

acid-based microarray approaches should help inform the development of protein microarrays. Despite the challenges inherent with protein microarrays, significant progress has been made towards providing working solutions that can deliver on the promise of this technology. This chapter will summarize recent efforts to achieve proteomic analyses through the use of the protein microarray platform.

Proteins can be arrayed on a microscale using a number of different formats, including on slides or beads, in wells of microtiter plates, and other configurations. In this chapter we focus on work using microarray slides, typically 1×3 inch microscope slides composed of a silica base material chemically derivatized to provide optimum protein attachment. The 2D planar surface provides a rigid support where protein interactions can take place, with specific information about the printed positions of the immobilized proteins being integral to the analysis. Analytical software tools are used to process the large data sets generated by this high-content format assay. The most significant advantages of using the protein microarray format result from the density of content displayed, the small amount of analytical material required, and its compatibility with both hardware and software tools already developed for DNA microarrays.

The microarray is best seen as an interaction surface for a large range of biologically active material, on which a mixture of solution and solid-phase molecules occurs. Small molecules, nucleic acids, peptides, purified proteins, antibodies, cell lysates, and body fluids have all been applied to the surface of protein microarrays, either adhered to the slide surface for display or applied to the functional surface as analytes in solution. Arrays are typically classified as being either "forward phase" arrays, which use purified immobilized molecules positioned on the slide surface to measure the solution phase, or "reverse phase" arrays, which have complex mixtures of proteins immobilized on the array surface for analysis. Assays are also classified as delivering either protein expression profiles or protein function profiles (enzymatic activities and binding interactions with proteins or small molecules are current areas of emphasis).

Using the microarray surface to study growing cells and analyze tissue sections further extends the range of studies available with microarrays. We will conclude with an update on these exciting applications.

2. Protein Microarray Format

Many methodologies study protein functions in a dilute solution phase, whereas the microarray format involves solution interactions with a surface. Both schemes contrast with the cellular environment where proteins normally function. Microarray formatted assays require proteins to retain structural and functional stability on the surface after printing. The limits of detection for assay signals and minimizing background because of nonspecific surface interactions are also critical factors for technological development using microarrays.

2.1. Protein Immobilization on the Microarray Surface

Immobilization of proteins onto specific locations on the surface of the slide is an essential feature of this technology. The base slide material is usually made of glass, silicon, or plastic. Some surface activation or coating is often used to prepare for specific interactions *(5–10)*. The nature of the solid support is important not only for its ability to attach the desired moiety, but also because its surface properties influence the degree of nonspecific binding. Assay development requires optimization using a variety of surface chemistries.

Proteins bind to the slide via adsorption, covalent bond formation, or binding affinity between capture agents on the surface and specific protein epitopes or epitope tags *(11–14)*. Additional factors such as the accessibility and surface density of available binding sites and the orientation of binding should be considered, although much work remains to be done before the impact of these issues on array performance is clearly understood *(5,15–17)*.

2.2. Protein Microarray Printing

Robotic arrayers originally developed for printing DNA microarrays are typically used to print high-content protein arrays. These commercially available machines must run under environmentally controlled conditions to enable reproducible protein printing. Performance requirements include the ability to operate at 4°C with humidity control and to dispense solutions that have considerable viscosity, conditions that preserve protein stability by preventing denaturation and dehydration during and after printing *(10,18)*.

Contact printing uses micromachined metal pins (either solid or quill type) to dispense samples by making direct contact with the slide surface. Quality control methods must be rigorously employed to ensure reproducible results, as individual pins can clog or deform over time *(19)*. Alternatively, noncontact printing can be performed using piezoelectric devices originally developed for printing ink, and subsequently adapted for printing nucleic acids. These printers can have better precision than contact printers, but provide lower throughput as they currently have fewer dispensers. New printing technologies are also being developed *(20)*.

2.3. Protein Microarray Assay Detection Methods

After applying the experimental analyte to the array surface, detection of the resulting interactions on the microarray surface is critical for assay development. Several widely used methods include the use of radioactivity, fluorescence, or chemiluminescence. Fluorescence is the technique most commonly used, as it can be formatted to allow for multiplex detection, avoids the risks and costs of using radioactivity, and makes use of commercially available hardware and software previously developed for nucleic acid microarrays.

Signal amplification can improve an assay's limits of detection. However, amplification methods that generate soluble products are not compatible with the microarray format, as the signal will diffuse away from the spot of origin. Fluorescent amplification of signals can be achieved using enzymatic tyramide signal amplification *(21,22)*. Rolling circle amplification (RCA) is another amplification technique, which utilizes enzymatic extension of a DNA primer–antibody conjugate followed by fluorescent oligonucleotide hybridization *(23,24)*. Recently RCA has been adapted such that two-color fluorescence can be used, with detection limits demonstrated in the low picogram per milliliter range *(25)*.

There has been a lot of recent effort to develop label-free detection methods. Two of the best known label-free methods are surface plasmon resonance and mass spectroscopy (MS) *(26,27)*. Although there have been recent improvements, significant throughput and sensitivity challenges remain to be overcome before these technologies can be used to monitor high-content arrays *(28,29)*. Many other label-free technologies continue to be developed *(30–37)*.

3. Using Protein Microarrays

Studies utilizing protein microarrays are similar to those using nucleic acid microarrays, in that comprehensive analysis is the ultimate goal. The analysis desired will depend on the type of analyte applied to the microarray. Blood serum and plasma, as well as other body fluids, are generally analyzed for antibody and antigen content. Clinical tissues and tissue cultures are usually interrogated for individual protein expression levels and degree of biological activity. Biochemical fractions, purified proteins, and chemical compounds are used to examine biochemical interactions with the displayed content. For this chapter, the discussion is separated into two general categories: analysis of protein content (often described as protein expression profiling) and analysis of protein function.

3.1. Protein Expression Profiling

Protein expression profiling requires the ability to measure a large sampling of specific protein content. Estimates of human protein diversity range from 100,000 to more than 1,500,000 different protein molecules present in a cell, resulting from domain shuffling and PTM of the 25,000–35,000 human genes *(38)*. Although truly global analysis of the human proteome is currently out of reach, protein microarrays have real potential for delivering protein expression profiles for specific areas of focus. Many tools have already been developed towards this end.

Significant challenges remain regarding the large number of discrete protein species, the huge dynamic range in protein concentration, and the spectrum of PTM observed, particularly when quantification is required. Using a single surface and detector to probe the range of protein abundance necessitates care in matching the probe to analyte. Similar issues have been encountered in the 2D gel/MS approach to protein expression profiling *(39)*. Microarrays techniques have to deal with the same dynamic range of protein concentrations, but they can be formatted to directly enrich low-abundance proteins. When the microarray is printed with purified proteins it is possible to normalize concentrations of proteins with low or high cellular abundance. Alternatively, capture antibodies enrich particular species directly on the array surface. Of course, one should also be aware that positive interactions on the array require validation and may not reflect what occurs in the cell, where proteins are at different concentrations and restricted to particular subcellular locations.

The following sections discuss the range of protein content currently used for protein expression profiling with microarrays. Sections are divided into forward phase arrays containing defined immobilized material or reverse phase arrays on which complex mixtures are immobilized.

3.1.1. Forward Phase Arrays

Forward phase arrays use a wide range of purified materials immobilized on the array surface in defined positions that allow for identification after interacting with the analyte. Examples include the use of immobilized antibodies, antigenic particles, proteins, nucleic acids, and carbohydrates *(14,40–42)*.

Generation of the purified material to be deposited on the microarray is one of the major bottlenecks in protein microarray technology. This "content" consists of recombinant proteins presented on the microarray for interactions or various capture agents used to detect specific proteins. Ideally, high-content arrays will allow for global analysis of the proteome or of relevant subproteomes. Current printing technology allows for 10,000–15,000 discrete features to be arrayed on a microscope slide, sufficient for displaying a significant portion of proteomic content. However, there is no general resource or shortcut for providing the thousands of individual species required, particularly when one desires purity and functionality. Additionally, important proteins classes, such as integral membrane proteins, do not have well-established methods for production and require further development. Some examples from the literature, categorized as either displaying antibodies or purified recombinant proteins (antigens and proteomes), are discussed below.

3.1.1.1. ANTIBODY ARRAYS

Antibody arrays containing surface-immobilized capture antibodies have the potential for highly parallel analysis that is limited only by the availability of antibodies having sufficient affinity and selectivity, as well as the degree of signal amplification performed. Current stocks of useful reagents fall far short of providing comprehensive coverage of the proteome, and development of relevant subsets of antibodies for microarrays is required *(43)*. Cytokines are one example of a biologically relevant subset, with the availability of highly specific antibodies against these secreted proteins making them an attractive target for microarray analysis. Schweitzer and coworkers *(23)* used antibodies against 75 cytokines, combined with RCA-based signal amplification, to analyze human dendritic cell responses. Time-dependent cytokine secretion patterns were measured, and both known and novel cytokine releases were seen. Importantly, extensive evaluation of antibody crossreactivity was required to maximize the number of truly specific antibodies that could be used simultaneously on the microarray. Recent work has similarly qualified up to 170 serum or plasma antibody features for RCA-based detection, after starting with 1200 individual reagents *(44,45)*. Other studies have described antibody arrays containing as many as 224 features, although without careful examination for antibody selectivity *(46)*.

Defining antibody specificity is crucial to analyzing highly multiplexed assays. Antibodies to several thousand proteins are currently available; however, it is not clear how many of these can be properly used on microarrays. Many studies have shown unexpected antibody crossreactivity *(47–49)*. Detailed characterization of several hundred to thousands of antibodies will require new analytical techniques *(50)*. The microarray format itself is ideal for analyzing the specificity of single antibodies in solution against a displayed protein slide *(51,52)*. Recently a study utilized a yeast proteome microarray containing >4000 full-length proteins to test the specificity of 11 commercially available antibodies and demonstrated that the majority displayed crossreactivity *(52)*. Others have suggested that only 5% of commercially available antibodies directed against intracellular targets are suitable because of a lack of specificity or affinity *(53)*. Certainly, a minority of antibodies have been fully tested for their usefulness on microarrays.

3.1.1.1.1. Direct vs Indirect Detection Methods. Direct methods for protein expression profiling require the protein analytes to be quantitatively fluorescent dye labeled or hapten tagged for dye labeling in a separate step. Two different sets of analytes that are to be compared are separately labeled with different tags and then mixed and applied onto the antibody array. Pairwise ratiometric comparisons of the bound species are performed that are conceptually very similar to DNA microarray methodologies or mass spectrometry methods such as DIGE or ICAT *(54–57)*. One-color methods have also been used, with a set of unlabeled experimental analytes competing against the labeled control analytes *(58)*; however, this approach may not be as sensitive as two-color methods *(58,59)*. Another means for optimizing results is to normalize the amount of antibody bound to each spot using an internal control secondary antibody *(60)*.

Indirect methods are similar to the traditional sandwich-assay format and use one antibody to capture unlabeled molecules, followed by a second antibody that provides labeling for detection. The second antibody is usually fluorescently labeled or provides RCA-based amplification and must bind to a different epitope than the capture antibody. Finding matching pairs of antibodies that meet the required performance criteria is a significant limiting issue in the field. Alternatively, the second antibody can display more generic immunoreactivity (for example, towards all phosphorylated tyrosine residues) allowing multiple captured species to be analyzed simultaneously *(61)*. One interesting alternative that might eliminate the complications arising from matching sandwich antibodies has been suggested by work using photoaptamers as the primary capture reagent and antibodies as the detection reagents *(62)*. If the protein characteristics which interact with an aptamer are significantly different than those recognized by an antibody, antibody-accessible epitopes will remain available in an immobilized antigen–aptamer complex.

Each of these formats results in qualitative results, whereas truly quantitative analysis requires calibration curves using purified antigens and would be operationally difficult for high-content assays. Limits of detection are typically in the low nanogram per milliliter range when using detection without amplification and can be in the pictogram per milliliter range when using amplification with RCA *(23,25)*.

3.1.1.1.2. Posttranslational Modification Antibodies. Cell biologists have been able to identify numerous functional changes correlating with enzymatic activity and cellular phenotypes using antibodies directed against various PTM of relevant proteins. Only a limited number of these antibodies have been used for microarray purposes, mostly with anti-phospho-tyrosine antibodies used as secondary reagents; however, additional generic and context-specific motif PTM antibodies should also be useful, with immunoprecipitating PTM antibodies particularly important capture reagents to develop for array content *(61,63)*. An alternative method for fluorescent detection is possible using a general phospho-specific dye reagent that has been developed *(64)*. Efforts using phospho-specific reagents are promising; however, one must be able to evaluate small fold-differences reliably, as the maximum fold-change in phosphorylation is often in the two- to fourfold range.

3.1.1.2. ANTIGEN ARRAYS

Analysis of serum antibodies provides an accessible window on many disease processes. Immune responses to pathogens, allergens, tumors, and to self can be analyzed by displaying these antigens on the array surface for serum profiling. One can screen clinical samples for diagnosis and tracking of disease and for discovery of immunogenic epitopes correlating with disease progression or therapeutic intervention.

Construction of these arrays requires obtaining purified antigens for immobilization. Manufacturing even small numbers of recombinant proteins can present challenges, but is generally straightforward. The first requirement is obtaining the appropriate source of DNA-encoding expressed open reading frames (ORFs). One can then produce recombinant proteins in bacteria, insect cells, or mammalian cells, or using in vitro transcription and translation techniques. Purification is usually provided through the use of epitope tags that are appended to the proteins. Obtaining larger numbers of purified antigens for high-content arrays is a significant challenge. In contrast to the production requirements necessary for functional studies of proteins, however, considerations of proper structure and function may not be as critical for arrays used strictly for analysis of serum immunoreactivity profiles.

3.1.1.2.1. Pathogen and Allergen Arrays. Arrays containing pathogenic proteins have been used for tracking of immune responses during exposure or vaccination. Recent work includes monitoring of both simian-human immunodeficiency virus and HIV-specific immune responses, as well as responses to vaccinia virus and viral meningitides *(65–68)*.

Allergen arrays can be used for diagnosis or for identifying novel immunogenic epitopes. Recently, 190 common allergens were purified and arrayed to profile patient sera for allergen-specific IgE *(69)*. The array assay showed good agreement with clinical allergen sensitivity assays. In another study, IgE profiling of subjects allergic to Brazil nuts identified a conformational epitope responsible for eliciting the allergic response *(70)*.

3.1.1.2.2. Autoantigen Arrays. Autoantigen arrays have the potential to diagnose and study autoimmune diseases, which result from aberrant activation of one's own immune system and the production of autoantibodies that attack self-molecules *(71,72)*. If specific autoantigens are already known, they can be printed on focused antigen arrays. Alternatively, autoantigens can be discovered by looking for reactivity towards broader proteomic microarray displays. Serum profiling using microarray technology should allow patients to be diagnosed and differentiated into clinically distinct populations within a disease group.

Robinson and coworkers *(40)* printed 196 distinct autoantigens previously identified from autoimmune rheumatic diseases and applied serum from various patients. No autoantibodies were present in sera from a healthy individual, whereas specific patterns of immunoreactivity were seen using sera from people having different diseases. A more focused group of antigens was used by this group to study a mouse model for multiple sclerosis, with results showing autoreactive B-cell responses that were predictive of disease activity *(73)*. More recently, Robinson's group used an array containing 225 peptides and protein antigens to analyze patients with rheumatoid arthritis, providing diagnostic information that allowed for patient stratification *(74)*. Quintana et al. *(75)* generated an array containing 266 different antigens, a subset of which was able to discriminate sensitive and resistant populations in a mouse model of Type 1 diabetes. Cahill and coworkers *(76,77)* utilized a human fetal brain cDNA library cloned into an *Escherichia coli* expression vector to produce high-content arrays using purified human fusion proteins. These workers used these arrays to discover eight autoantigens in serum from alopecia areata patients. They have also produced arrays from a mouse Th1 cDNA expression library, which was used to discover autoantibodies for a mouse model of systemic lupus erythematosus *(77,78)*.

3.1.1.2.3. Tumor Autoantigen Arrays. Cancer cells often overexpress normal or mutant proteins which can be recognized by the immune system. Analyzing a patient's blood or serum for immunoreactivity to tumor antigens may allow identification of cancer biomarkers and represents a promising emerging strategy for noninvasive diagnosis. As single biomarker correlations have not generally been predictive, efforts have focused on discovering diagnostic patterns from multiple potential biomarker sets *(79)*. Recently, serological analysis of recombinant cDNA libraries (SEREX) has used high-titer IgGs to identify antigens expressed by autologous cancers, with over 1000 SEREX-defined antigens currently defined *(80)*. Protein microarrays should also allow for potential cancer biomarker discovery and correlations between sera and disease, whereas providing greater ease of use and better antigen presentation at lower cost. Proteins or peptide sequences found to be tumor expressed can be immobilized on microarrays and used to examine serum for cancer-specific immune responses *(81,82)*.

Alternatively, "naïve" high-content protein microarrays can be used to discover cancer biomarkers. Using arrays containing 5000 human proteins, collaborative efforts are underway to utilize protein microarrays to profile serum from a variety of cancer patients, with the goal of identifying specific cancer serum profile biomarkers *(see* Fig. 1) *(72)*.

3.1.2. Reverse Phase Arrays

Reverse phase arrays contain complex mixtures of uncharacterized biological material immobilized in defined spots. By arraying thousands of mixtures derived from clinical material or from whole or fractionated cell lysates, one can assay for an individual component across all of the samples simultaneously under identical experimental conditions.

The complex protein mixtures are often spotted onto microarrays in a dilution series as an effective measure to deal with the large dynamic range of protein

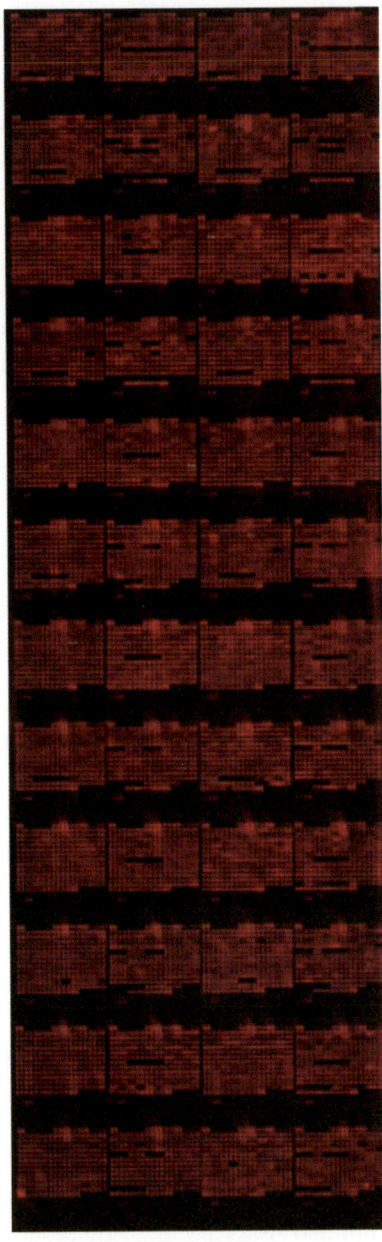

Fig. 1. Human ProtoArray™ v3: Protein Microarray Containing >5000 Purified Human Proteins. A microarray slide displaying 48 sub-arrays of epitope-tagged, purified human proteins (printed as duplicates) was incubated with an anti-GST antibody, detected with a fluorescently labeled (AlexaFluor647) anti-rabbit antibody, and scanned. Copyright © Invitrogen Corporation.

concentration and then probed with antibodies. This is essentially equivalent to an analytical dot blot, but having much greater throughput and using much less starting material. One group estimated the amount of neat protein required to be deposited on each spot as being equivalent to the amount derived from 20 cells, whereas another estimated the functional sensitivity for their approach to be about 5000 molecules per spot *(83,84)*. This technique is therefore well suited for studying rare cell populations, including those which are difficult to purify or grow, as well as clinical patient biopsies, and using methods such as laser capture microdissection *(21)*.

A major disadvantage for using reverse phase arrays is that application of a single analytical antibody detects only a single protein in each sample. Two-color fluorescent dye methods using a control antibody can facilitate quantitative results (e.g., comparing phosphorylated and nonphosphorylated signals) or can be used to double the assay throughput *(85)*. Novel technologies, such as quantum dots, may allow for further multiplexing *(86)*. Alternatively, complex serum samples containing unknown autoantibodies can be applied, with analysis looking for patterns of immmunoreactivity associated with the disease. Molecular identification of any correlating pattern would require additional analysis.

Examples of analyses using reverse phase arrays include a study using the standard NCI-60 cancer cell lines, with printed lysates probed with antibodies to profile the relative protein abundance of 52 proteins *(87)*. Another study analyzed follicular lymphoma samples (collected using laser capture microdissection) with 21 different antibodies, identifying potential prognostic markers *(88)*. Studies examining the changes in phosphorylation state of a number of signaling molecules have been performed using Jurkat cells and a variety of primary cells and tumor samples *(83,89)*. Hanash and coworkers *(90–92)* have printed hundreds to thousands of chromatography fractions for analysis from microdissected patient tumor biopsies and characterized autoantibody patterns for colon, prostate, and lung cancer patients. Janzi and coworkers *(93)* spotted over 2000 serum samples for analysis of serum IgA levels, with the results comparing well with those from other techniques.

3.2. Protein Function Profiling

Traditional biochemical methods are appropriate for focused assays, but cannot be applied to all of the proteins in an organism. The microarray slide format allows for highly parallel miniaturized assays that are limited only by the generation of content and by development of particular functional assays. In contrast to antigen arrays (*see* Section 3.1.1.2. above), the availability of full-length, correctly folded proteins is critical for measuring biochemical activity. Current assays either measure binding or enzymatic activity.

3.2.1. Binding Activity Profiling

3.2.1.1. PROTEIN–PROTEIN BINDING INTERACTION PROFILING

Protein microarrays are well configured for comparing binding interactions with a large number of immobilized molecules. Simply applying a protein or small molecule in solution to the arrayed components allows an examination of the binding profile across the entire slide content simultaneously and under uniform conditions. Microarrays are tools that compare to either yeast two-hybrid or phage display methods for studying binary binding interactions, with the distinct advantages of high-content display and the analytical tools previously developed for DNA microarray technology.

3.2.1.2.1. Domains. Protein domains mediate many cellular protein–protein interactions. These interactions are often regulated by PTM such as phosphorylation, methylation, and acetylation. Because domains are small and can fold independently from the rest of the protein, they are ideally suited for display on microarrays.

Bedford and coworkers *(94)* generated 212 individual protein domains (from more than 14 types of domains) as GST fusion molecules and arrayed them on nitrocellulose-coated slides. Biotinylated peptides known to bind to specific domains were synthesized and applied to the microarray for analysis. Unique binding profiles were found for each peptide corresponding to the normal domain interaction, with some methylation-specific interactions observed. In addition, these arrays were utilized for binding proteins from cell lysates, with binding profiles being similar to that shown with the peptides.

Keating and coworkers *(95)* produced purified peptides corresponding to 49 human bZIP domains as well as 10 yeast domains. A comprehensive matrix analysis of all pairwise interactions showed 14% having interactions (5.8% strong), with 90% of the interactions independent of which of the pair was immobilized. The demonstration of symmetrical binding is strong evidence of reliable interactions and is a useful means for validation. Known family-specific interactions were seen with a low false-negative rate, and novel interactions were identified in circadian clock and unfolded protein-response pathways.

Recently, MacBeath and coworkers *(96)* studied interactions of 61 phosphopeptides (from the EGFR receptor tyrosine kinase family) on a domain array containing 160 recombinant domains. Over 77,000 independent measurements yielded an interaction network that both confirmed 43 known interactions and suggested 116 novel interactions relevant to EGFR family signaling.

3.2.1.2.2. Proteins and proteomes. MacBeath and Schreiber were the first to show several protein–protein interactions on microarrays *(18)*. More recently, Blackburn and coworkers *(97)* showed the expected binding interactions between 50 arrayed isoforms of the transcription factor p53 and a fluorescently labeled known interactor (MDM2).

Zhu and Snyder *(14)* constructed the first proteome-scale microarrays, displaying approximately 70% of the *Saccharomyces cerevisiae* proteome (>4000 full-length purified proteins) oriented via His-tag binding onto nickel-coated slides. These arrays were used to perform protein–protein interaction profiling using biotinylated calmodulin and fluorescent streptavidin for detection. Expected interacting proteins were seen, and a number of novel interactors were identified and validated, allowing sequence alignments to define a putative calmodulin-binding motif. Binding interactions on yeast proteome arrays were also shown by Michaud and coworkers *(98,99)*.

Human proteome arrays are currently being developed and sold by Invitrogen Corporation. Satoh and coworkers *(100)* used a human proteome array containing over 1700 full-length proteins to identify 20 novel interactions with a 14-3-3 protein. Invitrogen currently has an array available which displays approximately 5000 full-length human proteins (Human ProtoArray™ v3, *see* Fig. 1). Examples of protein–protein interactions using these arrays are shown in Fig. 2.

3.2.1.3. PROTEIN–SMALL MOLECULE BINDING

Measuring the binding of small molecules to proteins is important both for understanding small molecules found in the cellular environment and for characterizing compounds used for drug development. Microarrays can display immobilized small molecule libraries, allowing them to be screened for interactions with particular target proteins. Alternatively, arrayed proteins can be analyzed for their ability to bind a solution-phase small molecule whose mechanism of action is unknown. Probing protein microarrays with such a compound provides an opportunity to discover the drug's molecular target. This approach also allows discovery of potential off-target binding.

Schreiber and coworkers *(101,102)* were the first to show microarray-based interactions between small molecule compounds and proteins, and this group went on to modify the slide chemistry to allow Diversity-Oriented Synthetic (DOS) libraries to be attached to the slide surface. This approach was used to bind a DOS-derived library containing over 3700 compounds, identifying molecules that directly bound to the fluorescent-tagged yeast transcriptional repressor Ure2 *(103)*. Subsequently Schreiber and coworkers *(104)* used a DOS library of 12,000 compounds to identify a single compound that specifically bound to the yeast transcription factor Hap3.

Labeled small molecules have also been used to bind to immobilized protein arrays. Zhu et al. *(14)* probed yeast proteome arrays with 5 different biotinylated phosphoinositide-containing liposomes, detecting 150 binding interactions. Importantly, a large number of interactions with membrane proteins were observed, suggesting the functionality of this class of proteins on microarrays. Similarly, Lahiri and coworkers *(105)* were able to immobilize G-protein coupled receptors in a functional form that could bind to fluorescently labeled peptide ligands.

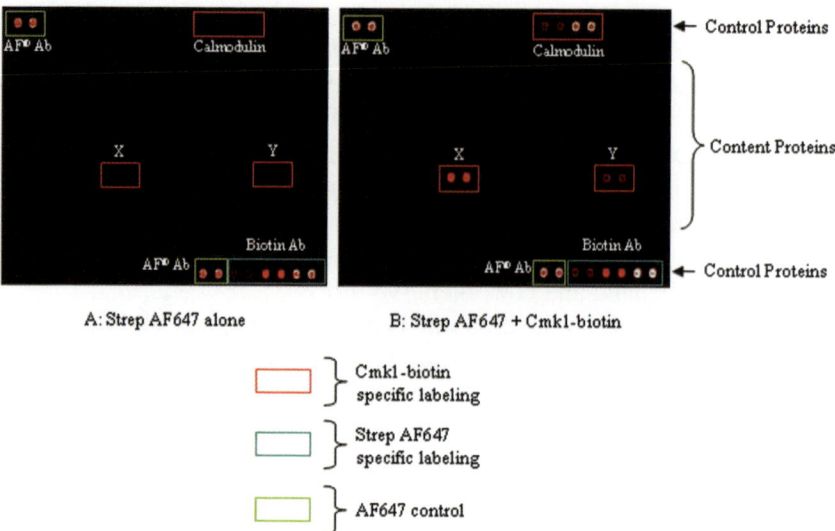

Fig. 2. One Subarray of Human Protoarray™ v3: Protein-Protein Interactions Using an Epitope-tagged Protein. Each image represents a single subarray of a Human ProtoArray™ v3 slide after probing with the designated reagents and scanning. The top and bottom rows contain control proteins printed in every sub-array, while the middle rows display the human protein content: A. AlexaFluor647-labeled Streptavidin detection reagent (Strep AF647) alone. B. Epitope-tagged calmodulin kinase (Cmk1-biotin) probing of the array, with Strep AF647 detection. Green boxes highlight printed AlexaFluor647-labeled antibody (AF®Ab). Blue boxes highlight a biotinylated antibody gradient (Biotin Ab) which binds only Strep AF647. Red boxes highlight a control protein (Calmodulin), and two content proteins (X and Y) which are detected only in the presence of Cmk1-biotin. Copyright © Invitrogen Corporation.

Several mechanism of action studies have been described. Schreiber and coworkers *(106)* used small molecule inhibitors of rapamycin to probe a yeast proteome microarray. Candidate binding proteins were identified that showed roles in TOR (target of rapamycin) signaling in follow-up studies. Michaud and coworkers *(107)* also used the yeast proteome microarray in studies to determine the mechanism of action for a small molecule identified as being neuroprotective in a cell-based screen. A biotinylated derivative of the small molecule was found to bind specifically to Sir2, a histone deacetylase known to be involved in regulating apoptosis and other cellular processes, allowing hypothesis-driven experiments to be pursued.

3.2.1.4. PROTEIN–NUCLEIC ACID BINDING

Protein microarrays can be used for binding studies with either DNA or RNA species. A focused microarray was used by Blackburn and coworkers *(97)* to show isoform-specific binding profiles for mutants of the cancer-associated transcription factor p53, with some degree of correlation seen with other data sets. Yeast proteome arrays have been used to show that approximately 5% of the proteins encoded by the yeast genome bind DNA *(14,108)*. Characterization of DNA–protein binding specificities has already provided valuable data for assessing transcriptional regulation of DNA by protein, and the microarray approach arrays is expected to be applied to gain understanding of recombination, splicing, replication, and structural regulation of nucleic acids.

3.2.2. Enzymatic Activity Profiling

Measuring enzymatic activity and inhibition is a focus for biologists and for drug development efforts. In this section, we will highlight work profiling enzyme activity using microarrays. Activity-based microarray assays can be categorized based on the type of material immobilized, i.e., immobilized substrate or immobilized enzyme slides. Additionally, there is a separate category using activity-based probes, which can be configured to work either as immobilized capture reagents or soluble probes.

3.2.2.1. SUBSTRATE ARRAYS

MacBeath and Schreiber *(18)* first demonstrated enzymatic activity on a microarray surface using solution-phase kinases to phosphorylate either immobilized peptide or protein substrates with radiolabeled ATP. The peptide substrate was seen to be specifically phosphorylated by PKA, and protein substrates were specifically phosphorylated by either CKII or ERK2. Many researchers have since developed array-based assays that utilize either peptide or protein substrates and have focused on kinase or hydrolase activity.

3.2.2.1.1. Peptide Substrate Arrays. Although appropriate for some applications, direct spotting of small peptides onto a solid support does not always allow for proper peptide–protein interactions with enzymes or detection antibodies *(109)*. Protocols that include blocking with BSA can also reduce peptide accessibility *(101)*. Therefore, fusion of peptide sequences to chemical or protein linkers is usually used to elevate the short peptide sequence away from the array surface, permitting better accessibility *(110)*. Reimer and coworkers *(111)* developed high-content peptide arrays for kinases assays using an N-terminal linker to immobilize a 1433 peptide molecule library. Radiometric kinase assays were performed using Abl kinase, and analysis yielded a consensus phosphorylation sequence prediction that was better than the existing computer algorithm. Subsequent work by this group, utilizing 2923 peptides and anti-phospho-tyrosine antibodies for detection, gave similar results to those found using the radioactive format *(112)*. Additional work profiled more than 13,000 peptides for their ability to be phosphorylated by CK2, with previously known phosphorylation sites identified.

Peptide arrays for measuring various hydrolases have also been developed. A common approach utilizes immobilized coumarin derivatives that have appended substrate moieties *(113,114)*. The conjugated substrate sequence quenches the normal fluorescence of coumarin until the substrate is hydrolyzed. This approach has been used to assay proteases, phosphatases, and epoxide hydrolases.

3.2.2.1.2. Protein Substrate Arrays. A focused microarray containing 50 isoforms of the transcription factor p53 has been used by Blackburn and coworkers *(97)* to show site-specific phosphorylation by soluble CKII using phospho-specific antibody detection.

Using high-content arrays allows a broader assessment of a kinase's substrate preferences, providing information for understanding signaling networks or discovering substrates to be used for drug development. A yeast proteome array, containing >4000 full-length purified proteins (representing approximately 70% of the entire proteome of the yeast *S. cerevisiae*), is commercially available from Invitrogen and has been used as a substrate array capable of being phosphorylated by exogenously applied kinases. Initial work demonstrated the use of these arrays to identify substrate phosphorylation by solution-phase PKA and showed dose-dependent inhibition by a PKA inhibitor *(98)*. More recently, Snyder and coworkers *(115)* have used these arrays to perform radiometric kinase substrate identification assays using 87 individual kinases, representing more than 70% of the yeast kinome. Nearly 4200 phosphorylations of 1325 yeast proteins were measured, with 73% of the identified substrates being phosphorylated by three or fewer kinases, suggesting that there is a high degree of selectivity for particular substrates. These data were used along with protein interaction and transcription factor-binding data sets to identify common regulatory modules that were not evident from the separate databases. This study represents the closest approach to a truly global analysis of potential phosphorylation reactions across a complete proteome. An example of human protein substrate phosphorylation reactions on microarrays can be seen in Fig. 3.

Another high-content array approach has been developed by Cahill and coworkers *(77)*, using proteins produced and purified under denaturing conditions in *E. coli*. Kersten and coworkers *(116)* utilized this method to generate nitrocellulose-coated microarrays containing 1690 *Arabidopsis* proteins as potential substrates for three kinases. A total of 48 MPK3 substrates were identified, 42 of which were validated in solution-phase reactions; 39 MPK6 substrates were identified, 26 of which were also MPK3 substrates. In contrast, PKA had three substrates which overlapped with MPK3.

Although much of the recent focus has been directly towards kinase and hydrolase activities, one can foresee interrogation of many other enzyme classes. In particular, biologically important ligase and transferase enzymes which have the ability to transfer labeled groups should work well on a microarray platform.

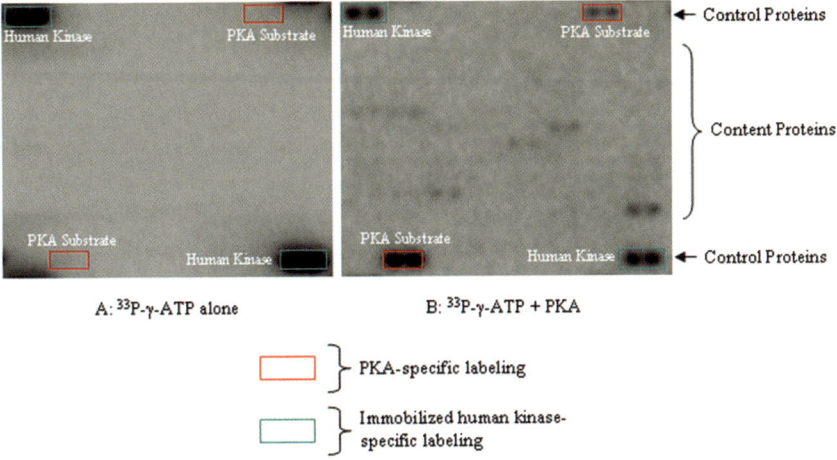

A: ³³P-γ-ATP alone   B: ³³P-γ-ATP + PKA

Fig. 3. One Subarray of Human Protoarray™ v3: Immobilized and Solution-Phase Kinase Activity on Microarrays. Each image represents a single subarray of a Human ProtoArray™ v3 slide after reaction with the designated reagents and scanning. The top and bottom rows contain control proteins printed in every sub-array, while the middle rows display the human protein content: A. Reaction using ^{33}P-γ-ATP only. B. Reaction using solution-phase PKA and ^{33}P-γ-ATP. Blue boxes highlight an immobilized human kinase (used as an array fiduciary control) showing kinase autophosphorylation activity. Red boxes highlight a PKA substrate used as a control, showing radiolabeling only with PKA in the reaction. Additional PKA-specific substrates are seen in the rows displaying the human protein content, but are not highlighted. Copyright © Invitrogen Corporation.

Methyltransferase activity has been shown on membranes, and acetylases, deacetylases, ubiquitin ligases, and other relevant enzymes should be amenable to substrate array analysis *(117)*.

3.2.2.2. ENZYME ARRAYS

Immobilizing enzymes on the array surface provides a format for interrogation of their activity, including profiling the effects of inhibitory small molecule compounds used for drug development. This approach holds great promise as a low-cost method for analyzing many enzymes simultaneously. However, one must consider that unless an alternative method is developed, a single buffer will be used for all of the enzymes on the array. One potential solution to this problem would be to use themed arrays, with enzyme classes grouped on separate arrays or isolated regions of one slide, allowing for optimized buffers to be used *(118)*.

Kinases clearly can function in an enzymatic manner when immobilized. Snyder and coworkers *(119)* used 119 yeast kinases, immobilized via a crosslinker to a silicone elastomer, in radiometric kinase reactions that showed autophospho-rylation for many of the kinases. Predki and coworkers *(98)* have demonstrated that immobilized kinases can retain enzymatic activity on a modified glass microarray surface, also by showing autophosphorylation. Figure 3 shows the autophosphorylation activity of a human kinase immobilized on the Human Protoarray™ v3.

Other enzyme classes have been shown to work while immobilized on microarrays. Sulfotransferase activity has been demonstrated on microar-rays, with the immobilized enzyme acting on the substrate in the bulk solution phase *(120)*. Stephanopoulos and coworkers *(121)* have shown activity for luciferase, nucleoside diphosphate kinase, and five enzymes in the trehalose synthesis pathway when immobilized on the microarray surface. Cytochrome P450 enzymes have also been shown to function on the array surface *(122)*.

An indirect method to show enzymes are functional on microarrays uses activity-based probes (*see* Section 3.2.2.3. below).

3.2.2.3. ACTIVITY-BASED PROTEIN PROFILING

Another category of interactions utilizes chemical agents that specifically bind the active site of enzymes. These chemicals can be used for protein profiling of active enzymes in a lysate or for biochemical analysis of enzyme–inhibitor interactions.

Cravatt and coworkers *(123,124)* developed an enzyme active-site-based approach, using a probe that broadly targets serine and metalloproteases, and used antibody arrays to capture specific enzymes for analysis. Addition of protease inhibitors strongly reduced signals for the respective targets, suggesting that the amount of bound enzyme reflected enzyme activity. Because there are more specific antibodies than specific mechanism-based inhibitors, this approach should provide more potential for high-content analysis than directly immo-bilizing the inhibitor on the array *(125)*. Initial studies with this approach look promising, but so far have only used a handful of probes and antibodies. Use of a labeled activity-based probe essentially replaces the labeling of cell lysates and secondary detection antibodies used during traditional antibody microarray protein profiling. Work developing probes for a wider range of enzymes should enable this approach to target a more significant portion of the druggable proteome *(126,127)*.

Enzyme activity assays of preselected molecules can also be performed with these reagents. Yao and coworkers *(128)* directly immobilized phosphatases, cysteine proteases, and serine hydrolases on the array surface and assayed for their biochemical activity using activity-based fluorescent probes. The same type of approach has also been used by Miyake and coworkers *(118)*, with slide-immobilized members of the cysteine protease family being interrogated by fluorescent activity-based probes. This platform was used to characterize the inhibitory profile of compounds from a chemical library by first preincubating the array with one of the compounds followed by interrogation with the fluorescent activity-based probe. The inhibitory activity is measured by the compound's ability to compete with the fluorescent activity-based probe for binding to the enzyme's active site.

3.2.2.4. Novel Array Assay Formats to Measure Enzyme Activity

Two novel methods have been recently described that measure enzymatic activity on microarrays. Both require the enzymes to have fluorogenic substrates, and both methods are also performed immediately after the enzyme is applied by specialized equipment onto the array. Despite these caveats, these approaches clearly work to assay enzymatic activity on microarrays.

The multiple spotting technique (MIST) used by Angenendt and coworkers utilizes two spotting steps, with fluorogenic substrates printed in the first round, followed by a second round of printing to place a spot of enzyme in reaction buffer right on top of the substrate. Enzymatic activity generates a fluorescent product for detection. Surface activity of horseradish peroxidase (HRP), alkaline phosphatase, beta-galactosidase, and cathepsin D was demonstrated *(129)*. Using this type of approach, several thousand completely different interactions (or smaller subsets) can conceivably be performed at once, a degree of multiplexing that is unmatched by any other format. However, as the reaction occurs immediately after the second printing, extended print runs would be problematic.

A different method of delivering an enzyme solution onto the arrayed substrates uses an aerosolized spray. The spray mist covers the spots without depositing a volume that would cause neighboring spots to merge, thus allowing the discrete spatial positioning of peptide substrates to be retained without chemical linking to the array surface *(130)*. Work has continued using aerosolized enzymes that have fluorogenic substrates including cysteine proteases, metalloproteases, and kinases *(131)*. Mixing potential enzyme inhibitors along with the printed substrate, followed by spraying the surface with an enzyme, allowed 6600 small molecules to be examined for enzyme inhibition on a single microarray. This approach has great promise, but requires specialized equipment and will work only for enzymes that can tolerate 10% glycerol in the reaction.

4. Cell Arrays

The microarray can also be exploited for assays that use living cells grown on the surface. Prepositioned small molecules and adhesion molecules can interact with the cells and effect their growth and behavior. Additionally, nucleic acid vectors localized on the array surface cause cells to undergo "reverse transfection", allowing both overexpression and RNAi knockdown studies to be performed on the microarray.

4.1. Peptide and Small Molecule Effects on Cells

Bhatia and coworkers *(132)* have devised methods for positioning living cells in hydrogel-encapsulated arrays. This lab has recently gone on to develop a method to deposit combinations of extracellular matrix (ECM) molecules onto the hydrogel-coated microarray surface, allowing for studies of the effects of ECM on cell differentiation *(133)*. Detection techniques included immunostaining of primary rat hepatocytes and the use of a beta-galactosidase reporter construct in mouse embryonic stem cells. The cells retained viability for as long as 7 days. This approach should be amenable for studying the effects of the deposited material (ECM, polysaccharides, proteoglycans, tethered growth factors, etc.) on cells and also to allow adhesion of particular cell types to specific locations on the array.

Clark and coworkers *(122)* have developed a metabolizing enzyme toxicology assay chip which can be used to assess the biological activity of cytochrome P450 enzyme-generated metabolites on a variety of cells. This system utilizes sol–gel encapsulated P450 isoforms arrayed on the slide surface. Prodrugs are deposited onto the spot, leading to P450-generated metabolites being formed. Finally a cell layer is stamped onto the surface, exposing them to the synthesized metabolites. Analysis of cell viability or growth yielded accurate assessments of cytotoxicity.

Stockwell and coworkers *(134)* have created microarrays printed with small molecules impregnated in a biodegradable polymer. Monitoring the growth of a monolayer of cells grown on top of the array for up to 7 days allowed for assessment of the effects of the slow release of compound. By growing cells already undergoing RNAi-mediated gene knockdown of seven target genes on top of microarrays containing 70 compounds, a synthetic lethality screen successfully identified the effects of combining the DNA antimetabolite macbecinII and knockdown of TSC2, an mTOR interactor.

4.1.2. Reverse Transfection Arrays

Another exciting application of the cell microarray format is to use nucleic acid vectors, localized on the array surface, to affect gene expression in cells grown on the microarray. First described as "reverse transfection" by Ziauddin and Sabatini *(135)* in 2001, these researchers printed various plasmid DNA spots on the array, followed by treatment with a lipid transfection reagent and culturing of cells on the entire slide surface. DNA was taken up by the cells immediately above the arrayed DNA spot, and the resulting effect was seen after several cell divisions.

Phenotypic effects on the cell layer, such as apoptosis or growth control can be assayed, as well as analysis of intracellular events such as alterations in Ca flux or subcellular relocalization of proteins *(135–137)*. Uhler and coworkers *(138)* used cotransfection with a reporter plasmid, enabled by cospotted protein factors that facilitated transfection, to study transcriptional regulation by the protein kinase PKA. Microarrays containing cells can be fixed and prepared for various staining techniques. A low-resolution analysis can be performed using standard microarray scanners; however, motorized microscopes may be needed for automated analysis of features requiring higher resolution.

Interactions of an overexpressed protein with labeled small molecules applied to the array can also be observed, thus allowing this approach to be used in a similar fashion as a protein microarray, with the reverse transfected cell providing localized production and display of the overexpressed protein. This approach for high-throughput protein expression and analysis may be of partic-ular utility for membrane proteins, which are difficult to produce in a purified form. Delehanty et al. *(139)* used this approach to display single-chain antibodies on the cell surface, showing apparent affinities within 10-fold relative to their soluble forms.

In addition, this approach can also be used to determine the effects of gene knockdown, such as through the use of printed RNAi vectors *(140–142)*. Genome-wide annotation by loss of function studies is now possible using available libraries of RNAi vectors *(143)*. As RNAi cell microarrays can hold several thousand spots of RNAi reagents, whole-genome screens can be carried out on a small number of slides.

Critical factors required for full development of cell microarrays include a wider range of biological readouts, high-throughput image analysis, and data storage and analytical tools. Although this technology is still being developed, the conservation of reagents and rare cell lines, as well as the uniform scoring and high-throughput nature of the cell microarray, is likely to motivate contin-ued development of this technology.

5. Tissue Microarrays

Tissue microarrays (TMA) are formally a form of reverse phase arrays, dis-playing uncharacterized mixtures of biological material in defined positions on the slide surface for analytical probing. First performed in 1998, TMA techno-logy currently allows for simultaneous analysis of up to 1000 tissue samples per slide *(144)*. Tissue cylinders of approximately 0.5 mm are taken from primary tissue blocks, and microtome slices are fixed onto the slide surface. Use of these small specimens maximizes the use of precious biopsy samples. Early concerns that this small sample size would not be representative of the larger tissue have largely been dispelled after numerous studies with cancer tumors showed good correlations with standard techniques *(145–148)*.

A variety of analytical techniques can currently be performed on the slide, including immunohistochemistry, fluorescence in situ hybridization, or RNA in situ hybridization. Because all tissues are analyzed simultaneously with the same batch of reagents, TMA studies provide an unprecedented degree of standardization, speed, and cost efficiency. Specialized equipment for automated processing is being developed and is commercially available; however, analysis of tissue samples is not as straightforward as with protein microarrays. This technology will allow precious tissue samples to be used for many more assays than standard methods and presents unique opportunities for sample archiving.

6. Summary

The field of protein microarrays is still emerging, but is poised to achieve its potential as a tool for dissecting global proteomic diversity. In just a few years, it has become clear that many of the questions regarding use of proteins on the microarray platform have been addressed, and that arraying proteins for detection on microscope slides is a feasible solution for performing proteomic studies. This highly parallel assay system uses small sample sizes and has been optimized to work on pre-existing analytical equipment currently used for high-throughput DNA and RNA analysis.

Production of quality content is critical for developing protein microarrays to meet their potential. With the ORFeome analysis ongoing, the compilation of sequence information encoding the human proteome will soon be complete, providing researchers with the full cDNA library of expressed proteins *(149)*. However, the roles of these genes will only be understood by carefully expressing and analyzing the PTM, binding interactions, and enzymatic functions of their encoded protein products. Defining and producing those proteins essential for human health will make it possible to routinely use the protein microarray format for basic research and drug development. Equal to the need for production of these proteins is the need for specific and sensitive detection reagents. Antibody, aptamer, and other protein-specific detection technologies are required to match the growth in purified protein content.

Appropriate use of protein microarrays requires a clear understanding of the goals of the analysis. Certainly, microarrays can provide unique information for discovery work in the biological sciences, although the clinical use of this format has not been developed. Using capture arrays to analyze a patient's immune responses to pathogens, allergens, tumors, and aberrant reactivity to self-molecules is likely to provide many new biomarkers for disease states. Proteome arrays can be used to assay for global binding interactions and substrate identification. Developments in enzyme and small molecule assays using microarrays should provide new opportunities for reducing costs in the drug development process.

In summary, there are a wide variety of applications possible when placing biological material on a standard microscope slide. In the long view, it is reasonable to see the nucleic acid microarray field as a subset of the biological microarray field, leading the way in development of the format's detection and analytical tools. With the addition of cell microarrays and TMA, the "other" microarrays are ready to accelerate drug development and human understanding.

Acknowledgments

Barry Schweitzer, PhD, Director of Invitrogen's Protein Microarray Center, is acknowledged for his editing of the manuscript. Greg Michaud, PhD, Senior Scientist at Invitrogen's Protein Microarray Center, is acknowledged for his contributions to the figures.

References

1. Lander, E. S., et al. (2001) Initial sequencing and analysis of the human genome. *Nature* **409**(6822), 860–921.
2. Venter, J. C., et al. (2001) The sequence of the human genome. *Science* **291** (5507), 1304–1351.
3. Gygi, S. P., et al. (1999) Correlation between protein and mRNA abundance in yeast. *Mol. Cell. Biol.* **19**(3), 1720–1730.
4. Mann, M. and Jensen, O. N. (2003) Proteomic analysis of post-translational modifications. *Nat. Biotechnol.* **21**(3), 255–261.
5. Stillman, B. A. and Tonkinson, J. L. (2000) FAST slides: a novel surface for microarrays. *Biotechniques* **29**(3), 630–635.
6. Rubina, A. Y., et al. (2003) Hydrogel-based protein microchips: manufacturing, properties, and applications. *Biotechniques* **34**(5), 1008–1014, 1016–1020, 1022.
7. Afanassiev, V., Hanemann, V., and Wolfl, S. (2000) Preparation of DNA and protein micro arrays on glass slides coated with an agarose film. *Nucleic Acids Res.* **28**(12), E66.
8. Cretich, M., et al. (2004) A new polymeric coating for protein microarrays. *Anal. Biochem.* **332**(1), 67–74.
9. Angenendt, P., et al. (2003) Next generation of protein microarray support materials: evaluation for protein and antibody microarray applications. *J. Chromatogr. A* **1009**(1–2), 97–104.
10. Kusnezow, W. and Hoheisel, J. D. (2003) Solid supports for microarray immunoassays. *J. Mol. Recognit.* **16**(4), 165–176.
11. Wacker, R., Schroder, H., and Niemeyer, C. M. (2004) Performance of antibody microarrays fabricated by either DNA-directed immobilization, direct spotting, or

streptavidin-biotin attachment: a comparative study. *Anal. Biochem.* **330**(2), 281–287.

12. Peluso, P., et al. (2003) Optimizing antibody immobilization strategies for the construction of protein microarrays. *Anal. Biochem.* **312**(2), 113–124.

13. Lue, R. Y., et al. (2004) Versatile protein biotinylation strategies for potential high-throughput proteomics. *J. Am. Chem. Soc.* **126**(4), 1055–1062.

14. Zhu, H., et al. (2001) Global analysis of protein activities using proteome chips. *Science* **293**(5537), 2101–2105.

15. Espina, V., et al. (2003) Protein microarrays: molecular profiling technologies for clinical specimens. *Proteomics* **3**(11), 2091–2100.

16. Tonkinson, J. L. and Stillman, B. A. (2002) Nitrocellulose: a tried and true polymer finds utility as a post-genomic substrate. *Front Biosci.* **7**, c1–c12.

17. Kusnezow, W. and Hoheisel, J. D. (2002) Antibody microarrays: promises and problems. *Biotechniques* Suppl. 14–23.

18. MacBeath, G. and Schreiber, S. L. (2000) Printing proteins as microarrays for high-throughput function determination. *Science* **289**(5485), 1760–1763.

19. Zhou, F. X., Bonin, J., and Predki, P. F. (2004) Development of functional protein microarrays for drug discovery: progress and challenges. *Comb. Chem. High Throughput Screen* **7**(6), 539–546.

20. Ringeisen, B. R., et al. (2002) Picoliter-scale protein microarrays by laser direct write. *Biotechnol. Prog.* **18**(5), 1126–1129.

21. Grubb, R. L., et al. (2003) Signal pathway profiling of prostate cancer using reverse phase protein arrays. *Proteomics* **3**(11), 2142–2146.

22. van Gijlswijk, R. P., et al. (1997) Fluorochrome-labeled tyramides: use in immunocytochemistry and fluorescence in situ hybridization. *J. Histochem. Cytochem.* **45**(3), 375–382.

23. Schweitzer, B., et al. (2002) Multiplexed protein profiling on microarrays by rolling-circle amplification. *Nat. Biotechnol.* **20**(4), 359–365.

24. Schweitzer, B., et al. (2000) Inaugural article: immunoassays with rolling circle DNA amplification: a versatile platform for ultrasensitive antigen detection. *Proc. Natl Acad. Sci. U. S. A.* **97**(18), 10,113–10,119.

25. Haab, B. B. and Zhou, H. (2004) Multiplexed protein analysis using spotted antibody microarrays. *Methods Mol. Biol.* **264**, 33–45.

26. Hodneland, C. D., et al. (2002) Selective immobilization of proteins to self-assembled monolayers presenting active site-directed capture ligands. *Proc. Natl Acad. Sci. U. S. A.* **99**(8), 5048–5052.

27. Grus, F. H., Joachim, S. C., and Pfeiffer, N. (2003) Analysis of complex autoantibody repertoires by surface-enhanced laser desorption/ionization-time of flight mass spectrometry. *Proteomics* **3**(6), 957–961.

28. Kyo, M., Usui-Aoki, K., and Koga, H. (2005) Label-free detection of proteins in crude cell lysate with antibody arrays by a surface plasmon resonance imaging technique. *Anal. Chem.* **77**(22), 7115–7121.

29. Yuk, J. S., et al. (2006) Analysis of protein interactions on protein arrays by a novel spectral surface plasmon resonance imaging. *Biosens. Bioelectron.* **21**(8), 1521–1528.

30. Cheran, L. E., et al. (2004) Protein microarray scanning in label-free format by Kelvin nanoprobe. *Analyst* **129**(2), 161–168.

31. Li, L., et al. (2002) In situ single-molecule detection of antibody-antigen binding by tapping-mode atomic force microscopy. *Anal. Chem.* **74**(23), 6017–6022.

32. Savran, C. A., et al. (2004) Micromechanical detection of proteins using aptamer-based receptor molecules. *Anal. Chem.* **76**(11), 3194–3198.

33. Blank, K., et al. (2003) A force-based protein biochip. *Proc. Natl Acad. Sci. U. S. A.* **100**(20), 11,356–11,360.

34. McKendry, R., et al. (2002) Multiple label-free biodetection and quantitative DNA-binding assays on a nanomechanical cantilever array. *Proc. Natl Acad. Sci. U. S. A.* **99**(15), 9783–9788.

35. Goh, J. B., et al. (2003) A quantitative diffraction-based sandwich immunoassay. *Anal. Biochem.* **313**(2), 262–266.

36. Zheng, G., et al. (2005) Multiplexed electrical detection of cancer markers with nanowire sensor arrays. *Nat. Biotechnol.* **23**(10), 1294–1301.

37. Striebel, H. M., et al. (2004) Readout of protein microarrays using intrinsic time resolved UV fluorescence for label-free detection. *Proteomics* **4**(6), 1703–1711.

38. Miklos, G. L. and Maleszka, R. (2001) Integrating molecular medicine with functional proteomics: realities and expectations. *Proteomics* **1**(1), 30–41.

39. Diamandis, E. P. (2004) Mass spectrometry as a diagnostic and a cancer biomarker discovery tool: opportunities and potential limitations. *Mol. Cell. Proteomics* **3**(4), 367–378.

40. Robinson, W. H., et al. (2002) Autoantigen microarrays for multiplex characterization of autoantibody responses. *Nat. Med.* **8**(3), 295–301.

41. Lueking, A., et al. (1999) Protein microarrays for gene expression and antibody screening. *Anal. Biochem.* **270**(1), 103–111.

42. Willats, W. G., et al. (2002) Sugar-coated microarrays: a novel slide surface for the high-throughput analysis of glycans. *Proteomics* **2**(12), 1666–1671.

43. Uhlen, M. and Ponten, F. (2005) Antibody-based proteomics for human tissue profiling. *Mol. Cell. Proteomics* **4**(4), 384–393.

44. Perlee, L., et al. (2004) Development and standardization of multiplexed antibody microarrays for use in quantitative proteomics. *Proteome Sci.* **2**(1), 9.

45. Shao, W., et al. (2003) Optimization of rolling-circle amplified protein microarrays for multiplexed protein profiling. *J. Biomed. Biotechnol.* **2003**(5), 299–307.

46. Kopf, E., Shnitzer, D., and Zharhary, D. (2005) Panorama Ab Microarray Cell Signaling kit: a unique tool for protein expression analysis. *Proteomics* **5**(9), 2412–2416.

47. Potgens, A. J., et al. (2002) Monoclonal antibody CD133-2 (AC141) against hematopoietic stem cell antigen CD133 shows crossreactivity with cytokeratin 18. *J. Histochem. Cytochem.* **50**(8), 1131–1134.

48. Fouraux, M. A., et al. (2002) Cross-reactivity of the anti-La monoclonal antibody SW5 with early endosome antigen 2. *Immunology* **106**(3), 336–342.

49. Hoglund, A. S., Jones, A. M., and Josefsson, L. G. (2002) An antigen expressed during plant vascular development crossreacts with antibodies towards KLH (keyhole limpet hemocyanin). *J. Histochem. Cytochem.* **50**(8), 999–1003.

50. Poetz, O., et al. (2005) Protein microarrays for antibody profiling: specificity and affinity determination on a chip. *Proteomics* **5**(9), 2402–2411.
51. Bangham, R., et al. (2005) Protein microarray-based screening of antibody specificity. *Methods Mol. Med.* **114,** 173–182.
52. Michaud, G. A., et al. (2003) Analyzing antibody specificity with whole proteome microarrays. *Nat. Biotechnol.* **21**(12), 1509–1512.
53. MacBeath, G. (2002) Protein microarrays and proteomics. *Nat. Genet.* **32**(Suppl), 526–532.
54. Haab, B. B., Dunham, M. J., and Brown, P. O. (2001) Protein microarrays for highly parallel detection and quantitation of specific proteins and antibodies in complex solutions. *Genome Biol.* **2**(2), RESEARCH0004.
55. Gygi, S. P., et al. (1999) Quantitative analysis of complex protein mixtures using isotope-coded affinity tags. *Nat. Biotechnol.* **17**(10), 994–999.
56. Sreekumar, A., et al. (2001) Profiling of cancer cells using protein microarrays: discovery of novel radiation-regulated proteins. *Cancer Res.* **61**(20), 7585–7593.
57. Unlu, M., Morgan, M. E., and Minden, J. S. (1997) Difference gel electrophoresis: a single gel method for detecting changes in protein extracts. *Electrophoresis* **18**(11), 2071–2077.
58. Barry, R., et al. (2003) Competitive assay formats for high-throughput affinity arrays. *J. Biomol. Screen* **8**(3), 257–263.
59. Yeretssian, G., et al. (2005) Competition on nitrocellulose-immobilized antibody arrays: from bacterial protein binding assay to protein profiling in breast cancer cells. *Mol. Cell. Proteomics* **4**(5), 605–617.
60. Olle, E. W., et al. (2005) Development of an Internally Controlled Antibody Microarray. *Mol. Cell. Proteomics* **4**(11), 1664–1672.
61. Gembitsky, D. S., et al. (2004) A prototype antibody microarray platform to monitor changes in protein tyrosine phosphorylation. *Mol. Cell. Proteomics* **3**(11), 1102–1118.
62. Bock, C., et al. (2004) Photoaptamer arrays applied to multiplexed proteomic analysis. *Proteomics* **4**(3), 609–618.
63. Nielsen, U. B., et al. (2003) Profiling receptor tyrosine kinase activation by using Ab microarrays. *Proc. Natl Acad. Sci. U. S. A.* **100**(16), 9330–9335.
64. Martin, K., et al. (2003) Quantitative analysis of protein phosphorylation status and protein kinase activity on microarrays using a novel fluorescent phosphorylation sensor dye. *Proteomics* **3**(7), 1244–1255.
65. Neuman de Vegvar, H. E., et al. (2003) Microarray profiling of antibody responses against simian-human immunodeficiency virus: postchallenge convergence of reactivities independent of host histocompatibility type and vaccine regimen. *J. Virol.* **77**(20), 11,125–11,138.
66. Arnaud, M. C., et al. (2004) Array assessment of phage-displayed peptide mimics of Human Immunodeficiency Virus type 1 gp41 immunodominant epitope: binding to antibodies of infected individuals. *Proteomics* **4**(7), 1959–1964.

67. Davies, D. H., et al. (2005) Vaccinia virus H3L envelope protein is a major target of neutralizing antibodies in humans and elicits protection against lethal challenge in mice. *J. Virol.* **79**(18), 11,724–11,733.

68. Steller, S., et al. (2005) Bacterial protein microarrays for identification of new potential diagnostic markers for Neisseria meningitidis infections. *Proteomics* **5**(8), 2048–2055.

69. Hiller, R., et al. (2002) Microarrayed allergen molecules: diagnostic gatekeepers for allergy treatment. *FASEB J.* **16**(3), 414–416.

70. Alcocer, M. J., et al. (2004) The major human structural IgE epitope of the Brazil nut allergen Ber e 1: a chimaeric and protein microarray approach. *J. Mol. Biol.* **343**(3), 759–769.

71. Ridgway, W. M., Weiner, H. L., and Fathman, C. G. (1994) Regulation of autoimmune response. *Curr. Opin. Immunol.* **6**(6), 946–955.

72. Mattoon, D., et al. (2005) Biomarker discovery using protein microarray technology platforms: antibody-antigen complex profiling. *Expert. Rev. Proteomics* **2**(6), 879–889.

73. Robinson, W. H., et al. (2003) Protein microarrays guide tolerizing DNA vaccine treatment of autoimmune encephalomyelitis. *Nat. Biotechnol.* **21**(9), 1033–1039.

74. Hueber, W., et al. (2005) Antigen microarray profiling of autoantibodies in rheumatoid arthritis. *Arthritis Rheum.* **52**(9), 2645–2655.

75. Quintana, F. J., et al. (2004) Functional immunomics: microarray analysis of IgG autoantibody repertoires predicts the future response of mice to induced diabetes. *Proc. Natl Acad. Sci. U. S. A.* **101**(Suppl 2), 14,615–14,621.

76. Lueking, A., et al. (2005) Profiling of alopecia areata autoantigens based on protein microarray technology. *Mol. Cell. Proteomics* **4**(9), 1382–1390.

77. Lueking, A., et al. (2003) A nonredundant human protein chip for antibody screening and serum profiling. *Mol. Cell. Proteomics* **2**(12), 1342–1349.

78. Gutjahr, C., et al. (2005) Mouse protein arrays from a TH1 cell cDNA library for antibody screening and serum profiling. *Genomics* **85**(3), 285–296.

79. Canto, E. I., Shariat, S. F., and Slawin, K. M. (2003) Biochemical staging of prostate cancer. *Urol. Clin. North Am.* **30**(2), 263–277.

80. Segal, N. H., et al. (2005) Antigens recognized by autologous antibodies of patients with soft tissue sarcoma. *Cancer Immun.* **5**, 4.

81. Cekaite, L., et al. (2004) Analysis of the humoral immune response to immunoselected phage-displayed peptides by a microarray-based method. *Proteomics* **4**(9), 2572–2582.

82. Wang, X., et al. (2005) Autoantibody signatures in prostate cancer. *N. Engl. J. Med.* **353**(12), 1224–1235.

83. Chan, S. M., et al. (2004) Protein microarrays for multiplex analysis of signal transduction pathways. *Nat. Med.* **10**(12), 1390–1396.

84. Liotta, L. A., et al. (2003) Protein microarrays: meeting analytical challenges for clinical applications. *Cancer Cell* **3**(4), 317–325.

85. Calvert, V., Tang, Y., Boveia, V., et al. (2004) Development of multiplexed protein profiling and detection using near infrared detection of reverse-phase protein microarrays. *Clin. Proteomics J.* **1**, 81–89.

86. Geho, D., et al. (2005) Pegylated, steptavidin-conjugated quantum dots are effective detection elements for reverse-phase protein microarrays. *Bioconjug. Chem.* **16**(3), 559–566.

87. Nishizuka, S., et al. (2003) Proteomic profiling of the NCI-60 cancer cell lines using new high-density reverse-phase lysate microarrays. *Proc. Natl Acad. Sci. U. S. A.* **100**(24), 14,229–14,234.

88. Gulmann, C., et al. (2005) Proteomic analysis of apoptotic pathways reveals prognostic factors in follicular lymphoma. *Clin. Cancer Res.* **11**(16), 5847–5855.

89. Belluco, C., et al. (2005) Kinase substrate protein microarray analysis of human colon cancer and hepatic metastasis. *Clin. Chim. Acta* **357**(2), 180–183.

90. Qiu, J., et al. (2004) Development of natural protein microarrays for diagnosing cancer based on an antibody response to tumor antigens. *J. Proteome Res.* **3**(2), 261–267.

91. Bouwman, K., et al. (2003) Microarrays of tumor cell derived proteins uncover a distinct pattern of prostate cancer serum immunoreactivity. *Proteomics* **3**(11), 2200–2207.

92. Nam, M. J., et al. (2003) Molecular profiling of the immune response in colon cancer using protein microarrays: occurrence of autoantibodies to ubiquitin C-terminal hydrolase L3. *Proteomics* **3**(11), 2108–2115.

93. Janzi, M., et al. (2005) Serum Microarrays for Large Scale Screening of Protein Levels. *Mol. Cell. Proteomics* **4**(12), 1942–1947.

94. Espejo, A., et al. (2002) A protein-domain microarray identifies novel protein-protein interactions. *Biochem. J.* **367**(Pt 3), 697–702.

95. Newman, J. R. and Keating, A. E. (2003) Comprehensive identification of human bZIP interactions with coiled-coil arrays. *Science* **300**(5628), 2097–2101.

96. Jones, R. B., et al. (2006) A quantitative protein interaction network for the ErbB receptors using protein microarrays. *Nature* **439**(7073), 168–174.

97. Boutell, J. M., et al. (2004) Functional protein microarrays for parallel characterisation of p53 mutants. *Proteomics* **4**(7), 1950–1958.

98. Merkel, J. S., et al. (2005) Functional protein microarrays: just how functional are they? *Curr. Opin. Biotechnol.* **16**(4), 447–452.

99. Michaud, G., Bangham, R., Salcius, M., and Predki, P. (2004) Functional protein microarrays for pathway mapping. *DDT: Targets* **3**(6), 238–245.

100. Satoh, J. I., Nanri, Y., and Yamamura, T. (2006) Rapid identification of 14-3-3-binding proteins by protein microarray analysis. *J. Neurosci. Methods.* **152**(1–2), 278–288.

101. MacBeath, G., Koehler, A., and Schreiber, S. (1999) Printing small molecules as microarrays and detecting protein–ligand interactions en masse. *J. Am. Chem. Soc.* **121**(34), 7967–7968.

102. Hergenrother, P. J., Depew, K., and Schreiber, S. L. (2000) Small molecule microarrays: Covalent attachment and screening of alcohol-containing small molecules on glass slides. *J. Am. Chem. Soc.* **122**(32), 7849–7850.

103. Kuruvilla, F. G., et al. (2002) Dissecting glucose signalling with diversity-oriented synthesis and small-molecule microarrays. *Nature* **416**(6881), 653–657.

104. Koehler, A. N., Shamji, A. F., and Schreiber, S. L. (2003) Discovery of an inhibitor of a transcription factor using small molecule microarrays and diversity-oriented synthesis. *J. Am. Chem. Soc.* **125**(28), 8420–8421.

105. Fang, Y., Frutos, A. G., and Lahiri, J. (2002) Membrane protein microarrays. *J. Am. Chem. Soc.* **124**(11), 2394–2395.

106. Huang, J., et al. (2004) Finding new components of the target of rapamycin (TOR) signaling network through chemical genetics and proteome chips. *Proc. Natl Acad. Sci. U. S. A.* **101**(47), 16,594–16,599.

107. Michaud, G., Samuels, M. L., and Schweitzer, B. (2006) Functional protein arrays to facilitate drug discovery and development. *IDrugs*, **9**(4), 266–272.

108. Hall, D. A., et al. (2004) Regulation of gene expression by a metabolic enzyme. *Science* **306**(5695), 482–484.

109. Wenschuh, H., et al. (2000) Coherent membrane supports for parallel microsynthesis and screening of bioactive peptides. *Biopolymers* **55**(3), 188–206.

110. Lee, S. J. and Lee, S. Y. (2004) Microarrays of peptides elevated on the protein layer for efficient protein kinase assay. *Anal. Biochem.* **330**(2), 311–316.

111. Rychlewski, L., et al. (2004) Target specificity analysis of the Abl kinase using peptide microarray data. *J. Mol. Biol.* **336**(2), 307–311.

112. Panse, S., et al. (2004) Profiling of generic anti-phosphopeptide antibodies and kinases with peptide microarrays using radioactive and fluorescence-based assays. *Mol. Divers* **8**(3), 291–299.

113. Zhu, Q., et al. (2003) Enzymatic profiling system in a small-molecule microarray. *Org. Lett.* **5**(8), 1257–1260.

114. Salisbury, C. M., Maly, D. J., and Ellman, J. A. (2002) Peptide microarrays for the determination of protease substrate specificity. *J. Am. Chem. Soc.* **124**(50), 14,868–14,870.

115. Ptacek, J., et al. (2005) Global analysis of protein phosphorylation in yeast. *Nature* **438**(7068), 679–684.

116. Feilner, T., et al. (2005) High Throughput Identification of Potential Arabidopsis Mitogen-activated Protein Kinases Substrates. *Mol. Cell. Proteomics* **4**(10), 1558–1568.

117. Lee, J. and Bedford, M. T. (2002) PABP1 identified as an arginine methyltransferase substrate using high-density protein arrays. *EMBO Rep.* **3**(3), 268–273.

118. Funeriu, D. P., et al. (2005) Enzyme family-specific and activity-based screening of chemical libraries using enzyme microarrays. *Nat. Biotechnol.* **23**(5), 622–627.

119. Zhu, H., et al. (2000) Analysis of yeast protein kinases using protein chips. *Nat. Genet.* **26**(3), 283–289.

120. Cha, T., Guo, A., and Zhu, X. Y. (2005) Enzymatic activity on a chip: the critical role of protein orientation. *Proteomics* **5**(2), 416–419.

121. Jung, G. Y. and Stephanopoulos, G. (2004) A functional protein chip for pathway optimization and in vitro metabolic engineering. *Science* **304**(5669), 428–431.

122. Lee, M. Y., et al. (2005) Metabolizing enzyme toxicology assay chip (MetaChip) for high-throughput microscale toxicity analyses. *Proc. Natl Acad. Sci. U. S. A.* **102**(4), 983–987.

123. Sieber, S. A., et al. (2004) Microarray platform for profiling enzyme activities in complex proteomes. *J. Am. Chem. Soc.* **126**(48), 15,640–15,641.

124. Liu, Y., Patricelli, M. P., and Cravatt, B. F. (1999) Activity-based protein profiling: the serine hydrolases. *Proc. Natl Acad. Sci. U. S. A.* **96**(26), 14,694–14,699.

125. Winssinger, N., et al. (2002) Profiling protein function with small molecule microarrays. *Proc. Natl Acad. Sci. U. S. A.* **99**(17), 11,139–11,144.

126. Speers, A. E. and Cravatt, B. F. (2004) Chemical strategies for activity-based proteomics. *Chembiochem.* **5**(1), 41–47.

127. Russ, A. P. and Lampel, S. (2005) The druggable genome: an update. *Drug Discov. Today* **10**(23–24), 1607–1610.

128. Chen, G. Y., et al. (2003) Developing a strategy for activity-based detection of enzymes in a protein microarray. *Chembiochem.* **4**(4), 336–339.

129. Angenendt, P., et al. (2005) Subnanoliter enzymatic assays on microarrays. *Proteomics* **5**(2), 420–425.

130. Gosalia, D. N. and Diamond, S. L. (2003) Printing chemical libraries on microarrays for fluid phase nanoliter reactions. *Proc. Natl Acad. Sci. U. S. A.* **100**(15), 8721–8726.

131. Ma, H., et al. (2005) Nanoliter homogenous ultra-high throughput screening microarray for lead discoveries and IC50 profiling. *Assay Drug Dev. Technol.* **3**(2), 177–187.

132. Albrecht, D. R., et al. (2005). *Lab Chip* **5**(1), 111–118.

133. Flaim, C. J., Chien, S., and Bhatia, S. N. (2005) An extracellular matrix microarray for probing cellular differentiation. *Nat. Methods* **2**(2), 119–125.

134. Bailey, S. N., Sabatini, D. M., and Stockwell, B. R. (2004) Microarrays of small molecules embedded in biodegradable polymers for use in mammalian cell-based screens. *Proc. Natl Acad. Sci. U. S. A.*, **101**(46), 16,144–16,149.

135. Ziauddin, J. and Sabatini, D. M. (2001) Microarrays of cells expressing defined cDNAs. *Nature* **411**(6833), 107–110.

136. Mishina, Y. M., et al. Multiplex GPCR assay in reverse transfection cell microarrays. *J. Biomol. Screen* **9**(3), 196–207.

137. Conrad, C., et al. (2004) Automatic identification of subcellular phenotypes on human cell arrays. *Genome Res.* **14**(6), 1130–1136.

138. Redmond, T. M., et al. (2004) Microarray transfection analysis of transcriptional regulation by cAMP-dependent protein kinase. *Mol. Cell. Proteomics* **3**(8), 770–779.

139. Delehanty, J. B., Shaffer, K. M., and Lin, B. (2004) Transfected cell microarrays for the expression of membrane-displayed single-chain antibodies. *Anal. Chem.* **76**(24), 7323–7328.

140. Silva, J. M., et al. (2004) RNA interference microarrays: high-throughput loss-of-function genetics in mammalian cells. *Proc. Natl Acad. Sci. U. S. A.* **101**(17), 6548–6552.

141. Vanhecke, D. and Janitz, M. (2004) High-throughput gene silencing using cell arrays. *Oncogene* **23**(51), 8353–8358.

142. Wheeler, D. B., et al. (2004) RNAi living-cell microarrays for loss-of-function screens in Drosophila melanogaster cells. *Nat. Methods* **1**(2), 127–132.

143. Paddison, P. J., et al. (2004) A resource for large-scale RNA-interference-based screens in mammals. *Nature* **428**(6981)**,** 427–431.

144. Kononen, J., et al. (1998) Tissue microarrays for high-throughput molecular profiling of tumor specimens. *Nat. Med.* **4**(7)**,** 844–847.

145. Zu, Y., et al. (2005) Validation of tissue microarray immunohistochemistry staining and interpretation in diffuse large B-cell lymphoma. *Leuk. Lymphoma* **46**(5)**,** 693–701.

146. Hoos, A., et al. (2002) Clinical significance of molecular expression profiles of Hurthle cell tumors of the thyroid gland analyzed via tissue microarrays. *Am. J. Pathol.* **160**(1)**,** 175–183.

147. Rubin, M. A., et al. (2002) Tissue microarray sampling strategy for prostate cancer biomarker analysis. *Am. J. Surg. Pathol.* **26**(3)**,** 312–319.

148. Hendriks, Y., et al. (2003) Conventional and tissue microarray immunohistochemical expression analysis of mismatch repair in hereditary colorectal tumors. *Am. J. Pathol.* **162**(2)**,** 469–477.

149. Brasch, M. A., Hartley, J. L., and Vidal, M. (2004) ORFeome cloning and systems biology: standardized mass production of the parts from the parts-list. *Genome Res.* **14**(10B)**,** 2001–2009.

6

Pharmacogenetic Issues in Biopharmaceutical Drug Development

Robert L. Haining

Abstract

Several converging factors, including the sequencing of the human genome, fine manufacturing and chemical synthesis processes, and the advent of targeted biotherapeutic molecules, make this an exciting era in personalized medicine. Pharmacogenetics, the term used to describe idiosyncratic responses to drug therapy having a genetic basis, has become much more than a few SNPs in some drug-metabolizing enzymes. With the advent of biopharmaceutical drugs such as monoclonal antibodies, cytokines, and peptidomimetics, the term "pharmacogenetics" must be interpreted in a much broader sense. Indeed, biopharmaceutical agents are often subject to traditional pharmacogenetic polymorphisms. But much more than that, pharmacogenetics encompasses the use of any and all genetic information available or obtainable from an individual and/or pathogenic organisms that can be put into use in clinical practice. By their very nature, many of the diseases in question are multigenic and do not lend themselves to simple pharmacogenetic analysis. Some key examples of these and other issues surrounding the use of genetic information to combat human diseases are presented.

Key Words: Pharmacogenetics; polymorphism; immunogenicity; biopharmaceuticals; drug development; antibodies; cytokines; subsets of disease.

1. Introduction and Chapter Scope

1.1. Introduction

With the completion of the human genome sequencing project, our understanding of the genetic basis underlying idiosyncratic reactions to small-molecule drug therapy has expanded exponentially. Combined with over four decades of research into the enzyme families responsible for drug metabolism, it is now common to see commercially available "gene chips" designed for the purpose of identifying common single nucleotide polymorphisms (SNPs) and other

From: *Biopharmaceutical Drug Design and Development*
Edited by: S. Wu-Pong and Y. Rojanasakul © Humana Press Inc., Totowa, NJ

genetic alterations which can result in a "poor metabolizer" phenotype because of missing or altered enzymatic activity. At this early stage, it is still unusual to perform genotyping studies prior to initiating drug therapy. In the coming decades, however, as gene chips and clinical correlations become established, it will become ever more common for a person to prospectively know what adverse reactions and diseases they are susceptible to. More so, our understanding of the genetic basis behind distinct subsets of a disease, and how an individual will respond to drug, opens the door to individualized therapy.

In the strictest sense, a true pharmacogenetic polymorphism, unlike an inborn error of metabolism, is defined as a phenotype which is only observed in a subset of the population following exposure to the drug. In the case of biopharmaceuticals, however, we must interpret pharmacogenetics in a much broader sense. That is, the advent of biopharmaceuticals opens the door to truly individualized medicine based on genetic makeup. Already certain cancers are treated using the patient's own immune cells, temporarily removed from the body for supplementation prior to reintroduction. Pharmacogenetics in terms of biopharmaceuticals may be taken to mean the use of a person's individual genetic makeup to tailor drug therapy. The distinction lies in the inclusion of genetic disorders and disease subset, which have a known cause and which may be treated with biopharmaceuticals.

1.2. Types of Biopharmaceutical Agents Considered

Biopharmaceuticals too is a term that may take on many meanings and is undoubtedly defined in several places throughout this work. From a biotechnology standpoint, a "biopharmaceutical" may be defined as any drug that is produced using biological methods as opposed to strictly chemical methods. The term can therefore include small molecules produced by biotechnological as opposed to chemical means. In the following article, however, I will restrict the definition of biopharmaceuticals largely to macromolecular drugs which mimic naturally occurring substances. In particular, I will limit discussion to proteinaceous biopharmaceuticals, up to and including peptidomimetics drugs which may or may not contain biologically cleavable peptide bonds. The reader interested in nucleic acid-based therapies such as RNAi will unfortunately be disappointed, as only one example is mentioned (asthma). The decision to limit discussion to protein-based drugs is based on the realization that the term "pharmacogenetics" becomes substantially blurred when referring to drugs that are, in fact, genetic material. The author recognizes, however, that nucleic acid-based therapies are a tremendous potential source of biopharmaceutical agents which open the door to truly individualized medicine based on genetic makeup. However, in the interests of highlighting more traditional pharmacogenetic aspects of biopharmaceutical agents, it was felt that inclusion of nucleic acids was beyond the current scope. In this chapter then, I shall consider several examples which illustrate some of the successes and failures in a pharmacogenetic approach to protein-based biopharmaceutical drug development and utilization.

2. Overarching Pharmacogenetic Issues Regarding Biopharmaceuticals

2.1. Which Polymorphisms Are Important? Clinical Correlations and Statistics

Beyond cataloguing the extent of genetic variability in the human population, many hurdles remain before we can put the information gleaned from the human genome sequencing project into clinical practice. Once the polymorphisms are identified, genetic information must then be obtained in a high-throughput manner so that enough data exist to confirm clinical correlations in a statistically meaningful manner. The enormous amount of information this entails means that the field of bioinformatics is critical to the success of pharmacogenetic applications *(1)*. Several online databases exist for this purpose, such as LocusLink (http://www.ncbi.nlm.nih.gov/LocusLink/), dbSNP (http://www.ncbi.nlm.nih.gov/SNP/), and RefSeq (http://www.ncbi.nlm.nih.gov/LocusLink/refseq.html). Go!Poly is a recent addition which extracts human gene-linked sequence variations of all common types (SNP, insertion–deletion, simple tandem repeat, and complex nucleotides variations) from public resources. Polymorphism data are then categorized into different genetic loci, and the reference sequences given by LocusLink are used for positioning *(2)*. Tools such as these along with many more clinical correlation studies are needed to realize the full potential of pharmacogenetics.

2.2. Immunogenicity and the HLA Locus

Without doubt, one of the largest and most consistent sources of interindividual variability in response to biopharmaceutical drugs lies in the patient's immune response. This can take the form either as a desired course of the drug pharmacology, as with cytokines and monoclonal antibodies (MAbs), or of an unwanted host response to the introduction of a foreign antigenic protein. By definition, a polymorphic pharmacogenetic response is idiosyncratic, i.e., not within the normal range of response. In the case of the immune system, many factors may contribute to natural, nongenetic variability including health, diet, smoking history, coadministered drugs, and so on. From what we have learned from the pharmacogenetics of nonbiopharmaceutical agents affecting the immune system, the classic case of polymorphism involves the major histocompatibility complex (MHC) proteins. These proteins, also known as human leukocyte antigens (HLA), are encoded by a cluster of genes residing at the *HLA* locus of human chromosome 6. They were named based on their heavy involvement in tissue graft rejection. Note the word "histocompatibility" literally means the ability of tissues to get along with each other. This process of self- vs nonself-recognition and subsequent rejection of nonself-tissue is mediated by cells of the immune system, i.e., human leukocytes, thus the name human leukocyte antigens and the nomenclature for the genetic locus encoding these proteins.

Early examples exist in which *HLA* mutations were found to be important in the pharmacogenetics of disease, one of the earliest being the role these antigens play in the treatment of rheumatoid arthritis (RA) *(3)*. Even before that, it was known that polymorphisms within the *HLA* locus were related to the development of complex diseases such as juvenile diabetes *(4)*. It is now known that several alleles make up the HLA locus, and that the complex pattern of inheritance of these alleles determines an individual's response to a particular antigen. More important to future drug development is an overall general understanding of why the *HLA* gene products are important in pharmacogenetic polymorphism. Such an understanding is crucial to our discussion of the real and possible ramifications of the use of biopharmaceutical agents. For starters, small-molecule drugs are themselves frequently reactive or may be bioactivated by human drug-metabolizing enzymes which are then capable of conjugation with normal cellular macromolecules, including proteins. This process can directly alter the functional properties of the protein, for example, in the case of drugs that are known suicide inhibitors of enzyme activity. Alternatively, the direct effect on the modified protein may only be marginal, yet the modification may introduce an antigenic determinant which is foreign to the host immune system in a process known as haptenization, which can then cause the host to elicit an immune response against the modified protein. In the worst cases, this immune response is not specific to the hapten in question but can extend to the normal protein as well causing unwanted side effects. Cells and organs which are most highly exposed to drug are naturally the most affected by haptenization, thus hepatotoxicities and/or disorders of red blood cells are often apparent in drug-induced autoimmune disorders *(5)*.

On a philosophical level, the job of the MHC proteins in a cell is to continually sample all of the antigens in the cell's environment, from both the inside and the outside, in order to make a determination regarding its own health that it is able to convey to the immune system. Should the cell be infected with a virus or vastly overexpress a given protein, this sampling in effect means the cell is continually offering itself up for examination and potential elimination by the immune system. It achieves this through the natural recycling processes that must occur in order for a cell to regenerate itself. That is, during its lifetime, oxidative damage and toxic insults continually damage cell components, often causing them to act in a defective or abnormal manner. These proteins and other cell components must therefore be replaced. In the case of proteins, this process often first involves ubiquitination (*see* below) in order to tag the proteins for degradation. Once tagged, they can be more efficiently shuttled about and hydrolyzed by cellular proteases into smaller "chunks" of amino acids. Peptides of 8–10 amino acids in length fit a binding pocket in the MHC protein, whose job is then to display this peptide antigen on the surface of the cell. A complementary binding protein found on the surface of immune cells is then able to recognize the displayed antigen. If it recognizes the displayed antigen as a part of a viral protein or an excess amount of normal protein (as in many cancers), a cascade of immunologic events leading to target cell death may be initiated. It is the specificity of the relationship between MHC protein and processed antigen which determines how tightly the two will bind and thus how strongly they will be held on the cell surface. This interaction is therefore unique to both antigen and MHC protein and is crucial for the recognition of this antigen as natural or foreign in origin.

As most biopharmaceuticals drugs approved to date are proteinaceous in nature (MAbs, cytokines, and enzyme replacement therapies), they too are subject to proteolytic processing and antigen display. Thus we can expect polymorphisms to arise in the display of novel antigens found in biopharmaceutical drugs. In the case of MAbs, the hope is for the MAb to stimulate a specific immune response without itself eliciting an immune response from the host. However, in practice, even very minor differences between recombinant immunoglobulin molecules and normal human antibodies will eventually be recognized and become limiting to use, the only real question appears to be "when?" The formation of neutralizing antibodies may make subsequent doses ineffectual at best or cause a severe hypersensitivity reaction to the drug. If the immune response can be effectively slowed, however, several courses of MAb therapy may be tolerated before the host response to foreign antigen (MAb) becomes limiting.

Complicating the above is the realization that other components of the biopharmaceutical formulation may also have the ability to trigger an immune response. For example, the agent may be contaminated by other protein antigens arising from the organism used to produce it. Unlike small-molecule drugs which lend themselves to entirely chemical synthesis methods, biopharmaceutical drugs are more practically synthesized in a "bioreactor"—a nonhuman, cell-based approach to the overexpression of a therapeutic protein, typically *Escherichia coli*, yeast, or plant-based bioreactors. Contaminant protein antigens can be extremely difficult to remove and under the proper genetic background conditions, idiosyncratic hypersensitivities may result. The clinical end result may then be the stoppage of treatment because of toxic side effects. In the case of a biopharmaceutical derived from a bacterial source, for example, the presence of contaminating bacterial proteins or other components can profoundly influence the overall immune response. An immune system primed by bacterial components into hyperactivity will more readily overact to the foreign antibody. A further potential contributor to immunogenicity lies in posttranslational modifications. Such alterations may be unique to the bioreactor cell type, potentially resulting in unexpected or otherwise unwanted, nonhuman epitopes, similar to the process of haptenization. Particularly in the case of generic versions soon developed from existing biopharmaceuticals which are destined to go off-patent in the coming years, the immunogenicity of the formulation itself becomes extremely important. Any such generic agents in development would be wise to take this into consideration. Along these lines, many biopharmaceutical drug developers are moving to the use of the plant bioreactors rather than yeast or bacterial versions which reduce some of these contamination and immunologic issues.

2.3. Ubiquitination

Another process common to many proteins within the cell during recycling and regeneration of cellular components is that of ubiquitination. One of the first steps in eliminating a normal protein gone bad is the recognition of instability. Often, cell components can be salvaged by refolding upon a "chaperonin" or so-called heat shock protein, so named because of their expression during times of cellular stress. However, beyond a certain point, the protein itself may be designated as unsalvageable and therefore relegated for parts. This is what ubiquitination does—marks the protein for destruction and parting out. However, polymorphisms can exist within the enzymes which carry out this process, resulting in interindividual variability in the ability to clear unwanted proteins. In the case of biopharmaceuticals, what this means is that for every novel protein or protein-like substance introduced into a human, part of its clearance pathway may involve ubiquitination, and this process is subject to genetic variation. Variations in such ubiquitin-related genes have been implicated in a number of diseases, such as cancer and viral infections. Hence, in addition to its potential for general defects in the clearance of biopharmaceutical agents, this pathway is itself a target for new drug discovery *(6)*.

2.4. Drug Targeting: Tissue Issues

Because of their typically macromolecular nature, biopharmaceutical drugs generally either are unable to effectively penetrate to the interior of cells or do not retain their antigen-binding characteristics in the cellular environment. Recently, a class of antibody-derived therapeutic molecules termed "intrabodies" has been investigated and found to possess the ability to enter cells *(7)*. This raises the possibility of monoclonal intrabodies targeted to internal cell components, a potentially very useful therapeutic approach. However, these drugs have yet to realize any real benefit in humans. Related to the inability of biopharmaceuticals to enter cells is the fact that they do not readily cross the blood–brain barrier. Despite an explosion in our understanding of the genetic basis behind many common neurological disorders and the identification of many new drug targets, the usefulness of biopharmaceuticals in this regard has therefore hitherto been extremely limited. For this reason, despite an intense personal interest in CNS disorders, discussion of the pharmacogenetics issues surrounding them has been omitted from the current work.

The fact is that most biopharmaceutical drugs must be administered by direct injection. Nonetheless, imaginative approaches are being explored whenever feasible, as illustrated by a new class of antiasthma drugs based on the concept of RNA interference. Because the target organ in this case is accessible from the atmosphere, an inhalant form of RNAi drug has been developed in which a solution of the RNA molecule drug is first dispersed into tiny droplets *(8)*. Though limited in this case to the lung, the ability to target a very specific gene product has obvious and exciting ramifications which may well be termed "pharmacogenetic". For example, an allelic variant of the Alzheimer's beta-amyloid precursor protein which is unstable and lends itself to aggregation and the carrier individual to Alzheimer's disease could potentially be shut off using an RNAi molecule specific for the allelic variant, whereas leaving the expression of the nondisease-related allele(s) untouched. This idea cannot be considered revolutionary in this era; the revolution required is getting the drug molecule to the site where it can be effective. Though experimentally proven feasible in rodents to introduce an RNAi molecule directly into the brain and have it produce a desired effect *(9,10)*, these techniques are not suited for human use. However, because of its tremendous promise, this is likely to be a hot area of research in the coming decades.

3. Selected Examples of Biopharmaceuticals Exhibiting Clinically Observable Pharmacogenetic Polymorphisms

3.1. Agammaglobulinemia

Despite the above discussions of MAbs, an even earlier clear-cut case of a pharmacogenetic polymorphism involving a biopharmaceutical agent has been in use in clinical practice for many years. That is, approximately 50 years ago, Ogden Bruton, considered the father of the study of genetically determined immune deficiency syndromes, discovered a young patient who was abnormally sensitive to bacterial infection (for a review, *see [11]*). Although the newly discovered antibiotic wonder-drugs (at that time) cured each infection, the boy would subsequently relapse with another infection. Eventually, periodic injection of nonspecific human gamma globulin was found to "cure" this patient.

The genetic basis for the low levels of circulating gamma globulin which characterizes this disorder was traced back to mutations on the X-chromosome, now known to be highly varied *(12)*, and this disorder became known as Bruton's agammaglobulinemia. However, other genetic loci can result in an identical phenotype as evidenced by forms of this disease which do not carry X-chromosome mutations *(13)*. Such autosomally inherited defects have been found in genes encoding the immunoglobulin mu heavy chain, the immunoglobulin alpha chain, and the lambda5 component of pre-B-cell receptors *(14)*. Thus, this polymorphic disorder, having distinct genetic subsets with the same phenotype, resulting in an idiosyncratic response to bacterial infection and antibiotic therapy and subsequently treated with a biopharmaceutical agent (IgG), may well be the first documented case of a pharmacogenetic polymorphism involving a biopharmaceutical agent.

3.2. Rituximab and ADCC Fc Polymorphisms

MAbs are front-line biopharmaceutical agents that have shown tremendous potential and growth. Originally of mouse origin, the first MAbs were fraught with complications centering on their unfortunate ability to generate an immune response in the host against the therapeutic agent itself. Technological advances *(15)*, including the use of chimeric and humanized MAbs, as well as the masking of epitopes and stabilization of the molecule through polyethylene glycol conjugation (pegylation), have overcome many of these obstacles to make MAbs viable therapeutic alternatives. Nevertheless, on widespread application in the general population, some of these nontraditional drugs have been shown to exhibit very traditional polymorphic multimodal behavior characteristic of a genetic effect.

Rituxan™ (rituximab) was the first MAb available in the United States for the treatment of cancer. It was approved in 1997 for the treatment of non-Hodgkin's lymphoma (NHL), a B-cell-proliferative defect, in particular low-grade or follicular NHL in which the causative cell type is shown to carry the CD20 cell surface marker. Already arguably a case of pharmacogenetics at work, given the subset of NHL known to respond to rituximab therapy (CD20 positive or CD20+), the more traditional pharmacogenetic point I wish to illustrate concerns the efficacy of Rituxan in that subset for which it is used. A bit of understanding regarding the pharmacological mechanism of rituximab is in order. Like many MAbs, the idea behind using a tight binding antibody is often to stimulate the body's own immune system into better recognizing the target antigen. In the case of CD20+ NHL, the binding of Rituxan to the CD20 surface antigen results in both complement-mediated cell lysis and antibody-dependent cellular cytotoxicity (ADCC), processes which rely entirely on the immune system's recognition of the antigen–antibody complex.

Despite its remarkable success in tumor response *(16–18)*, a certain subset of patients who are CD20+ still do not respond to rituximab. In 2003, a major cause of nonresponse was discovered in a study examining the effectiveness of rituximab in reducing the CD20+ B-cell population in systemic lupus erythematosus *(19)*. These authors examined genetic polymorphisms encoding the Fc receptors (FcR) on effector cells involved in the recognition of the antigen–antibody complexes required for drug efficacy. In particular, an isoform of the FcRIIIa receptor exists which is a poorer binder of the Fc portion of the rituximab molecule than the "normal" receptor. This receptor is found on the surface of natural killer cells, thus a poorer binding variant reduces the ability of these cells to carry out the ADCC necessary for drug effectiveness. Moreover, it was determined that a single amino acid change at position 158 of this receptor appears to be the major culprit in this polymorphic response.

Not to detract from the success story that is Rituxan, however, combination therapy with immune response stimulators appears able to overcome the limitations inherent in homozygotic carriers of the 158F FcRIIIa allele. In June 2004 at a meeting of the American Society of Clinical Oncology (ASCO), Chiron BioPharmaceuticals Corporation announced the initiation of a new Phase II study of ProleukinR (aldesleukin) interleukin-2 (IL-2) plus rituximab in rituximab-naive patients with low-grade NHL. Based on their data, a 31% response rate to rituximab combination therapy was achieved in carriers of the reduced-binding FcRIIIa allele. Clearly, other factors are at play, many of which undoubtedly have a genetic basis. Nonetheless, this is an exciting development in our understanding of the pharmacogenetic issues surrounding the effectiveness of biopharmaceutical agents. Because of its success, Rituxan is also now used widely for other CD20+ problematic cell disorders, including Waldenstrom's macroglobulinemia *(20,21)*, RA *(22,23)*, SLE *(19)*, posttransplant lymphoproliferative disorder *(24)*, Sjogren's syndrome *(25)*, opsoclonus-myoclonus syndrome *(26)*, and the list keeps growing *(27,28)*.

3.3. Herceptin and EGFR Polymorphisms

Like rituximab, the story of Herceptin, which goes by the generic name trastuzumab, is one without which any discussion of pharmacogenetic issues in biopharmaceutical drug development would be incomplete. Both agents recognize protein antigens found on the surface of cancerous cells in greater abundance than on normal cells. But this is where the similarities stop. Herceptin is a humanized antibody developed by Genentech BioOncology™ and approved for use in the United States in 1998. Unlike rituximab, the mechanism of action behind Herceptin is not the induction of cell killing via complement-mediated lysis or ADCC but rather the disruption of the normal cellular signaling processes which are involved in the abnormal rate of cell division characteristic of cancer.

In a certain subset of breast tumors, an overabundance of the human epidermal growth factor receptor, HER2, is associated with aggressive tumor growth and a poor response to traditional chemotherapy. Following Herceptin's discovery and widespread usage, breast tumors are now routinely screened for HER2 status. Originally done by immunohistochemical means, Genentech BioOncology™ codeveloped the DAKO HercepTest® to help identify those patients most likely to benefit from Herceptin therapy. Additional studies confirmed a chromosomal abnormality common to patients overexpressing HER2 which could then be used in a genetic screen for Herceptin sensitivity via fluorescence in situ hybridization (http://www.herceptin.com/).

The HER2 gene is considered an oncogene, i.e., one which is found in normal cells but which can then become associated with cancer progression. Early in the search for the underlying genetic basis for cancer, systems involving growth factors were of special interest for obvious reason. Thus it is perhaps not surprising that the family of genes encoding for receptors of these growth factors, the so-called epidermal growth factor receptors or EGFRs, are now prime targets of biopharmaceutical agents. The use of Herceptin in combination with traditional chemotherapy has shown a survival benefit in women with metastatic HER2-positive breast cancer when compared to women treated with chemotherapy alone *(29)*, a feat which very few therapies have been able to demonstrate in this type of breast cancer.

Interestingly, the pharmacogenetic aspects of Herceptin therapy are still being discovered, in large part because of its relative infancy compared with traditional chemotherapeutic agents. For example, a small percentage of patients administered Herceptin develop ventricular dysfunction and congestive heart failure. However, on coadministration of Herceptin with anthracyclines, this small percentage with heart problems jumps to a much higher level (~25%) *(30)*, thus it is important to monitor the cardiac function of patients undergoing Herceptin therapy. At this time, it is unknown why a certain subset of patients are subject to this polymorphic response, but the cause can and will almost certainly be traced back to genetics in the not-too-distant future.

3.4. Imatinib Mesylate (Gleevec) for Patients with bcr/abl-Positive Chronic Myelogenous Leukemia

Of the examples examined herein, Gleevec is arguably the least "biopharmaceutical" of them all, being much more like a small molecule than a protein. Nonetheless, given the amide bond and activity of this drug as an antagonist of a protein phosphorylation enzyme, it may still be characterized as a peptidomimetic agent, defined as any compound that mimics the biological activity of a peptide but not necessarily containing enzymatically cleavable peptide bonds (http://www.chemicool.com/definition/peptidomimetic.html). Definitions aside, the clear-cut case of pharmacogenetic polymorphism makes it an attractive story for presentation here. As a cytochrome P450 3A4 substrate, polymorphisms within this enzyme are likely to affect its pharmacokinetics *(31)*. However, this is not where our current story lies.

Notably, the genetic alteration behind the pharmacological action of Gleevec is relatively large as opposed to comparatively minor SNPs or exon deletions. In this instance, portions of chromosomes 9 and 22 in fact undergo a reciprocal translocation, i.e., whole chunks of these two human chromosomes are swapped. As is the case with many cancers, such chromosomal rearrangements are not at all uncommon. However, in this instance, the "new" chromosome 22 encodes for an aberrantly regulated enzyme with tyrosine kinase activity from the newly formed *BCR-ABL* oncogene. Unfortunately, this arrangement allows certain white blood cell types to divide uncontrollably resulting in chronic myelogenous leukemia (CML), among other disorders. This tyrosine kinase is connected through cell-signaling pathways involving phosphoinositide 3-kinase *(32)*, among others, which directly results in the cancerous phenotype. Thus Gleevec is the first drug to directly target a cancer-causing protein.

Though truly a pharmacogenetic polymorphism in that only those leukemia patients who have this chromosomal rearrangement (termed the "Philadelphia chromosome") are likely to benefit from Gleevec therapy in the first place, this is not where the pharmacogenetic polymorphism story of Gleevec ends. That is, 60–70% of patients treated in the blast phase of the disease (characterized by rapid proliferation and numerous circulating blast cells) become resistant to the drug, whereas resistance is only rarely seen during treatment in chronic phase patients. As reported by Gorre and others *(33)*, one cause for this resistance is a further mutated form of the *Bcr-Abl* gene that is less able to bind Gleevec.

3.5. Kineret (Anti-IL-1) for Rheumatoid Arthritis

Kineret (generic name anakinra) is a biopharmaceutical agent with tremendous potential and power in inflammatory disorders such as RA which targets the inflammatory mediator interleukin I (IL-I). This cytokine has numerous functions, not the least of which is the induction of a local loss in proteoglycans, a class of cell surface proteins conjugated to complex sugars. This loss of proteoglycans is mediated normally through the binding of IL-I to the IL-I receptor IL-IR1, in turn leading to the cartilage degradation characteristic of RA. IL-I is found in the joints of RA patients at much higher levels than normal *(34)*, too high for the naturally occurring receptor antagonist IL-IRa to cope *(35,36)*. Thus, the strategy behind Kineret is to add exogenous IL-I receptor antagonist in an effort to stop the damaging activities caused by excess IL-I *(37)* Approved in November 2001, Kineret is the first such drug to employ this strategy and can be used alone or in combination with strategies which employ an antitumor necrosis factor (anti-TNF) approach *(38)*. Anakinra itself differs from native IL-IRa only in the addition of a methionine residue at the N terminus to its normally 152 amino acid length as a result of the *E. coli* expression system used to produce it. For this reason, persons sensitive to proteins produced in *E. coli* are advised to avoid this product (http://www.kineretrx.com/) and this known sensitivity will present extra concerns for future generic equivalents.

Soon after its introduction it was noted that the response to Kineret/anti-TNF therapy varied substantially between individuals, leading to the search for a genetic polymorphism behind this response. And indeed, much progress has been made in understanding the result of polymorphism in genes of the IL variety, found on human chromosome 2. For illustrative purposes of the pharmacogenetic aspects of biopharmaceutical agents, we focus on a point mutation at position +4845 of the IL-1A which was associated with this response. As reported at the 66th Annual Scientific Meeting of the American College of Rheumatology, the overall response rate of 91 patients to anakinra was 48% (44/91 patients). However, carriers of the rarer allele at IL1A (+4845) responded 63.4% of the time vs only 26.3% in noncarriers *(39)*. Perhaps even more important in terms of pharmaceutical development, Bansback et al. *(40)* showed that PG testing was economically feasible and practical to consider prior to initiation of therapy against RA.

3.6. Human Growth Hormone Therapy and Receptor Gene Polymorphism

Human growth hormone (GH) is a protein normally expressed in the human pituitary gland throughout a person's lifetime in response to signals from the hypothalamus region of the brain. Of the many other hormones and growth factors found in the body, GH is the major regulator of growth. In childhood, this translates into stimulation of bone elongation and tissue growth. In adulthood, GH continues to be important by regulating not only bone density but also cholesterol levels, roles extremely important in preventing osteoporosis and heart disease, respectively (http://www.gene.com/gene/products/education/biotherapeutics/growthhormone.jsp). For children with short stature because of GH deficiency (GHD), GH replacement therapy with the biopharmaceutical equivalent is now standard therapy.

Studies of interindividual variability to GH therapy have revealed many numerous variables but only partially explain the clinical observations. As with other examples herein, an obvious place to look for a pharmacogenetic response to GH would be in the receptor(s) which mediate its action. Not surprisingly, a report in 2005 from Jorge et al. *(41)* reveals an association between response to GH and a polymorphism within the *GHR* gene. These authors found a significant correlation between patients carrying at least one copy of an exon-3-deleted allelic version of *GHR* and those without to growth velocity in 58 children in their first year of GH replacement therapy, as well as a correlation to adult height in 44 patients who had been receiving GH therapy indiscriminate of genotype after 7.5±3.0 yr of therapy. Apparently, the growth abnormalities in those individuals who do not have two normal copies of the *GHR* gene (exon-3 inclusive) may partially be explained by GHR polymorphism.

4. The New Pharmacogenetics Paradigm: Host/Pathogen Interactions

Like disease susceptibility genes within our own genomes, a large source of genetic variability arises because of the genomes carried within the infectious organisms which are external causes of disease. Application of pharmacogenetic knowledge to the treatment of infectious diseases thus requires consideration of both the host and pathogen genomes *(42)*. With our increased knowledge in the field of comparative genomics, scientist are becoming ever more aware that pharmacogenetic polymorphism with the use of biopharmaceuticals and small-molecule drugs alike is not an isolated phenomenon. Genetic variability exists at all levels of the disease process as well as the intervention process. Often the task lies in sorting out the complexities behind genetic vs environmental variables and clinically important vs useful vs unimportant differences in DNA sequence between individuals. The following are instances in which PG knowledge has been applied to explain and understand observed polymorphisms in response to drug therapy against infectious organisms.

4.1. Response of Hepatitis C Virus to Interferon Therapy

Interferon (IFN) is a human cytokine that is used as a biopharmaceutical agent for a number of diseases involving the immune system and, as is seemingly the norm, exhibits a polymorphic response. Within a population, responders and nonresponders are the most often recognized outcome. All hepatitis C virus (HCV) treatment protocols currently in use are based on IFN products, generally in combination with nucleoside analog viral inhibitors owing to the low sustained responses observed with IFN alone. At least five IFN preparations have been approved in the United States for the treatment of chronic hepatitis C in adults, including interferon alpha-2a (Roferon-A; Hoffmann-La Roche), inteferon alpha-2b (Intron-A; Schering-Plough), interferon alfacon-1 (Infergen; Intermune), peginterferon alpha-2b (Peg-Intron; Schering-Plough), and peginterferon alpha-2a (Pegasys; Hoffmann-La Roche). The latter are examples of biopharmaceutical agents improved via conjugation of a polyethylene glycol moiety to the drug in question. Advantages of the pegylated forms include higher stability and lower immunogenicity, making them more suited for long-term therapy with less frequent injections required.

But the pharmacogenetic story does not begin here. In the first place, most of us are already familiar with the fact that hepatitis is not a single human disease but rather a broad classification of disease within a given organ (the liver), exhibiting shared symptoms but having varied causes. Thus it only needs pointing out because it is so familiar and in fact old, the reader may not recognize it as a case of the "new pharmacogenetics": disease subsets which can be targeted individually for maximum pharmacological response. One cause of human hepatitis is infection with the blood-borne pathogen HCV. The next level of pharmacogenetics of this story lies in the subsets of HCV. That is, through today's technology, it is not at all beyond reason (though easily beyond cost/benefit in most cases) to individually sequence the entire genome of an infectious organism; thus for smaller genomes typically carried by viruses such as HCV, several subtypes are recognized in the causation of human liver disease.

4.1.1. Hepatitis C Virus Type 1 and IFN/Ribavirin Combination Therapy

In 2003, Yee et al. reported on a pharmacogenetic response to interferon-alpha (IFN-a) treatment in a subset of Caucasian hepatitis sufferers carrying the HCV type 1 virus. Within the HCV1 subtype, these authors examined two SNPs, one at nucleotide position −318 in a viral gene promoter region and another at +49 in a coding region (exon 1) of that gene. The gene in question is called the cytotoxic T-lymphocyte (CTL) antigen 4 gene and, as its name implies, encodes a protein recognized by CTLs. Patients infected with HCV1 were first separated into two groups of equal size, strong responders (SR) vs nonresponders (NR), following treatment with IFN-a in combination with ribavirin. SRs were found to have a much higher frequency of the +49G polymorphism both alone and in haplotype combination with the −318C promoter polymorphism than NRs. Importantly, these associations persisted after multivariable analysis. In addition, no relationship was found with nontype 1 HCV viruses. Thus in the HCV1 subtype, it appears that two SNPs in a single gene carried by this virus largely determine the response to IFN-a combination therapy *(43)*.

4.1.2. Hepatitis C Virus Type 2b and IFN Monotherapy

More recently, Tanabe et al. reported on real-time pharmacogenetic polymorphisms of the viral genome in carriers of the hepatitis C virus 2b subtype (HCV2b) in response to IFN monotherapy in a Japanese population. That is, some viral genomes have long been known to undergo rapid mutation during the course of an infection in order to evade the host immune response. These authors sequenced the complete genome of the HCV2b virus in a group of hepatitis patients before, during, and after IFN treatment over a period of years to scan for all genetic changes that occurred during the infection and subsequent treatment. In particular, they were interested in those alterations that resulted *not* in a sustained viral response (sVR), but rather in a sustained biochemical response (sBR) but with persistent viral loads. A set of five such patients, along with a set of five patients exhibiting no measurable response at all (NR), was examined for comparison. Interestingly, these investigators found that the overall substitution rate of amino acids in the full-length HCV genome was higher in the sBR group than in the NR group. Furthermore, these authors were able to pinpoint the major proteins affected by these genetic coding changes and found that the amino acid changes cluster to those regions of proteins, in particular the NS4A antigen, which are normally recognized by the HLA class I proteins of the human immune system. The flip side of the coin mentioned in Section 1, in which *HLA* genetic polymorphisms result in variable susceptibility to hypersensitivities and other disorders, in this case, the viruses in the sBR group, appears to have undergone more rapid mutation in the antigens which cause them to be recognized by the immune system via HLA, thus evading it *(44)*.

4.2. HIV

Owing to the intensive investigations over the last decades, a comprehensive review of the pharmacogenetic issues surrounding HIV-AIDS progression and treatment would fill volumes. Thus I have little hope of serving the topic justice here. Beyond the variability within host and pathogen genomes as with HCV discussed above, the treatment of HIV involves multiple drugs of differing classes administered simultaneously. Overlapping resistances to one or more classes of drug are not uncommon. At least one commercially available test is designed to screen HIV genomic mutations known to confer resistance to specific antiretroviral drugs. The TRUGENE™ HIV-1 Genotyping Kit and OpenGene™ DNA Sequencing System were approved by the FDA in 2001 as an integrated system. Using this system, a patient's blood sample is simply sent to one of the scores of laboratories qualified to carry out the test, which is continually updated with new pharmacogenetic polymorphisms as they are discovered and verified.

Some of the more important traditional pharmacogenetic polymorphisms that have appeared following widespread application of highly active antiretroviral therapy (HAART) include differences in drug metabolism through cytochrome P450, as exemplified by CYP2D6, and clearance through the p-glycoprotein drug transporter (*mdr* gene product) *(45)*. Other polymorphisms have become apparent within the HIV pathogen which relate to small-molecule pharmacogenetics. For example, strains of HIV more prominent in Africa (HIV-A and HIV-C) have been shown to carry HIV-protease genes containing mutations within the protease inhibitor binding site which decrease their binding potential to protease inhibitor drugs as compared to the HIV-B strains to which the drugs were developed *(46)*. However, HAART therapy appears to be just as effective in African patients carrying non-B strains of HIV *(47)*.

In addition, known pharmacogenetic polymorphisms in human HLA genotypes influence potentially fatal hypersensitivity to some AIDS drugs such as abacavir (reviewed by Martin) *(48)*. In particular, patients carrying the HLA-B*5701 allele are over 100-fold more likely to exhibit this response *(49)*.

Some of the earliest applications of biopharmaceutical agents aimed at HIV-AIDS followed the discovery of the gp120/CD4 link, in which it was learned of the specific binding requirement between viral coat glycoprotein (gp120) and a surface marker on white blood cells that initiated the entry of the HIV genome into the host cell. Early strategies to block this interaction via MAbs and/or soluble CD4 antigens were met with much hope but little success. However, one of the currently hot drug targets in anti-HIV therapy involves the human chemokine 5 coreceptor (CCR5), a surface molecule found on human T cells that is used in conjunction with CD4 by the HIV coat proteins to recognize and enter their host. Numerous strategies are being developed that attempt to disrupt this interaction, including small molecules and biopharmaceuticals alike. At least three new small-molecule drugs are in phase II or III clinical trials at this time designed to specifically disrupt the CCR5–virus interaction by binding to CCR5 *(50)*. And already, a number of pharmacogenetic issues have arisen surrounding the potential clinical application of CCR5 inhibitors. Notably, an allele of CCR5 exists in the human population containing a 32 nucleotide basepair deletion as compared to the more common full-length gene. Homozygotic carriers of this allele (CCR5-delta32) have been shown to be resistant to HIV infection *(51)*, apparently because of an inability of HIV to then utilize this binding protein for entry into CD4+ T cells. The protective effect afforded by the CCR5-delta33 allele has lately been subjected to intense scrutiny in terms of evolutionary pressure, natural selection, and geographic spread *(52–54)*.

Recently, an HIV-1 clonal isolate was found to be resistant to several small-molecule inhibitors targeting entry via CCR5, yet was still unable to enter CCR5-delta32 homozygous cells. In addition, MAbs against CCR5 were able to block viral entry, and a specific inhibitor for an alternate corecognition protein (CXCR4) was ineffectual, indicating that the viral isolate was in fact still utilizing the CCR5 protein as an entry point. Not surprisingly, amino acid changes in these "escape mutants" were traced back to the HIV gp120 surface protein which binds to CCR5 + CD4 (or CXCR4 + CD4) *(55)*, illustrating the flip side of pharmacogenetic complications in antiviral therapy—real time changes in pathogen DNA sequence.

Finally, at least two novel biopharmaceutical approaches against HIV are being tested to disrupt the CCR5 interaction. In one approach, a bifunctional fusion protein containing a CCR5 binding domain and the Fv portion of an anti-CD4 antibody are combined in order to simultaneously disrupt both proteins required (CD4 and one of several chemokine coreceptors, CCR5 being one) for

HIV entry into the host cell. Using in vitro tests, the fusion protein was shown to be superior to the use of either protein binding agent alone *(56)*. The other approach utilizes a sort of molecular evolution of anti-CCR5 peptides through exon shuffling of a phage display library. Using this approach, a peptide with an IC-50 of 5 μM was generated which will need further testing to determine its therapeutic potential *(57)*.

4.3. Influenza Vaccination

The last example presently examined of the role viral genomics plays in pharmacogenetics is influenza, which despite its familiarity and perhaps lack of fear-invoking potential has in fact been one of the most devastating illnesses throughout history. We are of course familiar with the influenza epidemics which sweep the globe every year, making noses run and fevers rise. But these symptoms are mild in comparison to the influenza's true potential and only so because of our long history of exposure to the virus—we are survivors. Like other viruses, influenza mutates rapidly in the host, hence the need to develop strain-specific vaccines each year. But these mutations (termed antigenic drift) are minor in comparison—just enough to evade the immune system for yet another year and cause worldwide misery—enough even to kill the weak and elderly. Every so often, a more substantial change occurs in the architecture of the influenza virus, making it one that very few individuals have natural immunity to. The result is a pandemic—millions of deaths worldwide, and the next one could be just around the corner. Prudent governments worldwide are making preparations for combating the avian flu virus in an attempt to ensure it is not the source of the next human pandemic. Therefore, biopharmaceuticals used to combat or prevent influenza infection will undoubtedly be key players for decades and centuries to come.

Neutralizing antibodies against two influenza virus coat proteins, hemagglutinin and neuraminidase, are the body's chief defense against influenza. These antigenic determinants are the basis for influenza nomenclature, H1N5, for example, corresponding to the particular combination of these two coat proteins carried by a given strain. Likewise, vaccination strategies generally rely on the injection of a combination of these two proteins to induce antibodies against viral coat proteins prior to actual viral exposure. However, even when all non-genetic variables are taken into account, some individuals still do not mount an efficient immune response following vaccination. In other words, there is an apparent pharmacogenetic polymorphism in response to vaccines, as with other drugs and biopharmaceuticals, resulting in a class of nonresponders separating out from the main patient group.

Given the role of the HLA (Section 2.1.) in the display of peptides as a requisite step in the development of immunity, Lambkin et al. decided to look for polymorphisms within the *HLA* locus which might explain the lack of influenza vaccination response. That is, perhaps some individuals carry changes within their *HLA* Class II DNA sequences which encode for HLA/MHC proteins which disallow the efficient binding of peptides derived from the hemagglutinin and neuraminidase of the vaccine, thus preventing the desired antibody response to these proteins. What they found is evidence indicating that the HLA-DRB1*0701 allele is overexpressed in those individuals who fail to mount a neutralizing antibody response following influenza vaccination *(58)*. Other vaccine strategies have been shown to be subject to similar *HLA* polymorphisms, and future strategies would be wise to take this into consideration. Much more work needs to be done before we will fully understand the relationship between the protein antigenic structures and the HLA binding determinants which control this response before we can hope to prevent the next pandemic.

4.4. Bacterial Genomes: Future Prospects

Antibiotics today seemingly have as much potential for harm as good, as resistances to whole classes of drugs become widespread. Though no specific biopharmaceutical agents have been approved that take advantage of knowledge gained through comparative genomics, the potential of this arena makes it one that cannot be left without mentioning here. Gene expression profiling and the rapid DNA analysis technique of pyrosequencing *(59)* are just some of the mechanisms making it practical to predict the protein/antigen expression patterns of pathogenic microorganisms based on genetic sequence *(42)*. By targeting the right protein or combination of proteins and knowing in advance which ones are likely to be overexpressed by a pathogenic organism, we can better hope to eliminate unwanted bacteria while leaving the normal intestinal flora intact. With the inclusion of genomic information from the patient in the clinic, antibiotic therapy tailored to both host and pathogen may be achieved. When this becomes a reality, the term pharmacogenomics will truly apply.

5. Conclusions

In the preceding pages, I have attempted to collate and present some of the more well-known pharmacogenetic polymorphisms which occur with the use of biopharmaceutical agents, as well as to highlight some of the important overriding issues involved. Though we have gained much knowledge in the last 10 yr, the use of pharmacogenetic information in drug development and clinical practice is still in its infancy. Large-molecule drugs are becoming ever more practical; however, new approaches are still needed to overcome the limitations imposed by current delivery protocols. As evidenced by the growing body of literature, an understanding of pharmacogenetics is crucial for all aspects of drug discovery and clinical practice of biopharmaceutical agents. Unfortunately, a quick jump from "bench to bedside" or "genes to drugs" cannot be expected: the early history of biopharmaceutical drug development has made us realize that the biology of disease can be extremely complex. Only with an insight into the organism as a whole can we hope to properly target new drug entities. This then becomes the arena of pharmacogenomics and systems biology approaches—the studies of system- and organism-wide change that takes place in response to a drug or disease process. However, our ability to handle large amounts of information is ever increasing, and the arena of pharmacogenetics promises to expand with it.

References

1. Geraghty, D. E., Fortelny, S., Guthrie, B., et al. (2000) Data acquisition, data storage, and data presentation in a modern genetics laboratory. *Rev. Immunogenet.* **2**(4), 532–540.
2. Zhang, G., Zhang, S., Chen, W., et al. (2001) Go!Poly: A gene-oriented polymorphism database. *Hum. Mutat.* **18**(5), 382–387.
3. Stastny, P. (1983) Rheumatoid arthritis: relationship with HLA-D. *Am. J. Med.* **75**(6A), 9–15.
4. Ludwig, H., Schernthaner, G., and Mayr, W. R. (1977) The importance of HLA genes to susceptibility in the development of juvenile diabetes mellitus. A study of 93 patients and 68 first degree blood relations. *Diabetes Metab.* **3**(1), 43–48.
5. Vesell, E. S. (1984) Pharmacogenetic perspectives: genes, drugs and disease. *Hepatology* **4**(5), 959–965.
6. Wong, B. R., Parlati, F., Qu, K., et al. (2003) Drug discovery in the ubiquitin regulatory pathway. *Drug Discov. Today* **8**(16), 746–754.
7. Stocks, M. R. (2004) Intrabodies: production and promise. *Drug Discov. Today* **9**(22), 960–966.
8. Dreyfus, D. H., Matczuk, A., and Fuleihan, R. (2004) An RNA external guide sequence ribozyme targeting human interleukin-4 receptor alpha mRNA. *Int. Immunopharmacol.* **4**(8), 1015–1027.
9. Bhargava, A., Dallman, M. F., Pearce, D., and Choi, S. (2004) Long double-stranded RNA-mediated RNA interference as a tool to achieve site-specific silencing of hypothalamic neuropeptides. *Brain Res. Brain Res. Protoc.* **13**(2), 115–125.
10. Akaneya, Y., Jiang, B., and Tsumoto, T. (2005) RNAi-induced gene silencing by local electroporation in targeting brain region. *J. Neurophysiol.* **93**(1), 594–602.
11. Hitzig, W. H. (2003) The discovery of agammaglobulinaemia in 1952. *Eur. J. Pediatr.* **162**(5), 289–304.
12. Vihinen, M., Kwan, S. P., Lester, T., et al. (1999) Mutations of the human BTK gene coding for bruton tyrosine kinase in X-linked agammaglobulinemia. *Hum. Mutat.* **13**(4), 280–285.
13. Morell, A., Skvaril, F., Radl, J., Dooren, L. J., and Barandun, S. (1975) IgG - subclass abnormalities in primary immunodeficiency diseases. *Birth Defects Orig. Artic. Ser.* **11**(1), 108–111.
14. Iglesias Alzueta, J. and Matamoros Flori, N. (2001) [Common variable immunodeficiency. Review]. *Allergol Immunopathol. (Madr)* **29**(3), 113–118.

15. Carter, P. (2001) Improving the efficacy of antibody-based cancer therapies. *Nat. Rev. Cancer* **1**(2), 118–129.

16. Anderson, D. R., Grillo-Lopez, A., Varns, C., Chambers, K. S., and Hanna, N. (1997) Targeted anti-cancer therapy using rituximab, a chimaeric anti-CD20 antibody (IDEC-C2B8) in the treatment of non-Hodgkin's B-cell lymphoma. *Biochem. Soc. Trans.* **25**(2), 705–708.

17. Maloney, D. G., Grillo-Lopez, A. J., Bodkin, D. J., et al. (1997) IDEC-C2B8: results of a phase I multiple-dose trial in patients with relapsed non-Hodgkin's lymphoma. *J. Clin. Oncol.* **15**(10), 3266–3274.

18. Maloney, D. G., Grillo-Lopez, A. J., White, C. A., et al. (1997) IDEC-C2B8 (Rituximab) anti-CD20 monoclonal antibody therapy in patients with relapsed low-grade non-Hodgkin's lymphoma. *Blood* **90**(6), 2188–2195.

19. Anolik, J., Sanz, I., and Looney, R. J. (2003) B cell depletion therapy in systemic lupus erythematosus. *Curr. Rheumatol. Rep.* **5**(5), 350–356.

20. Dimopoulos, M. A., Zervas, C., Zomas, A., et al. (2002) Treatment of Waldenstrom's macroglobulinemia with rituximab. *J. Clin. Oncol.* **20**(9), 2327–2333.

21. Gertz, M. A., Rue, M., Blood, E., Kaminer, L. S., Vesole, D. H., and Greipp, P. R. (2004) Multicenter phase 2 trial of rituximab for Waldenstrom macroglobulinemia (WM): an Eastern Cooperative Oncology Group Study (E3A98). *Leuk. Lymphoma* **45**(10), 2047–2055.

22. Leandro, M. J., Edwards, J. C., and Cambridge, G. (2002) Clinical outcome in 22 patients with rheumatoid arthritis treated with B lymphocyte depletion. *Ann. Rheum. Dis.* **61**(10), 883–838.

23. Moore, J., Ma, D., Will, R., Cannell, P., Handel, M., and Milliken, S. (2004) A phase II study of Rituximab in rheumatoid arthritis patients with recurrent disease following haematopoietic stem cell transplantation. *Bone Marrow Transplant.* **34**(3), 241–247.

24. Choquet, S., Leblond, V., Herbrecht, R., et al. (2006) Efficacy and safety of rituximab in B-cell post-transplant lymphoproliferative disorders: results of a prospective multicentre phase II study. *Blood* **107**(8), 3053–3057.

25. Pijpe, J., van Imhoff, G. W., Spijkervet, F. K., et al. (2005) Rituximab treatment in patients with primary Sjogren's syndrome: an open-label phase II study. *Arthritis Rheum.* **52**(9), 2740–2750.

26. Pranzatelli, M. R., Tate, E. D., Travelstead, A. L., and Longee, D. (2005) Immunologic and clinical responses to rituximab in a child with opsoclonus-myoclonus syndrome. *Pediatrics* **115**(1), e115–e119.

27. Eisenberg, R. and Looney, R. J. (2005) The therapeutic potential of anti-CD20 "What do B-cells do?" *Clin. Immunol.*.

28. Keystone, E. (2005) B cell targeted therapies. *Arthritis Res. Ther.* **7**(Suppl 3), S13–S18.

29. Baselga, J. (2001) Herceptin alone or in combination with chemotherapy in the treatment of HER2-positive metastatic breast cancer: pivotal trials. *Oncology* **61** (Suppl 2), 14–21.

30. Seidman, A., Hudis, C., Pierri, M. K., et al. (2002) Cardiac dysfunction in the trastuzumab clinical trials experience. *J. Clin. Oncol.* **20**(5), 1215–1221.
31. Peng, B., Lloyd, P., and Schran, H. (2005) Clinical pharmacokinetics of imatinib. *Clin. Pharmacokinet.* **44**(9), 879–894.
32. Tseng, P. H., Lin, H. P., Zhu, J., et al. (2005) Synergistic interactions between imatinib mesylate and the novel phosphoinositide-dependent kinase-1 inhibitor OSU-03012 in overcoming imatinib mesylate resistance. *Blood* **105**(10), 4021–4027.
33. Gorre, M. E., Ellwood-Yen, K., Chiosis, G., Rosen, N., and Sawyers, C. L. (2002) BCR-ABL point mutants isolated from patients with imatinib mesylate-resistant chronic myeloid leukemia remain sensitive to inhibitors of the BCR-ABL chaperone heat shock protein 90. *Blood* **100**(8), 3041–3044.
34. Deleuran, B. W., Chu, C. Q., Field, M., et al. (1992) Localization of interleukin-1 alpha, type 1 interleukin-1 receptor and interleukin-1 receptor antagonist in the synovial membrane and cartilage/pannus junction in rheumatoid arthritis. *Br. J. Rheumatol.* **31**(12), 801–809.
35. Chomarat, P., Vannier, E., Dechanet, J., et al. (1995) Balance of IL-1 receptor antagonist/IL-1 beta in rheumatoid synovium and its regulation by IL-4 and IL-10. *J. Immunol.* **154**(3), 1432–1439.
36. Firestein, G. S., Boyle, D. L., Yu, C., et al. (1994) Synovial interleukin-1 receptor antagonist and interleukin-1 balance in rheumatoid arthritis. *Arthritis Rheum.* **37**(5), 644–652.
37. Hannum, C. H., Wilcox, C. J., Arend, W. P., et al. (1990) Interleukin-1 receptor antagonist activity of a human interleukin-1 inhibitor. *Nature* **343**(6256), 336–340.
38. Bresnihan, B. (1999) Treatment of rheumatoid arthritis with interleukin 1 receptor antagonist. *Ann. Rheum. Dis.* **58**(Suppl 1), I96–198.
39. Camp, N. J., Cox, A., di Giovine, F. S., McCabe, D., Rich, W., and Duff, G. W. (2005) Evidence of a pharmacogenomic response to interleukin-1 receptor antagonist in rheumatoid arthritis. *Genes Immun.* **6**(6), 467–471.
40. Bansback, N. J., Regier, D. A., Ara, R., et al. (2005) An overview of economic evaluations for drugs used in rheumatoid arthritis: focus on tumour necrosis factor-alpha antagonists. *Drugs* **65**(4), 473–496.
41. Jorge, A. A. L., Marchisotti, F. G., Montenegro, L. R., Carvalho, L. R., Mendonca, B. B., and Arnhold, I. J. P. (2005) Growth hormone (GH) pharmacogenetics: influence of GH receptor exon 3 retention or deletion on first-year growth response and final height in patients with severe GH deficiency. *J. Clin. Endocrinol. Metab.* (doi:10.1210/jc.2005-2005).
42. Hayney, M. S. (2002) Pharmacogenomics and infectious diseases: impact on drug response and applications to disease management. *Am. J. Health Syst. Pharm.* **59**(17), 1626–1631.
43. Yee, L. J., Perez, K. A., Tang, J., van Leeuwen, D. J., and Kaslow, R. A. (2003) Association of CTLA4 polymorphisms with sustained response to interferon and ribavirin therapy for chronic hepatitis C virus infection. *J. Infect. Dis.* **187**(8), 1264–1271.

44. Tanabe, Y., Nagayama, K., Enomoto, N., et al. (2005) Characteristic sequence changes of hepatitis C virus genotype 2b associated with sustained biochemical response to IFN therapy. *J. Viral Hepat.* **12**(3), 251–261.

45. Fellay, J., Marzolini, C., Meaden, E. R., et al. (2002) Response to antiretroviral treatment in HIV-1-infected individuals with allelic variants of the multidrug resistance transporter 1: a pharmacogenetics study. *Lancet* **359**(9300), 30–36.

46. Velazquez-Campoy, A., Vega, S., and Freire, E. (2002) Amplification of the effects of drug resistance mutations by background polymorphisms in HIV-1 protease from African subtypes. *Biochemistry* **41**(27), 8613–8619.

47. Frater, A. J., Beardall, A., Ariyoshi, K., et al. (2001) Impact of baseline polymorphisms in RT and protease on outcome of highly active antiretroviral therapy in HIV-1-infected African patients. *Aids* **15**(12), 1493–1502.

48. Martin, A. M., Nolan, D., Gaudieri, S., Phillips, E., and Mallal, S. (2004) Pharmacogenetics of antiretroviral therapy: genetic variation of response and toxicity. *Pharmacogenomics* **5**(6), 643–655.

49. Mallal, S., Nolan, D., Witt, C., et al. (2002) Association between presence of HLA-B*5701, HLA-DR7, and HLA-DQ3 and hypersensitivity to HIV-1 reverse-transcriptase inhibitor abacavir. *Lancet* **359**(9308), 727–732.

50. Pharmd, V. I. (2005) Human Immunodeficiency Virus (HIV) Entry Inhibitors (CCR5 Specific Blockers) in Development: Are They the Next Novel Therapies? *HIV Clin. Trials* **6**(5), 272–277.

51. Bogner, J. R., Lutz, B., Klein, H. G., Pollerer, C., Troendle, U., and Goebel, F. D. (2004) Association of highly active antiretroviral therapy failure with chemokine receptor 5 wild type. *HIV Med.* **5**(4), 264–272.

52. Galvani, A. P. and Novembre, J. (2005) The evolutionary history of the CCR5-Delta32 HIV-resistance mutation. *Microbes Infect.* **7**(2), 302–309.

53. Sabeti, P. C., Walsh, E., Schaffner, S. F., et al. (2005) The case for selection at CCR5-Delta32. *PLoS Biol.* **3**(11), e378.

54. Novembre, J., Galvani, A. P., and Slatkin, M. (2005) The geographic spread of the CCR5 Delta32 HIV-resistance allele. *PLoS Biol.* **3**(11), e339.

55. Marozsan, A. J., Kuhmann, S. E., Morgan, T., et al. (2005) Generation and properties of a human immunodeficiency virus type 1 isolate resistant to the small molecule CCR5 inhibitor, SCH-417690 (SCH-D). *Virology* **338**(1), 182–199.

56. Mack, M., Pfirstinger, J., Haas, J., et al. (2005) Preferential Targeting of CD4-CCR5 Complexes with Bifunctional Inhibitors: A Novel Approach to Block HIV-1 Infection. *J. Immunol.* **175**(11), 7586–7593.

57. Vyroubalaova, E. C., Hartley, O., Mermod, N., and Fisch, I. (2006) Identification of peptide ligands to the chemokine receptor CCR5 and their maturation by gene shuffling. *Mol. Immunol.* **43**(10), 1573–1578. Epub 2005 Nov 8.

58. Lambkin, R., Novelli, P., Oxford, J., and Gelder, C. (2004) Human genetics and responses to influenza vaccination: clinical implications. *Am. J. Pharmacogenomics* **4**(5), 293–298.

59. Clarke, S. C. (2005) Pyrosequencing: nucleotide sequencing technology with bacterial genotyping applications. *Expert Rev. Mol. Diagn.* **5**(6), 947–953.

7

Development and Applications of Transgenics in Biotechnology and Medicine

Wagner Dos Santos and Helen L. Fillmore

Abstract

The possibility of expressing foreign genes in mammals and plants by gene transfer has opened new dimensions in the genetic manipulation of these organisms. The use of transgenic animals as an experimental system for the study of gene regulation, genetic modeling of diseases, and testing of novel therapies or as a way to produce important bioactive drugs has provided great advances in agriculture and medicine. Transgenic technology has been used successfully to generate animals exhibiting features associated with human diseases or genetic disorders such as hemoglobinopathies, diabetes, cystic fibrosis (CF), Huntington's and Alzheimer's diseases providing significant advances in understanding the development and pathophysiological aspects of these diseases. Plants have been generated to produce therapeutic proteins such as antibodies, blood products, cytokines, growth factors, hormones, and a variety of human and veterinary vaccines. In this chapter, we discuss the technology associated with the generation of transgenic animals, the new developments, and applications where transgenics have proven invaluable and promising.

Key Words: Transgenic; gene transfer; molecular farming; biotechnology; gene expression.

1. Introduction

Transgenic technology, which has the ability to introduce functional genes into animals, is a powerful and dynamic tool for dissecting complex biological processes. The potential applications and questions that can be addressed by using this technology are vast, encompassing scientific spectrum ranging from biomedical and biological mechanisms to production of bioactive drugs. Here we aim to give an overall discussion on transgenic technology development, the techniques involved, and its impact and applications in biotechnology, medicine, and pharmacy.

From: *Biopharmaceutical Drug Design and Development*
Edited by: S. Wu-Pong and Y. Rojanasakul © Humana Press Inc., Totowa, NJ

2. Background

2.1. The Origins of Transgenic Technology

The term transgenic was used for the first time by Gordon & Ruddle (1982) *(1,2)* to describe animals harboring new genes within their genomes. Now the term is more generally applied to the characterization of certain variants of species whose genome has been altered by the transfer of genes. Although several steps in the development of transgenic technology had been performed before 1970s, transgenesis was not widely recognized until the pioneer work by Palmiter et al. *(3)* in 1982. In this work they introduced the human growth hormone gene into mouse zygotes using a pronuclear microinjection method and the resulting transgenic offspring demonstrated a dramatic change in growth. Along with this work, other laboratories reported success at gene transfer *(4–8)*. Following these studies the transgenic technology was further improved and reached a status of extreme importance in biomedical and biopharmaceutical research allowing advances over general cell culture techniques. The manufacture of large quantities of complex bioactive proteins like hormones or growth factors for therapeutic purposes is only one example of a wide range of different applications that can be realized by transgenic technology. For this purpose, foreign DNA is introduced into fertilized oocytes or embryos of mice, rats, and other mammals *(9–11)*.

2.2. Methods Used for Gene Transfer

Transgenic methodologies that are currently utilized in laboratories have been pioneered using the mouse model. Today the mouse continues to serve as a starting point for implementing gene transfer procedures and is the standard for optimizing experimental efficiencies for other species. The production of transgenic mice has been paramount for the development of animal biotechnology. A close look in the early events leading to the first genetically engineered mice demonstrate that the procedure for DNA microinjection was described 25 years ago *(12,13)*. Since then, advances and numerous strategies have been made for producing genetically engineered animals which extend from mechanistic procedures (DNA microinjection, embryonic stem cell- or retrovirus-mediated transfer) to molecular (cloning) techniques.

There are essentially two ways of generating animals with the capacity to transmit a genetic element through the germline to their offspring. These are (1) injection of DNA into the pronucleus of a newly fertilized egg and (2) genetic manipulation of embryonic stem cells. Both methods have been employed to generate transgenic mice.

The goal in transgenic technology is to deliberately insert a gene into a host genome. This gene is prepared by recombinant DNA methodology and includes not only the DNA sequence of the gene itself but also other sequences that help the gene incorporate into the host DNA, as well as sequences needed for the gene to be expressed by the cells of the host. There are many considerations for DNA preparation for gene transfer technologies and include either the use of genomic DNA or cDNA or inclusion of a polyadenylation signal. Another consideration depending on the application is the selection of a desired promoter that can, for example, drive tissue/cell-specific expression of the transgene.

2.2.1. Pronuclear Injection

Pronuclear injection has been the major method used for generating transgenic animals. The pronucleus is the nucleus of either an egg cell or a sperm during the process of fertilization. Usually the male pronucleus is used for the injection, as it is normally larger than the female pronucleus. Fertilized eggs are collected from superovulated donors. The egg is held securely with a glass pipette, and the DNA solution is injected into a pronucleus by insertion of a fine glass injection needle. The injected DNA is integrated into the genome of 10–40% of surviving embryos. The injected zygotes are then transferred into foster mothers and allowed to develop to term. Carrier transgenic animals can pass the transgene through the germline as stable genetic information. In this way, transgenic mice, rats, pigs, and other animals have been generated. The advantage of this method is that large fragments of DNA can be injected into the pronucleus, thus allowing complete genes with their associated regulatory regions to be introduced into the zygote.

Many criteria considered important for successful production of transgenic mice using pronuclear injection were defined two decades ago *(14)*. Some considerations are: (1) linear DNA fragments integrate with greater efficiency than supercoiled DNA and the DNA fragment size or length does not affect integration frequency; (2) use of a low ionic strength microinjection buffer consisting of 10 mM Tris, pH 7.4, with 0.1–0.3 mM EDTA provides good results; (3) the DNA concentration between 1.0 and 2.0 ng/µL appears to be the most efficient range to produce transgenic mice (DNA integration and development of microinjected eggs to term); (4) linear DNA fragments with blunt ends have the lowest chromosomal integration frequency; (5) injection of DNA into the male pronucleus is slightly more efficient than injection into the female pronucleus; and (6) nuclear injection of foreign DNA is dramatically more efficient than cytoplasmic injection.

2.2.2. Embryonic Stem Cell-Mediated Gene Transfer

Embryonic stem (ES) cells are derived from a preimplantation-stage embryo, usually at the 3.5-d blastocyst stage. Early embryos are flushed from the uterine horns and maintained in cell culture medium in order to harvest ES cells from the inner cell mass of the blastocysts. Foreign DNA is added to the ES cells and transformation methods are used to promote the incorporation of the foreign

DNA to the host genome. Successfully transformed cells are selected and injected into the host inner cell mass of a blastocyst and then implanted into a pseudopregnant mouse. The transformed ES cells will combine with the inner cell mass component and contribute to the developing embryo *(15)*. Offspring have somatic tissue composed of both ES cell-derived cells and host blastocyst-derived cells. Offspring DNA is tested for the presence of the foreign DNA. Typically 10–20% of the offspring will be heterozygous for the gene. These mice can be mated and screened for homozygosity for the transgene.

Retroviruses have also been utilized in the production of transgenic animals. Retroviruses carrying recombinant genes have been used in gene transfer for a wide range of purposes since the early 1980s *(16–19)*. The major driving force stimulating the development of this procedure was the desire for a highly efficient gene transfer method for potential gene therapy applications in human diseases. Soon this technology was applied to the production of transgenic animals. By using this method in 4- to 16-cell stages of mouse embryos it is possible to produce mosaic transgenic mice, where not all somatic cells will contain the proviral insert. When preimplantation embryos are exposed to a retrovirus, a proportion of embryonic cells will stably integrate proviral sequences into their genome, usually as one copy per cell. Jaenisch *(20,21)* showed that adult mice derived after Moloney murine leukemia virus (M-MuLV) infection could transmit integrated proviral sequences through the germline. Unfortunately, preimplantation mouse embryos are not permissive for M-MuLV expression because of its promoter, the long terminal repeat promoter (LTR). Genes driven by this promoter are not expressed in the embryo. On integration, the provirus is subject to *de novo* methylation *(22,23)*, effectively shutting off proviral transgene expression even in cell lineages derived from the original infected cells. This problem has been circumvented at some extent by the use of an internal promoter to drive the expression of the transgene. Such a promoter can be aimed at providing either ubiquitous expression of a transgene as in the case of herpes simplex virus thymidine kinase promoter *(24)* or cell-specific expression as reported for β-globin promoter driving expression in hematopoietic tissues *(25)*. The major advantages of retroviral infection as a method of gene transfer include the fact that the recombinant proviral sequence transferred is integrated stably into the genome of the recipient cell as a single, randomly located integrant with predictable molecular structure and no subsequent cytopathy. Another advantage relates to the fact that the efficiency of gene transfer is very high compared to other nonviral methods. However, some limitations on the use of retroviruses to generate transgenic mice exist. One of them is the packaging limits of the virus, which restrict the insert size to approximately 9 kb. If the insert is much larger than this, the viral RNA cannot be packaged into the viral capsid. Also, the production of mosaic transgenic animals may occur depending on the developmental stage at which the cells are infected by the retrovirus and the number of embryonic cells infected. Therefore, only when the provirus is transmitted through the germline, the animal can truly be considered a transgenic.

3. Transgenic Technology Applications

Transgenic technology in biomedicine and pharmacy has opened a new era for animal model creation and drug testing. The successful development of transgenic animal models for human diseases has led to remarkable break-throughs that have significantly influenced approaches to the diagnosis, treatment, and intervention of human diseases. Moreover, transgenic animal models have clarified and shed light on our understanding of disease mechanisms and the onset and course of pathology associated with the disease.

For several reasons, the transgenic mouse has been the most commonly used animal model. Some of these reasons include the availability of extensive information for particular strains, well-developed techniques in handling the gametes, embryos, and surrogates, inexpensive and relatively limitless supplies, and short generation times. The transgenic mice will continue to play a critical role in the development of models for human diseases. However, transgenic technology has not been limited only to mice but also been extended to a variety of species including rats, rabbits, swine, ruminants (sheep, goats, and cattle), poultry, and fish. Furthermore, not only have animals been genetically engineered, but plants have also become more and more an attractive model for generating transgenics aimed at the development of pharmaceutical products or production of new varieties more resistant to infection *(26,27)*. All the applications this exciting technology has offered as well as all the transgenic animal models are out of scope of this chapter. Therefore among the numerous applications that transgenic technology has been successful, some will be discussed below in different categories.

3.1. Determination of Normal Gene Function

Deregulation of transgene expression in a whole animal environment has proven to be a useful tool to assess the normal function of a gene product. For example, deregulation of *c-fos* expression under the control of the metallothionein I promoter interfered with normal bone development, suggesting a role for this protein in bone modeling *(28)*. On the other hand, under the control of the H-2K promoter, expression of *c-fos* specifically interfered with thymus development even though the transgene was also expressed in other tissues *(29)*. Transgenic mice with inappropriate expression of the v-*mos* proto-oncogene develop neuro-pathological changes in the brain, suffer progressive limb paralysis, and show aberrant eye lens fiber differentiation *(30,31)*. However, it is worthwhile to note that when a transgene is expressed in many tissues, including those where the endogenous version of the gene is not normally expressed, it is questionable whether abnormalities in growth and development are directly related to the normal function of the gene.

The activity of an individual gene may also be neutralized by the transfer of a gene construct which contains a structural gene encoding an antisense RNA for the corresponding endogenous transcript. Intracellular hybridization of antisense RNA with the mRNA (sense RNA) encoded by the endogenous gene under study either effectively inhibits translation or leads to the synthesis of functionally impaired protein fragments. The efficiency of the inhibition of the endogenous gene expression primary depends on the synchronous or overlapping expression of the antisense transgene and the endogenous gene. It also depends on the relative excess of the antisense transcript because only this will guarantee that some of the endogenous RNA molecules of the structural gene in question will indeed be bound to the antisense RNA.

Integration of exogenous DNA fragments, such as transgenes or viruses, can induce a mutation in the gene the function of which is to be determined. The gene function characterization should be easy to determine because the exogenous DNA fragment has a known sequence or structure and can therefore act as a "molecular tag". These so-called insertional mutants have been generated by the methods discussed earlier to generate transgenes. Probing the mutant genome with the exogenous sequence has greatly facilitated the cloning and identification of the gene involved in each case.

3.2. Animal Models of Human Diseases and Disorders

Undoubtedly the most important use of transgenic animals has been in the identification of genes associated with human diseases and the understanding of the role of these genes in pathology associated with the diseases. Transgenic animals have been generated that show some of the physiological and pathological changes associated with human genetic disorders or diseases. These animals can be used to study disease progression or to test potential pharmaceuticals for pharmacologic potential. Behringer and coworkers (32) demonstrated that α- and β-globin genes could be correctly coexpressed in erythroid tissues of transgenic mice setting the foundation for attempts to generate mouse models of various hemoglobinopathies such as sickle cell disease and thalassemias. Expression of the sickle hemoglobin transgene in β-thalassemic mice produced partially anemic animals with erythrocytes presenting sickle shape when subjected to low oxygen tension (33,34).

Diabetes mellitus is another disease that has been studied in transgenic animal models. The overexpression of class I histocompatibility antigen (H-2K) in the pancreatic β-cells of transgenic mice induces insulin-dependent diabetes without an autoimmune response *(35)*. In another study, overexpression of calmodulin, a Ca^{2+}-binding protein involved in signal transduction, induces early onset of diabetes within hours after birth *(36,37)*.

Alzheimer's disease (AD) is characterized clinically by progressive memory loss that leads eventually to dementia. The neuropathology of AD involves neuronal and synaptic loss and also the development of two lesions: extracellular senile plaques, which are composed mostly of amyloid formed from the amyloid β peptide, and intraneuronal neurofibrillary tangles, composed of hyperphosphorylated forms of the microtubule-associated protein tau (MAPT) *(38)*. Several genes have been implicated in AD in humans, most notably, those encoding βA4 precursor protein (APP), presenilin 1 (PSEN1) and presenilin 2 (PSEN2). The first mouse models that developed amyloid plaque pathology were generated by expressing human APP containing mutations associated with early-onset AD. The first published AD transgenic mice *(39)* named PDAPP overexpressed a minigene construct encoding a V717F mutant form of the amyloid β precursor protein. These animals developed a robust amyloid plaque pathology by 6–9 months. The Tg2576 model, which overexpresses a human APP cDNA transgene with the K670M/N671L double mutation, developed amyloid plaque pathology in an age-dependent manner and was also shown to have correlative memory deficits as determined by Morris water-maze testing *(40–42)*. Since the reports showing that APP mutant transgenic mice develop amyloid plaques, additional novel models have been developed. Notable is the result of crossing APP mutant mice with PSEN1 transgenics (named PSAPP mice). These mice dramatically accelerate amyloid deposition because of the increase in Ab42 production mediated by PSEN1 mutations. The pathology in these transgenic mice include diffuse amyloid deposits and dense fibrillar plaques that resemble the senile plaques in human AD *(41,43,44)*. Despite the robust amyloid deposition observed in these models none of them develop a widespread neuronal loss or MAPT pathology. This observation may reflect the limitations of using rodent system to reproduce a human disease process that takes several decades and primarily involves higher cognitive function *(45)*.

Gene transfer technology has also been applied to generate mouse models of human genetic disorders. Several genes involved in human genetic syndromes have been mapped and the causative mutation identified. This has enabled engineering of analogous mutations into the murine homologue of that specific gene. One of the first mouse loci to be mutated was the hypoxanthine phosphoribosyltransferase (*hprt*) gene in an attempt to create an animal model for Lesch–Nyhan syndrome, a sex-linked recessive disease causing neurological and behavioral problems *(46,47)*. Surprisingly, HPRT-deficient mice are relatively normal and show no major metabolic or neurological characteristics associated with the syndrome. The lack of a phenotype in these mutant mice appears to be caused by the purine metabolism differences between rodents and humans. HPRT is the key enzyme for purine salvage in humans whereas adenine phosphoribosyltransferase

(APRT) is more important in rodents. However, the administration of an APRT inhibitor to HPRT-deficient mice induces persistent self-injurious behavior, which is one of the behavior alterations associated with the Lesch–Nyhan syndrome *(48)*. Although this model was not completely perfect, this work was a landmark in showing that human disorders may be produced in animals.

A model for Gaucher's disease, characterized by a lysosomal storage disorder, was generated by disrupting the glucocerebrosidase gene. This disease results from an autosomally inherited deficiency of the enzyme glucocerebrosidase or beta-D-glucosyl-*N*-acylsphingosine glucohydrolase *(49)*. Mice homozygous for the mutation have less than 4% of normal glucocerebrosidase activity, fail to degrade the sphingolipid glucocerebroside, and die within 24 hr of birth. In an attempt to generate a mouse model for arteriosclerosis, Maeda and coworkers tried to inactivate genes involved in lipid metabolism. Apolipoprotein A-I (apoA-I) is the major protein complexed with HDL in mammals and also participates in cholesterol metabolism. In humans, mutations of the gene encoding apoA-I are correlated with predisposition to arteriosclerosis. Mice lacking apoA-I protein show a marked reduction of plasma HDL cholesterol *(50)*. Because a reduction in plasma HDL levels in humans is associated with an increased risk of arteriogenesis it was expected that these animals would develop arteriosclerotic plaques with age. Apolipoprotein E (ApoE) is a glycoprotein that forms aggregates apart from low-density lipoprotein and is also involved in lipid metabolism. It functions primarily as a ligand for specific receptor containing particles to be removed from the circulatory system by the liver for further processing. Mice lacking ApoE protein have been generated. These animals show elevated cholesterol levels and develop spontaneous arterial lesions that gradually occlude the coronary and pulmonary arteries *(51)*.

The gene mutation associated with the genetic neurodegenerative disorder Huntington's disease (HD) has been known for over a decade but still no effective treatment is available. Attempts to generate transgenic models with the aim to develop novel therapeutic strategies have been made. The first transgenic mouse models for HD, named R6/1 and R6/2, were developed in 1996 *(52)*. These models were followed by many new HD transgenic lines of mice differing in the type of mutation expressed, portion of the protein included in the transgene, promoter employed, and level of expression of the mutant protein *(53,54)*.

CF is the most common autosomal recessive genetic disorder among Caucasians and is characterized by several symptoms, including elevated salt levels in sweat, hyperaccumulation of mucus in the airways and gastrointestinal tract, pancreatic enzyme insufficiency, deregulated absorption of intestinal contents, intestinal obstructions, and male sterility. CF transmembrane conductance regulator (*cftr*) gene was identified as the gene responsible for this genetic disorder.

Successful generation of mice carrying a disrupted *cftr* gene has been reported almost simultaneously *(55–58)*. These mice obtained by different procedures presented different phenotypes but essentially correlated to CF.

The list of human genetic diseases reproduced in transgenic mice continues to grow as well as improvement of old models to resemble more the clinical and pathological symptoms.

3.3. Oncogenes

Proto-oncogenes are important during normal development; however, when they are mutated and become oncogenes they may induce uncontrolled proliferation that is characteristic of cancer. Transgenic animals expressing activated oncogenes that predictably develop specific types of tumors are very useful in many fields of cancer research including pathogenesis studies, testing of carcinogenicity of certain compounds, and screening of anticancer compounds.

Oncogene transgenics may be used for studying the effects and consequences of its expression in animals. Both viral and cellular oncogenes have been used for generating transgenic mice. Analysis of such mice has increased our understanding of the mechanisms of oncogene function during normal and malignant development at the molecular level.

A variety of viral oncogenes, including the large T-antigen gene of SV40 virus, polyoma virus large and middle T genes, bovine papilloma virus, human JC and BK viruses, human T-cell leukemia virus tat oncogene, and human hepatitis B virus as well as cellular oncogenes such as *ras*, *myc*, and *abl* have been used for generating transgenic mice. One of the first transgenics, the so-called Onco mouse, was created by Leder et al. *(59)*. This mouse strain was generated by using a mammary tumor virus LTR/c-myc fusion gene and was expressed in a wide variety of tissues. Transgenic females expressing activated *ras* oncogene by the mammary tumor virus (MMTV) promoter and under hormonal control typically develop breast cancer on sexual maturation. Transgenic mice expressing an MMTV-TGFα fusion gene developed mammary gland hyperplasia culminating in adenocarcinoma development *(60)*. This tumor was shown to be accelerated and more dramatic after treatment of these transgenic with 7,12-dimethylbenzanthracene (DMBA) *(61)*, a chemical carcinogen used to induce tumors in mice.

SV40 viral-induced oncogenesis has also been the subject of much research. Use of SV40 (Simian virus) genes and the murine metallothionein promoter to produce transgenic mice caused papillomas and carcinomas of the choroid plexus *(62)*. Further investigations have identified a 72 bp element of the SV40 promoter as the crucial sequence for the genesis of these tumors *(63)*. Transgenic mice expressing SV40 T antigen in the retina develop heritable ocular tumors with histological, ultrastructural, and immunohistochemical features identical to those of human retinoblastoma, an autosomal recessive eye malignancy *(64)*.

4. Biotechnology

4.1. Transgenic Animals in the Production of Therapeutic Proteins

The possibility of expressing foreign genes in mammals by gene transfer has opened new dimensions in the genetic manipulation of animals. The techniques developed in mice were soon applied to larger animals offering the prospect of completely new breeding strategies and other novel applications focused on the use of animals as bioreactors and extending the transgenic technology to the so-called gene or molecular farming.

The major goal of gene farming is the production of recombinant proteins in the milk of transgenic animals *(65)*. The production of human pharmaceuticals in farm animals has been gaining application and becoming more popular particularly after the development of the first mice to produce a human drug tPA (tissue plasminogen activator) to treat blood clots in 1988 *(66)*. The strategy used to achieve these objectives is still being used today and consists of coupling the DNA gene for the protein drug of interest with a DNA signal (promoter sequence) directing production in the mammary gland. The new gene, although present in every cell of the animal, functions only in the mammary gland, so the protein drug is made only in the milk. Because the mammary gland and milk are essentially "outside" the main life support systems of the animal, there is virtually no danger of disease or harm to the animal in making the "foreign" protein drug. Production of recombinant proteins in the milk of transgenic farm animals has been well documented for goats *(67,68)*, rabbits *(69,70)*, pigs *(71)*, cattle *(9,72)*, and sheeps *(73–75)*.

Applications of gene transfer into farm animals from the biotechnology point of view fall basically into three groups: (1) the improvement of production efficiency and quality of animal products, (2) the production of new proteins of high quality, and (3) the creation of animal models for human diseases and organs for xenotransplantation. The first group is the most obvious application not only for economic reasons but, more important, for satisfying the ever-growing requirement for food as the word's population increases exponentially. The second group of applications that has been used in farm animals is the synthesis of proteins that is impossible or very difficult to produce at high purity, as raw materials for industrial processing; pharmaceuticals for human and veterinary medicine such as vaccines, growth factors, blood coagulation factors, antibodies, and so on; enzymes; or nutrients.

Another aspect that is potentially of great importance for medicine is the genetic alteration of animal organs so that they can be used for transplantation (xenotransplantation) into humans without being rejected by the recipient. Although much progress has been achieved in gene farming the methods of gene transfer in large mammals are still expensive. However, once technical optimizations are overcome, substantial reductions in cost should be expected.

4.2. Transgenic Plants as Pharmaceutical Factories

So far we have been focusing exclusively on gene transfer applied to animals; however, this technology has also been applied to generate transgenic plants. Plants have been used for medicinal purpose since the earliest stages of civilization. The active ingredients of many plants have now been identified, and close to one quarter of prescription drugs are still of plant origin. Gene transfer to generate transgenic organisms has extended the use of plants beyond its original boundaries.

The first recombinant plant-derived pharmaceutical protein was human serum albumin, initially produced in 1990 in transgenic tobacco *(76)*. Sixteen years have passed and many proteins produced in transgenic plants are on the market, and proof of concept has been established for the production of many therapeutic proteins including mammalian antibodies *(77,78)*, blood substitutes, cytokines, growth factors *(79,80)*, hormones, and vaccines *(81–88)*. Furthermore, several plant-derived pharmaceutical products for the treatment of human diseases are approaching commercialization including recombinant gastric lipase for the treatment of CF (Meristem Therapeutics), antibodies for the prevention of dental caries (Planet Biotechnology Inc.) and the treatment of non-Hodgkin's disease (Large Scale Biology Corp.). There are also several veterinary vaccines in the pipeline. Dow AgroSciences (Indianapolis, IN) recently announced their intention to produce plant-based vaccines for animal health industry.

Historically, bacteria were often the protein expression system of choice and yeast cells or baculovirus-infected insect cell systems were of lesser importance *(89,90)*. Whereas bacteria are an inexpensive, convenient production system, they are incapable of most of the posttranslational modifications necessary for the activity of many mammalian proteins. This limitation and the cost of expression of proteins in mammalian cells prompted the exploration of plants as cheap, safe, and efficient alternative. The successful expression of functional antibodies in plants represented a significant breakthrough showing that plants had the potential to produce complex mammalian proteins of medical importance. By analogy to the production of insulin in bacteria, which became the first recombinant protein to be approved for therapeutic use, the production of antibodies in plants had the potential to make large amounts of safe, inexpensive antibodies available.

Plant expression systems became attractive because they offer significant advantages over the classical expression systems. First, they have higher eukaryote protein synthesis pathway, very similar to animal cells with only minor differences in protein glycosylation *(91)*. In contrast, bacteria cannot produce full size antibodies or perform most of the important mammalian posttranslational modifications. Second, proteins produced in plants accumulate to high levels *(92)* and plant-derived antibodies are functionally equivalent to those produced by hybridomas *(77)*. Third, concerns about contamination of expressed proteins with human or animal pathogens (HIV, hepatitis viruses) or the copurification of blood-borne pathogens and oncogenic sequences are entirely avoided by using plants.

Transgenic plants producing high levels of safe, functional recombinant proteins can be cultivated on an agricultural scale and require only a virus-infected or transgenic plant, water, mineral salts, and sun light. The ease with which plants can be manipulated and grown in single cell suspension culture or scaled up for field-scale production is a great advantage over the more commonly used microbial methods, mammalian culture, and even transgenic animal technology.

Plant transformation involves the chromosomal integration of a heterologous gene, a process which is becoming straightforward. There are still technical and logistical hurdles to be overcome, such as developing efficient transformation for all major crop species. Developing plant lines expressing recombinant proteins is time intensive and expensive. Approximately 8–12 wk are needed for transgenic plants to be available, but the time required depends on the plant species. Though this is slower than some classical expression systems, the development of transient expression is rapid and results on protein expression can be obtained in days. This makes transient expression suitable for verifying that the gene product is functional before moving on to large-scale production in transgenic plants.

There are three major transient expression systems used to deliver a gene to plant cells: delivery of projectiles coated with "naked DNA" by particle bombardment, infiltration of intact tissue with recombinant agrobacteria (agroinfiltration), and infection with modified viral vectors. The overall level of transformation varies between these three systems. Particle bombardment usually reaches only a few cells and for transcription the DNA has to reach the cell nucleus *(93)*. Agrofiltration targets many more cells than particle bombardment and the T-DNA harboring the gene of interest is actively transferred into the nucleus with the aid of several bacterial proteins. A viral vector can systemically infect most cells in a plant. Transcription of the introduced gene in RNA viruses is achieved by viral replication in the cytoplasm, which transiently generates many transcripts of the gene of interest. However, when long-term production of recombinant proteins, such as antibodies, is necessary; stable transgenic plants are undoubtedly the most attractive strategy. The generation of transgenic plants uses two principal technologies: *Agrobacterium* sp. mediated gene transfer to dicots, such as tobacco and pea *(94,95)*, or biolistic delivery of genes to monocots, such as wheat and corn *(93)*. *Agrobacterium* sp. has a restricted host range and does not efficiently infect monocots but is the most widely used technique for dicot transformation. For transforming plants, the gene of interest is cloned into a binary vector that can be moved between *Escherichia coli* and *Agrobacterium* sp. The transformed *Agrobacterium* sp. itself delivers the target gene into the host cell genome. Transformation is followed by selection of cells with stably integrated copies of the target gene by following a selectable resistance gene that is introduced in the expression vector. The quantity of recombinant protein that can be harvested after success-ful transformation and selection is only limited by the number of hectares that can be planted with transgenic.

In conclusion, the number of mammalian proteins expressed in plants is expand-ing and include antibodies, plasma proteins, human enzymes, and recombinant vaccines. Another application of transgenic technology in plants is the production of vaccine antigens and edible vaccines (85), which may be an important technology in the future. There are still technical challenges that need to be overcome but plants may become a leading expression system for production of pharmaceutically important, commercially valuable proteins.

5. Concluding Remarks

The application of gene transfer techniques has provided us with new insights in developmental biology and the principles underlying tissue-specific gene expression. It has also furnished oncologists, immunologists, and medical geneticists with a plethora of important useful and detailed information. Conside-ring animal production for modeling diseases, gene transfer has acquired an important status and shown to be a promising technique for improving our understanding on gene function, mechanisms, and pathology of many diseases including cancer as well as improvement of performance and quality of animal products. In addition, gene transfer has also allowed the development of new production systems for pharmaceutically important proteins (gene farming). However, production of human pharmaceuticals in farm animals still has many technical barriers to overcome, although most researchers in the field agree that these technical difficulties will be eventually resolved. As a production method, animal farming is entirely unprecedented and as expected must undergo significant evaluation by the Food and Drug Administration. Human drugs purified from animal milk or blood are required to have exceptional levels of safety testing before animal and human health concerns are addressed to the satisfaction of consumers. The future of these applications needs to be balanced with concerns on issues such as animal welfare and biotechnology's redefinition and the relationship between humans and animals. Genetic engineering and transgenic animal research are essentially human endeavors to improve the availability, quality, and safety of drugs; to enhance human health; and to improve animal health. Animal breeding has gone on for centuries, but the ability to change the DNA of the animal brings breeding to a revolutionary new level.

References

1. Gordon, J. W. and Ruddle, F. H. (1982) Germ line transmission in transgenic mice. *Prog. Clin. Biol. Res.* **85**(Pt B), 111–124.
2. Gordon, J. W. and Ruddle, F. H. (1981) Integration and stable germ line transmission of genes injected into mouse pronuclei. *Science* **214**, 1244–1246.
3. Palmiter, R. D., Brinster, R. L., Hammer, R. E., et al. (1982) Dramatic growth of mice that develop from eggs microinjected with metallothionein-growth hormone fusion genes. *Nature* **300**, 611–615.
4. Brinster, R. L., Chen, H. Y., Trumbauer, M., Senear, A. W., Warren, R., and Palmiter, R. D. (1992) Somatic expression of herpes thymidine kinase in mice following injection of a fusion gene into eggs. *Biotechnology* **24**, 411–419.
5. Costantini, F. and Lacy, E. (1981) Introduction of a rabbit beta-globin gene into the mouse germ line. *Nature* **294**, 92–94.
6. Harbers, K., Jahner, D., and Jaenisch, R. (1981) Microinjection of cloned retroviral genomes into mouse zygotes: integration and expression in the animal. *Nature* **293**, 540–542.
7. Wagner, T. E., Hoppe, P. C., Jollick, J. D., Scholl, D. R., Hodinka, R. L., and Gault, J. B. (1981) Microinjection of a rabbit beta-globin gene into zygotes and its subsequent expression in adult mice and their offspring. *Proc. Natl Acad. Sci. U. S. A.* **78**, 6376–6380.
8. Wagner, E. F., Stewart, T. A., and Mintz, B. (1981) The human beta-globin gene and a functional viral thymidine kinase gene in developing mice. *Proc. Natl Acad. Sci. U. S. A.* **78**, 5016–5020.
9. Wall, R. J., Kerr, D. E., and Bondioli, K. R. (1997) Transgenic dairy cattle: genetic engineering on a large scale. *J. Dairy Sci.* **80**, 2213–2224.
10. Gordon, J. W., Scangos, G. A., Plotkin, D. J., Barbosa, J. A., and Ruddle, F. H. (1980) Genetic transformation of mouse embryos by microinjection of purified DNA. *Proc. Natl Acad. Sci. U. S. A.* **77**, 7380–7384.
11. Hammer, R. E., Pursel, V. G., Rexroad, C. E., Jr., et al. (1985) Production of transgenic rabbits, sheep and pigs by microinjection. *Nature* **315**, 680–683.
12. Lin, T. P. (1966) Microinjection of mouse eggs. *Science* **151**, 333–337.
13. Lin, T. P. (1967) Micropipetting cytoplasm from the mouse-egg. *Nature* **216**, 162–163.
14. Brinster, R. L., Chen, H. Y., Trumbauer, M. E., Yagle, M. K., and Palmiter, R. D. (1985) Factors affecting the efficiency of introducing foreign DNA into mice by microinjecting eggs. *Proc. Natl Acad. Sci. U. S. A.* **82**, 4438–4442.
15. Bradley, A., Evans, M., Kaufman, M. H., and Robertson, E. (1984) Formation of germ-line chimaeras from embryo-derived teratocarcinoma cell lines. *Nature* **309**, 255–256.
16. Wei, C. M., Gibson, M., Spear, P. G., and Scolnick, E. M. (1981) Construction and isolation of a transmissible retrovirus containing the src gene of Harvey murine sarcoma virus and the thymidine kinase gene of herpes simplex virus type 1. *J. Virol.* **39**, 935–944.
17. Joyner, A. L. and Bernstein, A. (1983) Retrovirus transduction: segregation of the viral transforming function and the herpes simplex virus tk gene in infectious

Friend spleen focus-forming virus thymidine kinase vectors. *Mol. Cell Biol.* **3,** 2191–2202.

18. Joyner, A. L. and Bernstein, A. (1983) Retrovirus transduction: generation of infectious retroviruses expressing dominant and selectable genes is associated with in vivo recombination and deletion events. *Mol. Cell Biol.* **3,** 2180–2190.

19. Miller, A. D., Jolly, D. J., Friedmann, T., and Verma, I. M. (1983) A transmissible retrovirus expressing human hypoxanthine phosphoribosyltransferase (HPRT): gene transfer into cells obtained from humans deficient in HPRT. *Proc. Natl Acad. Sci. U. S. A.* **80,** 4709–4713.

20. Jaenisch, R., Harbers, K., Schnieke, A., et al. (1983) Germline integration of moloney murine leukemia virus at the Mov13 locus leads to recessive lethal mutation and early embryonic death. *Cell* **32,** 209–216.

21. Jaenisch, R., Harbers, K., Jahner, D., Stewart, C., and Stuhlmann, H. (1983) Expression of retroviruses during early mouse embryogenesis. *Haematol. Blood Transfus.* **28,** 270–274.

22. Jahner, D., Stuhlmann, H., Stewart, C. L., et al. (1982) De novo methylation and expression of retroviral genomes during mouse embryogenesis. *Nature* **298,** 623–628.

23. Jahner, D., Stewart, C. L., Stuhlmann, H., Harbers, K., and Jaenisch, R. (1983) Retroviruses and embryogenesis: de novo methylation activity involved in gene expression. *Cold Spring Harb. Symp. Quant. Biol.* **47**(Pt 2)**,** 611–619.

24. Stewart, C. L., Schuetze, S., Vanek, M., and Wagner, E. F. (1987) Expression of retroviral vectors in transgenic mice obtained by embryo infection. *EMBO J.* **6,** 383–388.

25. Soriano, P., Cone, R. D., Mulligan, R. C., and Jaenisch, R. (1986) Tissue-specific and ectopic expression of genes introduced into transgenic mice by retroviruses. *Science* **234,** 1409–1413.

26. Benvenuto, E. and Tavladoraki, P. (1995) Immunotherapy of plant viral diseases. *Trends Microbiol.* **3,** 272–275.

27. Tavladoraki, P., Benvenuto, E., Trinca, S., De Martinis, D., Cattaneo, A., and Galeffi, P. (1993) Transgenic plants expressing a functional single-chain Fv antibody are specifically protected from virus attack. *Nature* **366,** 469–472.

28. Ruther, U., Garber, C., Komitowski, D., Muller, R., and Wagner, E. F. (1987) Deregulated c-fos expression interferes with normal bone development in transgenic mice. *Nature* **325,** 412–416.

29. Ruther, U., Muller, W., Sumida, T., Tokuhisa, T., Rajewsky, K., and Wagner, E. F. (1988) c-fos expression interferes with thymus development in transgenic mice. *Cell* **53,** 847–856.

30. Khillan, J. S., Oskarsson, M. K., Propst, F., Kuwabara, T., Vande Woude, G. F., and Westphal, H. (1987) Defects in lens fiber differentiation are linked to c-mos over-expression in transgenic mice. *Genes Dev.* **1,** 1327–1335.

31. Propst, F., Cork, L. C., Kovatch, R. M., Kasenally, A. B., Wallace, R., and Rosenberg, M. P. (1992) Progressive hind limb paralysis in mice carrying a v-Mos transgene. *J. Neuropathol. Exp. Neurol.* **51,** 499–505.

32. Behringer, R. R., Ryan, T. M., Reilly, M. P., et al. (1989) Synthesis of functional human hemoglobin in transgenic mice. *Science* **245,** 971–973.

33. Ryan, T. M., Townes, T. M., Reilly, M. P., et al. (1990) Human sickle hemoglobin in transgenic mice. *Science* **247,** 566–568.

34. Greaves, D. R., Fraser, P., Vidal, M. A., et al. (1990) A transgenic mouse model of sickle cell disorder. *Nature* **343,** 183–185.

35. Allison, J., Campbell, I. L., Morahan, G., Mandel, T. E., Harrison, L. C., and Miller, J. F. (1988) Diabetes in transgenic mice resulting from over-expression of class I histocompatibility molecules in pancreatic beta cells. *Nature* **333,** 529–533.

36. Epstein, P. N., Overbeek, P. A., and Means, A. R. (1989) Calmodulin-induced early-onset diabetes in transgenic mice. *Cell* **58,** 1067–1073.

37. Ribar, T. J., Epstein, P. N., Overbeek, P. A., and Means, A. R. (1995) Targeted over-expression of an inactive calmodulin that binds Ca2+ to the mouse pancreatic beta-cell results in impaired secretion and chronic hyperglycemia. *Endocrinology* **136,** 106–115.

38. Selkoe, D. J. (1991) The molecular pathology of Alzheimer's disease. *Neuron* **6,** 487–498.

39. Games, D., Adams, D., Alessandrini, R., et al. (1995) Alzheimer-type neuropathology in transgenic mice overexpressing V717F beta-amyloid precursor protein. *Nature* **373,** 523–527.

40. Ashe, K. H. (2001) Learning and memory in transgenic mice modeling Alzheimer's disease. *Learn. Mem.* **8,** 301–308.

41. Ashe, K. H. (2000) Synaptic structure and function in transgenic APP mice. *Ann. N. Y. Acad. Sci.* **924,** 39–41.

42. Westerman, M. A., Cooper-Blacketer, D., Mariash, A., et al. (2002) The relationship between Abeta and memory in the Tg2576 mouse model of Alzheimer's disease. *J. Neurosci.* **22,** 1858–1867.

43. Borchelt, D. R., Ratovitski, T., van Lare, J., et al. (1997) Accelerated amyloid deposition in the brains of transgenic mice coexpressing mutant presenilin 1 and amyloid precursor proteins. *Neuron* **19,** 939–945.

44. Holcomb, L., Gordon, M. N., McGowan, E., et al. (1998) Accelerated Alzheimer-type phenotype in transgenic mice carrying both mutant amyloid precursor protein and presenilin 1 transgenes. *Nat. Med.* **4,** 97–100.

45. McGowan, E., Eriksen, J., and Hutton, M. (2006) A decade of modeling Alzheimer's disease in transgenic mice. *Trends Genet.* **22,** 281–289.

46. Hooper, M., Hardy, K., Handyside, A., Hunter, S., and Monk, M. (1987) HPRT-deficient (Lesch–Nyhan) mouse embryos derived from germline colonization by cultured cells. *Nature* **326,** 292–295.

47. Kuehn, M. R., Bradley, A., Robertson, E. J., and Evans, M. J. (1987) A potential animal model for Lesch–Nyhan syndrome through introduction of HPRT mutations into mice. *Nature* **326,** 295–298.

48. Wu, C. L. and Melton, D. W. (1993) Production of a model for Lesch–Nyhan syndrome in hypoxanthine phosphoribosyltransferase-deficient mice. *Nat. Genet.* **3,** 235–240.

49. Tybulewicz, V. L., Tremblay, M. L., LaMarca, M. E., et al. (1992) Animal model of Gaucher's disease from targeted disruption of the mouse glucocerebrosidase gene. *Nature* **357,** 407–410.

50. Williamson, R., Lee, D., Hagaman, J., and Maeda, N. (1992) Marked reduction of high density lipoprotein cholesterol in mice genetically modified to lack apolipoprotein A-I. *Proc. Natl Acad. Sci. U. S. A.* **89,** 7134–7138.

51. Plump, A. S., Smith, J. D., Hayek, T., et al. (1992) Severe hypercholesterolemia and atherosclerosis in apolipoprotein E-deficient mice created by homologous recombination in ES cells. *Cell* **71,** 343–353.

52. Mangiarini, L., Sathasivam, K., Seller, M., et al. (1996) Exon 1 of the HD gene with an expanded CAG repeat is sufficient to cause a progressive neurological phenotype in transgenic mice. *Cell* **87,** 493–506.

53. Sathasivam, K., Hobbs, C., Mangiarini, L., et al. (1999) Transgenic models of Huntington's disease. *Philos. Trans. R. Soc. Lond. B Biol. Sci.* **354,** 963–969.

54. Menalled, L. B. and Chesselet, M. F. (2002) Mouse models of Huntington's disease. *Trends Pharmacol. Sci.* **23,** 32–39.

55. Snouwaert, J. N., Brigman, K. K., Latour, A. M., et al. (1992) An animal model for cystic fibrosis made by gene targeting. *Science* **257,** 1083–1088.

56. Ratcliff, R., Evans, M. J., Doran, J., Wainwright, B. J., Williamson, R., and Colledge, W. H. (1992) Disruption of the cystic fibrosis transmembrane conductance regulator gene in embryonic stem cells by gene targeting. *Transgenic Res.* **1,** 177–181.

57. Ratcliff, R., Evans, M. J., Cuthbert, A. W., et al. (1993) Production of a severe cystic fibrosis mutation in mice by gene targeting. *Nat. Genet.* **4,** 35–41.

58. Dorin, J. R., Dickinson, P., Alton, E. W., et al. (1992) Cystic fibrosis in the mouse by targeted insertional mutagenesis. *Nature* **359,** 211–215.

59. Leder, A., Pattengale, P. K., Kuo, A., Stewart, T. A., and Leder, P. (1986) Consequences of widespread deregulation of the c-myc gene in transgenic mice: multiple neoplasms and normal development. *Cell* **45,** 485–495.

60. Halter, S. A., Dempsey, P., Matsui, Y., et al. (1992) Distinctive patterns of hyperplasia in transgenic mice with mouse mammary tumor virus transforming growth factor-alpha. Characterization of mammary gland and skin proliferations. *Am. J. Pathol.* **140,** 1131–1146.

61. Coffey, R. J., Jr., Meise, K. S., Matsui, Y., Hogan, B. L., Dempsey, P. J., and Halter, S. A. (1994) Acceleration of mammary neoplasia in transforming growth factor alpha transgenic mice by 7,12-dimethylbenzanthracene. *Cancer Res.* **54,** 1678–1683.

62. Brinster, R. L., Chen, H. Y., Messing, A., van Dyke, T., Levine, A. J., and Palmiter, R. D. (1984) Transgenic mice harboring SV40 T-antigen genes develop characteristic brain tumors. *Cell* **37,** 367–379.

63. Palmiter, R. D., Chen, H. Y., Messing, A., and Brinster, R. L. (1985) SV40 enhancer and large-T antigen are instrumental in development of choroid plexus tumours in transgenic mice. *Nature* **316,** 457–460.

64. Windle, J. J., Albert, D. M., O'Brien, J. M., et al. (1990) Retinoblastoma in transgenic mice. *Nature* **343,** 665–669.

65. Van Brunt, J. (1990) Transgenics primed for research. *Biotechnology (N. Y.)* **8,** 725–728.

66. Pittius, C. W., Hennighausen, L., Lee, E., et al. (1988) A milk protein gene promoter directs the expression of human tissue plasminogen activator cDNA to the mammary gland in transgenic mice. *Proc. Natl Acad. Sci. U. S. A.* **85,** 5874–5878.

67. Denman, J., Hayes, M., O'Day, C., et al. (1991) Transgenic expression of a variant of human tissue-type plasminogen activator in goat milk: purification and characterization of the recombinant enzyme. *Biotechnology (N. Y.)* **9,** 839–843.

68. Ebert, K. M., Selgrath, J. P., DiTullio, P., et al. (1991) Transgenic production of a variant of human tissue-type plasminogen activator in goat milk: generation of transgenic goats and analysis of expression. *Biotechnology (N. Y.)* **9,** 835–838.

69. Buhler, T. A., Bruyere, T., Went, D. F., Stranzinger, G., and Burki, K. (1990) Rabbit beta-casein promoter directs secretion of human interleukin-2 into the milk of transgenic rabbits. *Biotechnology (N. Y.)* **8,** 140–143.

70. Limonta, J. M., Castro, F. O., Martinez, R., et al. (1995) Transgenic rabbits as bioreactors for the production of human growth hormone. *J. Biotechnol.* **40,** 49–58.

71. Wall, R. J., Pursel, V. G., Shamay, A., McKnight, R. A., Pittius, C. W., and Hennighausen, L. (1991) High-level synthesis of a heterologous milk protein in the mammary glands of transgenic swine. *Proc. Natl Acad. Sci. U. S. A.* **88,** 1696–1700.

72. van Berkel, P. H., Welling, M. M., Geerts, M., et al. (2002) Large scale production of recombinant human lactoferrin in the milk of transgenic cows. *Nat. Biotechnol.* **20,** 484–487.

73. Clark, A. J., Ali, S., Archibald, A. L., et al. (1989) The molecular manipulation of milk composition. *Genome* **31,** 950–955.

74. Wright, G., Carver, A., Cottom, D., et al. (1991) High level expression of active human alpha-1-antitrypsin in the milk of transgenic sheep. *Biotechnology (N. Y.)* **9,** 830–834.

75. Carver, A., Wright, G., Cottom, D., et al. (1992) Expression of human alpha 1 antitrypsin in transgenic sheep. *Cytotechnology* **9,** 77–84.

76. Sijmons, P. C., Dekker, B. M., Schrammeijer, B., Verwoerd, T. C., van den Elzen, P. J., and Hoekema, A. (1990) Production of correctly processed human serum albumin in transgenic plants. *Biotechnology (N. Y.)* **8,** 217–221.

77. Hiatt, A., Cafferkey, R., and Bowdish, K. (1989) Production of antibodies in transgenic plants. *Nature* **342,** 76–78.

78. Hiatt, A. (1990) Antibodies produced in plants. *Nature* **344,** 469–470.

79. Magnuson, N. S., Linzmaier, P. M., Gao, J. W., Reeves, R., An, G., and Lee, J. M. (1996) Enhanced recovery of a secreted mammalian protein from suspension culture of genetically modified tobacco cells. *Protein Expr. Purif.* **7,** 220–228.

80. Magnuson, N. S., Linzmaier, P. M., Reeves, R., An, G., HayGlass, K., and Lee, J. M. (1998) Secretion of biologically active human interleukin-2 and interleukin-4 from genetically modified tobacco cells in suspension culture. *Protein Expr. Purif.* **13,** 45–52.

81. Arakawa, T., Yu, J., Chong, D. K., Hough, J., Engen, P. C., and Langridge, W. H. (1998) A plant-based cholera toxin B subunit-insulin fusion protein protects against the development of autoimmune diabetes. *Nat. Biotechnol.* **16,** 934–938.

82. Arakawa, T., Chong, D. K., and Langridge, W. H. (1998) Efficacy of a food plant-based oral cholera toxin B subunit vaccine. *Nat. Biotechnol.* **16,** 292–297.

83. Haq, T. A., Mason, H. S., Clements, J. D., and Arntzen, C. J. (1995) Oral immunization with a recombinant bacterial antigen produced in transgenic plants. *Science* **268,** 714–716.

84. Kapusta, J., Modelska, A., Figlerowicz, M., et al. (1999) A plant-derived edible vaccine against hepatitis B virus. *FASEB J.* **13,** 1796–1799.

85. Walmsley, A. M. and Arntzen, C. J. (2003) Plant cell factories and mucosal vaccines. *Curr. Opin. Biotechnol.* **14,** 145–150.

86. Walmsley, A. M. and Arntzen, C. J. (2000) Plants for delivery of edible vaccines. *Curr. Opin. Biotechnol.* **11,** 126–129.

87. Fischer, R., Stoger, E., Schillberg, S., Christou, P., and Twyman, R. M. (2004) Plant-based production of biopharmaceuticals. *Curr. Opin. Plant Biol.* **7,** 152–158.

88. Twyman, R. M., Schillberg, S., and Fischer, R. (2005) Transgenic plants in the biopharmaceutical market. *Expert. Opin. Emerg. Drugs* **10,** 185–218.

89. Taticek, R. A., Lee, C. W., and Shuler, M. L. (1994) Large-scale insect and plant cell culture. *Curr. Opin. Biotechnol.* **5,** 165–174.

90. Skerra, A. (1993) Bacterial expression of immunoglobulin fragments. *Curr. Opin. Immunol.* **5,** 256–262.

91. Cabanes-Macheteau, M., Fitchette-Laine, A. C., Loutelier-Bourhis, C., et al. (1999) *N*-Glycosylation of a mouse IgG expressed in transgenic tobacco plants. *Glycobiology* **9,** 365–372.

92. Verwoerd, T. C., van Paridon, P. A., van Ooyen, A. J., van Lent, J. W., Hoekema, A., and Pen, J. (1995) Stable accumulation of Aspergillus niger phytase in transgenic tobacco leaves. *Plant Physiol.* **109,** 1199–1205.

93. Christou, P. (1995) Particle bombardment. *Methods Cell Biol.* **50,** 375–382.

94. Fraley, R. T., Rogers, S. G., Horsch, R. B., et al. (1983) Expression of bacterial genes in plant cells. *Proc. Natl Acad. Sci. U. S. A.* **80,** 4803–4807.

95. Horsch, R. B., Rogers, S. G., and Fraley, R. T. (1985) Transgenic plants. *Cold Spring Harb. Symp. Quant. Biol.* **50,** 433–437.

8

Viral, Nonviral, and Physical Methods for Gene Delivery

Jingjiao Guan, Xiaogang Pan, L. James Lee, and Robert J. Lee

Abstract

Gene transfer is an emerging therapeutic modality for a wide spectrum of diseases. Its clinical adoption is, however, limited by the lack of safe and efficient gene delivery methods. Three classes of methods are currently under evaluation. The first class consists of genetically modified viruses, which include retroviruses, adenoviruses, adeno-associated viruses, and several others. These vectors are relatively efficient. However, their clinical application is associated with significant safety concerns, such as oncogenesis and acute inflammatory response. The second class is nonviral vectors, which are composed of synthetic components. These include complexes of DNA with lipids, polymers, or their combination. Many nonviral vector formulations, which incorporate functional components to facilitate nuclease protection, cellular/tissue targeting, endosomal release, and nuclear localization, have been investigated, mostly in vitro. These efforts have resulted in incremental advances in gene transfer efficiency, requiring further improvements for clinical applications. The third class of methods is based on the use of physical energy or force. Examples are gene gun, electroporation, and magnetofection. These methods are suitable for locoregional gene delivery. In this chapter, we will provide an overview of the state-of-the-art gene transfer methods, their strengths and weaknesses, and challenges and opportunities in this critical area of research, which will, to a large extent, determine the future prospect of gene therapy in the clinic.

Key Words: Gene therapy; gene delivery; viral vector; nonviral vector.

1. Introduction to Gene Therapy

Gene therapy can be defined as the treatment of diseases through gene transfer. Genes are deoxyribonucleic acid (DNA) sequences encoding proteins, which are the building blocks of all living species and are involved in all biological processes. As a result, a single defective gene can lead to devastating disorders such as sickle cell anemia, hemophilia, and cystic fibrosis. Indeed, substitution of a defective or missing gene with a healthy one is the original and most

From: *Biopharmaceutical Drug Design and Development*
Edited by: S. Wu-Pong and Y. Rojanasakul © Humana Press Inc., Totowa, NJ

straightforward goal of gene therapy. Furthermore, common diseases such as cancer, neurological disorders, arthritis, cardiovascular diseases, and diabetes all have a genetic basis.

Exogenous genetic materials must be transferred into cells in gene therapy, for the production of therapeutic proteins. In contrast to therapeutic protein delivery, production of proteins from transgenes can last from days to the entire life span of a person. This is desirable for treating many chronic diseases. In addition, transgenes are expressed inside the cells whereas exogenous proteins have difficulty entering cells because of their high molecular weight. Gene therapy is thus potentially superior to protein therapy when the site of action is intracellular, e.g., for elicitation of cellular immunity during immunotherapy (1). Suicide gene therapy is also based on the transgene expression within cancer cells (2).

Despite potential benefits, gene therapy remains in an early stage of development 15 yr after the first human clinical trial (3). A limiting factor for clinical success of gene therapy is the lack of efficient and safe methods for the delivery of genes into target cells. This is exemplified by the lack of success in gene therapy for monogenic diseases. Depending on how DNA is introduced, gene therapy can be divided into ex vivo and in vivo. For ex vivo gene therapy, cells are removed from the patient, genetically modified, and reintroduced into the patient. For in vivo gene therapy, DNA is delivered directly into the body. In both cases, the therapeutic DNA must enter the cells by passing through a series of barriers. For ex vivo gene transfer, the cellular (endosomal) membrane, nucleases, and the nuclear envelope constitute formidable barriers to transgene expression. To achieve long-term gene expression, integration of the transgenes into the host chromosome is usually required, which is relatively inefficient and might cause unintended mutagenesis. For in vivo gene transfer, additional barriers must be overcome. To reach the target cells, DNA may have to pass through various tissues such as skin, blood vessel, and extracellular matrix, in which macromolecules and nanoparticles, such as DNA vector, have extremely low diffusivity. DNA is also prone to degradation by nucleases in extracellular matrix and in plasma, and gene vectors are subjected to clearance by the reticuloendothelial system (RES). Various gene delivery strategies have been developed to overcome these barriers. DNA is frequently formulated into vectors, with properties designed to enhance gene transfer via more efficient interaction with the target cells and tissues. The vectors are classified as viral and nonviral based on their composition. Viral vectors are viruses engineered for gene delivery and nonviral vectors are constructed from synthetic materials. A third strategy is based on the use of external energy or force to shuttle the genetic materials across biological barriers. All methods using this strategy are thus called physical methods. In the following sections, viral and nonviral vectors and physical methods will be separately reviewed.

2. Viral Vectors

2.1. Introduction

Viruses are submicron particles that can infect cells and self-replicate. It contains nucleic acid, either DNA or RNA, encoding viral proteins and regulatory sequences. The nucleic acid is enclosed in a coat that possess multiple functions, including protecting the nucleic acid against enzymatic degradation, targeting cells, and mediating viral entry into the cytosol. From the point of view of gene delivery, a virus is a naturally evolved machine for gene delivery into cells. To use viruses for the delivery of therapeutic DNA, parts of the viral genome are replaced with therapeutic genes. Viral vectors are engineered to be replication deficient so that they do not cause diseases, except in the case of oncolytic viruses, in which viral replication is part of the therapeutic strategy. Many viruses have been used for the production of viral vectors with different gene delivery characteristics, including retrovirus, adenovirus, adeno-associated virus (AAV), herpes simplex virus (HSV), and vaccinia virus. These are briefly reviewed below.

2.2. Retrovirus

A retrovirus is a spherical particle approximately 100 nm in diameter. It contains a lipid bilayer envelope and a protein capsid enclosing two copies of single-stranded RNA and enzymes, including reverse transcriptase, protease, and integrase. The viral genome consists of two noncoding long terminal repeats (LTRs) at each end that are important for regulation of gene expression. Between the LTRs are the genes encoding viral proteins. These sequences can be deleted for the incorporation of therapeutic genes.

Retroviral vectors enter cells via receptor-mediated endocytosis. In the cytoplasm, the outer protein capsid is shed and the transgene undergoes reverse transcription by viral reverse transcriptase to form a double-stranded DNA intermediate, which can then be integrated into the host genome facilitated by the viral integrase. The integration results in sustained transgene expression, desirable for treating hereditary and chronic diseases. However, the integration site randomly favors active genes and may lead to insertional mutagenesis by accidentally disrupting a tumor suppressor gene or activating an oncogene *(4,5)*.

Retroviruses are historically classified into three subgroups, onco-retrovirus, lentivirus, and spumavirus. Vectors based on onco-retroviral viruses, represented by murine leukemia virus (MLV), are extensively used in gene therapy strategies requiring long-lasting gene expression. A rare success story in the history of gene therapy research is the apparent cure of patients with X-linked severe combined immune deficiency using onco-retroviral vectors *(6)*. However, the application of the onco-retroviral vectors is limited because they can only transfect dividing cells and have a relatively small packaging capacity of approximately 8 kb. Lentiviral vectors have recently attracted considerable interest because they can efficiently transfect nondividing cells and have a larger packaging capacity of up to 18 kb *(7,8)*. Numerous in vitro and in vivo studies have demonstrated efficient transduction and stable expression of either reporter genes or therapeutic genes delivered by lentiviral vectors in various cells such as endothelial cells *(9,10)*, neural stem cells *(11)*, embryonic stem cells *(12)*, and hematopoietic stem cells *(13)*, and tissues such as liver *(10,14,15)*, artery *(16)*, and brain *(11,17)*. A major concern for the clinical use of the lentiviral vectors is safety because the most widely used lentiviral vectors are derived from the human immunodeficiency virus type 1 (HIV-1), which causes acquired immunodeficiency syndrome. One strategy to address this issue is to construct vectors based on nonprimate lentiviruses that are not infectious to humans *(18)*. Use of self-inactivating vectors, in which the viral enhancer and promoter sequences are blunted, also significantly improves the safety profile of the lentiviral vectors *(19)*.

2.3. Adenovirus

An adenovirus is a nonenveloped particle approximately 70–100 nm in diameter with a linear, double-stranded DNA of approximately 36 kb encapsulated in a protein capsid. Adenoviruses bind to cells via high-affinity interaction between the viral capsid proteins and cell surface receptors. Following the binding, the virions enter the cells via endocytosis, escape from the endosome into the cytoplasm, move towards and reach the nucleus, and eventually deliver the viral DNA into the host nucleus, where the viral DNA remains largely episomal *(20)*. Adenoviruses can efficiently infect a variety of cell types independent of cell cycle. Production of adenoviral vectors at high titer is relatively easy. Given these desirable features, adenoviral vectors are widely used in gene therapy studies. First generation of adenoviral vectors was constructed by deleting a small portion of the viral genome and incorporation of transgenes and regulatory sequences of up to 8 kb. Major limitations of these vectors include severe cytotoxicity and immunological reactions, rapid clearance of the vectors and transfected cells by the immune system, and relatively small DNA payload capacity. As a result, these vectors are only useful if long-term gene expression is not required or induction of immune responses is the therapeutic goal. By deleting the entire viral genome except the inverted terminal repeats (ITRs) and the packaging signal, a new generation of adenoviral vectors, helper-dependent adenoviral (HD-Ad) vectors (also referred to as high-capacity adenoviral and "gutless" adenoviral vectors), has been developed *(21)*. The vectors have a packaging capacity of up to 35 kb, allowing for the incorporation of multiple genes containing regulatory sequences. The safety profiles of the vectors are also significantly improved by avoiding the immunological responses provoked by the proteins encoded by the deleted viral genes. It has been shown that a single intravenous injection of HD-Ad vector completely and stably corrected a genetic disease, hypercholesterolemia, in mice for their entire natural life span up to 2.5 yr *(22)*. Another study using HD-Ad shows stable expression of a transgene encoding for dystrophin, a large protein for treating Duchenne's muscular dystrophy, in mice for a year *(23)*.

2.4. Adeno-associated Virus

AAV is a nonpathogenic human virus. It is nonenveloped, approximately 20 nm in diameter, and contains a linear, single-stranded DNA genome of approximately 4.7 kb. AAV genome contains two ITRs at each end, encoding the origin of replication and the packaging signal. Between the ITRs are two AAV-specific genes, rep and cap. Rep encodes proteins which control viral replication, structural gene expression, and integration into the host genome, and cap encodes capsid structural proteins. Wild-type AAV can integrate viral genes at a specific site in human chromosome 19 *(24)*. This ability is highly desirable for long-term gene therapy to reduce the risk of insertional mutagenesis, but AAV-derived vectors do not retain this ability because all viral genes are replaced by foreign sequences because of the very limited packaging size of AAV *(25)*. The size constraint can, however, be partially overcome by coadministration of two AAV vectors carrying different DNA sequences which can then recombine to form a functional transgene cassette *(26)*. Used as gene delivery vectors, AAV infects both dividing and nondividing cells of various types with low toxicity and immunogenicity. They have thus been used in numerous gene therapy studies for treating various diseases such as lung diseases *(27)*, hemophilia *(28)*, muscular dystrophy *(29)*, neurological disorders *(30)*, ophthalmic diseases *(31)*, and rheumatoid arthritis *(32)*.

2.5. Herpes Simplex Virus

HSV is an enveloped, double-stranded linear DNA virus. It contains a large genome size of 152 kb of which up to 40 kb can be replaced by transgene cassettes *(33)*. This large carrying capacity allows for the incorporation of multiple small genes or a single very large gene with requisite regulatory sequences. HSV can infect a wide variety of mammalian cell types and has a natural preference for neurons independent of cell cycle. It can travel retrograde along neuronal axons from the periphery to the central nervous system. After entering the nuclei of the host cells, the nonintegrated viral genome can assume a latent state without damaging the neurons. HSV-mediated gene therapy has been used to treat cancer *(34)*, chronic pain *(35)*, and neurological disorders and injury *(36,37)*. A major disadvantage of HSV vectors is their potential pathogenicity.

2.6. Vaccinia Virus

Vaccinia virus, also called poxvirus, was used as vaccines to eradicate small-pox worldwide. Among all viruses, it has been used most extensively in humans and has an excellent record of safety and effectiveness. Vaccinia virus has a double-stranded linear DNA genome of 192 kb, allowing incorporation of at least 25 kb of transgenic sequence *(38)*. It can infect almost all human cell types with high efficiency independent of cell cycle. Moreover, vaccinia virus is highly immunogenic, rendering it particularly attractive for immunotherapy against cancer. P53 tumor suppressor gene, cytokine genes, and gene encoding tumor-specific antigen have been delivered using vaccinia viral vectors and demonstrated therapeutic effectiveness in both animal studies and human clinical trials *(39–42)*.

2.7. Summary

Viral vectors are engineered viruses carrying therapeutic genes. A variety of viral vectors have been developed with different characteristics applicable to different conditions. They are currently used in a majority of gene transfer studies and have shown promising therapeutic effectiveness in numerous animal studies and human clinical trials. However, use of the viral vectors is restricted by a number of factors. Most notable among them are cytotoxicity, immune responses, and insertional mutagenesis that have been observed in human clinical trials. Others include the limited packaging capacity and difficulty and high cost associated with the vector production. These limitations have led to the use of synthetic materials for the construction of nonviral vectors that will be reviewed in the following section.

3. Nonviral Vectors

3.1. Introduction

Nonviral vectors are constructed from synthetic materials. As a result, they are free of some problems associated with viral vectors, e.g., potential infectivity, limited packaging size, and unwanted immunogenicity caused by viral proteins. Moreover, unlike viral vectors that must be produced from existing viruses, use of nonviral vectors offers an opportunity for constructing a system from the scratch by rational design. However, viruses consist of multiple components constructed in highly sophisticated structures that specifically evolved for overcoming various physiological barriers for gene delivery. It is not surprising that low efficiency is the major drawback of current nonviral vectors. A key issue in the development of nonviral vectors is thus the design strategy for overcoming various physiological barriers.

The functionalities of a nonviral vector come from the materials used for the construction of the vector. Among various types of materials used, polymers and lipids are the two major classes. Based on the material (or materials) used, the vectors are divided into lipid based, polymer based, and combined lipid/ polymer based. These systems are included in the following sections.

3.2. Design Strategies for Overcoming Physiological Barriers

A gene delivery system must overcome a series of barriers from the point of administration to the nuclei of a target cell. Overcoming these barriers is the main design goal of nonviral vectors.

3.2.1. Extracellular Barriers

Nuclease degradation is a major barrier to in vivo gene delivery. To overcome this obstacle, DNA can be condensed into small and compact structure through electrostatic interaction between negatively charged phosphates on the DNA backbone and positively charged cationic lipids or polymers. Polycationic polymer-condensed DNA usually exhibits toroidal or spherical structures with size ranging from 20 to several hundred nanometers *(43)* and can remain stable for hours in the presence of DNase *(44)*. Monovalent cationic lipid-condensed DNA has been shown to exhibit a "spaghetti-meatball" structure *(45)*. The condensation process is believed to be kinetically controlled and is highly dependent on the properties of cationic lipids/polymers and the ionic strength of the buffer. It is important to note, however, that excessively high affinity of DNA with vector components may hinder intracellular release of DNA, which is required for gene transcription *(46)*.

Another barrier to gene delivery is clearance of the vector by phagocytic cells of the RES *(47)*. This can be partially overcome by coating the surface of the vector with hydrophilic polymers such as polyethylene glycol (PEG) *(48)*, oligosaccharides *(49)*, or proteins *(50)*, which also enhances the colloidal stability of vector particles. However, this may reduce the gene delivery efficiency by inhibiting endosomal escape of the vector *(51)*.

Another barrier to effective gene therapy is the lack of target cell selectivity. The requirement for specific cell targeting in gene therapy varies by therapeutic applications. In the case of hemophilia, gene delivery to specific cells is not required as long as sufficient levels of secreted therapeutic proteins can be produced. In most other disease types, gene delivery to cells of interest, such as cancer cells, is required for therapeutic effectiveness. Several strategies have been evaluated to increase cellular selectivity and uptake. A number of membrane-bound receptors, such as EGFR *(52)*, HER-2 *(53)*, folate receptor *(54)*, have been explored for targeted gene delivery via receptor-mediated endocytosis.

3.2.2. Intracellular Barriers

Generally, nonviral vectors are internalized via either nonspecific pinocytosis *(55)* or a specific receptor-mediated endocytosis pathway, followed by endosomal escape, then eventually reach the nucleus and release the DNA. Nonviral vectors must overcome numerous intracellular barriers in this process. Only a small fraction of internalized vectors can reach their eventual target *(56)*. Thus, a thorough understanding of intracellular trafficking of nonviral vectors is essential for rational design of vectors.

3.2.2.1. INTERNALIZATION AND ENDOSOMAL ESCAPE

In the internalization pathway, vectors are engulfed in the endocytic vesicles that are en route to lysosomes filled with digestive enzymes. Escape from the endosome has been demonstrated as one of the major barriers for effective gene delivery. The vectors travel from early endosomes to late endosomes with an acidic pH *(57)*. Only a small portion of DNA complex can escape into cytoplasm and reach the nucleus. It has also been suggested that DNA can be released from vectors during the endosomal escape process, leaving naked DNA in the cytoplasm *(58)*.

Several strategies have been applied for improving endosomal escape. Incorporation of a helper lipid, such as dioleoyl phosphatidylethanolamine (DOPE), a cone-shape lipid favoring transition to a non-bilayer lipid phase, in the vector formulation could promote DNA release by facilitating endosomal disruption *(59)*. In addition, attaching a fusogenic peptide to the vector has been shown to increase gene transfection efficiency *(60,61)*. The cationic polymer polyethyleneimine (PEI) can facilitate endosomal release by acting as a "proton-sponge". PEI has a large amount of secondary and tertiary amines with pK_a in the range of endosomal pH *(20)*. Accumulation of protons and counterions eventually leads to osmotic swelling of endosome and rupture of the endosomal membrane *(62,63)*. Some special fusogenic liposomes and pH-sensitive liposomes have also been designed for facilitating endosome escape, with varying degree of success *(64)*.

3.2.2.2. TRANSPORT IN THE CYTOPLASM

On escaping the endosome, the vectors or released DNA must travel through cytoplasm to gain access to the nucleus, where transcription would take place. Vectors face two major challenges in the cytoplasm: the diffusion barrier and the enzymatic barrier. The cytoplasm is rich with proteins, microtubules, and many other organelles. All of them can hinder vector diffusion and decrease gene transfection efficacy. Diffusion of DNA in cytoplasm is also size dependent — the diffusion coefficient of DNA larger than 2000 base pairs in length is only less than 1% of that of DNA in water, and DNA larger than 3000 base pairs is essentially immobile in cytoplasm *(65)*. There is also no evidence of active transport of DNA in the cytoplasm to date, thus it is a reasonable consideration to use smaller and more condensed DNA. Besides the physical diffusion barrier, the richness of nucleases in the cytoplasm poses another significant barrier as well *(66)*.

3.2.2.3. NUCLEAR LOCALIZATION

The ultimate destination for delivered DNA inside the cell is the nucleus where the gene can be transcribed. Nature has enclosed the nucleus with a nuclear envelope that displays tightly regulated pores. Unlike the virus that has evolved methods to transport its gene into the nucleus of the host cell, the mechanism of nuclear transport of DNA by nonviral vector is still poorly characterized, though some possible routes have been proposed for transporting DNA into the nucleus. The nuclear pore complex allows the passage of small particles with sizes less than 9 nm in its closed state and can facilitate transport of particles with sizes less than 26 nm in its open state *(67)*. Clearly, typical nonviral vectors cannot pass through the nuclear pore by simple diffusion.

A more widely proposed mechanism of nuclear transport is that DNA gains access to the nucleus during cell mitosis in which the nuclear membrane is broken down. Transfection efficiency of polymer or lipid-based systems in cells undergoing division (S or G2 phase) was 30 to 500-fold higher than that in cells entering the cell cycle (G1 phase) *(68)*. Therefore, nuclear translocation for nondividing cells seems to be a formidable obstacle.

The nuclear localization sequence (NLS), a short cationic peptide, is responsible for the nuclear localization of many proteins. Covalently linking NLS to DNA or noncovalent association of NLS to DNA complex has been shown to enhance their nuclear translocation *(69,70)*. It is quite possible that the positively charged vector serves itself as a NLS to target the nucleus, but it may only work to some extent because of its low nuclear transport ability *(71–73)*.

3.2.2.4. IMMUNOLOGICAL FACTORS—THE ROLE OF CPG MOTIF

Contrary to early expectations, nonviral vectors frequently induce a highly potent inflammatory response. It has been found that both systemic and local administrations of lipoplex evoke rapid activation of the innate immune system that induces a large amount of proinflammatory cytokines, such as tumor necrosis factor-α (TNF-α), interferon-γ (IFN-γ), and interleukins *(74,75)*. These inflammatory cytokines inactivate transgene expression because of inhibition of transcription and destabilization of mRNA *(76)*. This may be caused by unmethylated CpG motifs in plasmid DNA, which is produced in bacteria *(77,78)*. Further understanding of nonviral vector-induced immune response is important for improving vector design for its minimization.

3.3. Lipid-Based Vectors

Lipid-based systems (liposomes, micelles, etc.) are attractive for drug delivery because of their favorable characteristics, such as being biodegradable, nontoxic, and easy to scale-up. Conventional liposomes were used for gene delivery in as early as the early 1980s. In 1987 Felgner et al. *(79)* demonstrated efficient in vitro gene delivery by cationic liposomes. To date, numerous studies have been performed to investigate lipid-based gene delivery vectors.

DNA can either be entrapped into or complexed with cationic liposomes to form a complex known as lipoplex. Based on the structural differences, cationic lipids can be categorized into three groups: single-chain lipids, double-chain lipids, and cholesterol-based lipids. Single-chain cationic lipids are generally less efficient than the other two. A majority of cationic lipids for gene delivery belong to double-chain lipids. Ever since the pioneering development of *N*-[1-(2,3-dioleyloxy)propyl]-*N,N,N*-trimethylammonium chloride (DOTMA), a glycerol backbone lipid, many other glycerol or nonglycerol-based cationic lipids have been developed, such as DOTAP, DMRIE, and so on *(79)*. The first effective cholesterol-based cationic lipid, DC-Chol, was designed and synthesized by Huang and coworker *(80)*. Many studies have since suggested that there is little structure–activity correlation of cationic lipids between in vitro and in vivo systems *(81)*. Thus, the design of cationic lipids still largely relies on a trial-and-error approach.

Toxicity encountered by lipoplexes is mostly associated with the excess of positively charged lipids. Highly positively charged lipoplexes are generally more toxic to cells. In addition, toxicity associated with nonspecific gene delivery by lipoplexes (as well as polymer-based vectors) has to be considered.

Lipoplexes have been investigated in both preclinical experiments and clinical trials. They have been shown to transfect endothelial cells of lung of mice with reasonable efficiency by intravenous administration. A clinical trial carried out in late 1990s showed gene expression after lipoplex delivery of cystic fibrosis transmembrane conductance regulator (CFTR) gene to the nose of patients with cystic fibrosis *(82–84)*. However, a number of limitations for lipoplexes will need to be overcome for therapeutic applications of these vectors, such as toxicity at high doses and low transfection efficiency.

3.4. Polymer-Based Vectors

Cationic polymers (polycations) designed for gene delivery are comprised of a variety of positively charged DNA-binding moieties, such as amines (primary, secondary, tertiary, or quaternary), amidines, and so on. The structure of polymers can be divided into linear, branched, or dendritic. Polymers are also designed for overcoming specific barriers to gene delivery, such as poor stability, biocompatibility, and inefficient endosomal escape. These polymers associate with DNA to form complexes known as polyplexes *(63)*.

3.4.1. First Generation—Polylysine, Polyethyleneimine, and Polyamidoamine (PAMAM)

Polylysine (PLL), polyethyleneimine (PEI), and polyamidoamine (PAMAM) are among the first polycations used in nonviral gene delivery systems *(62,85)*. Positively charged PLL can efficiently condense DNA to form polyplexes. However, these polyplexes have poor transfection efficiency because of the lack of efficient endosomal escape *(86,87)*. Addition of chloroquine, an endosomolytic agent, improved the gene transfer activity *(88)*. Conjugation of PLL with targeting moieties, such as antibody *(89)*, folate *(90)*, peptide *(91)*, and transferrin *(92)*, have shown to significantly enhance in vitro and in vivo transfection efficiency. PLL has relatively high cytotoxicity: the PLL polyplexes kill 40–60% cells in vitro *(93)*. Conjugation of PEG or introduction of degradable bonds to PLL has been employed to reduce cytotoxicity *(94)*.

PEI has branched and linear forms and is among the most effective and widely used polycations in gene delivery since its introduction as a gene delivery agent in 1995 *(62)*. PEI can effectively condense DNA and has endosomolytic activity because of the high density of amines and the associated proton buffering capacity *(62)*. PEI-based polyplexes have also been designed for targeting specific cells by coupling with different ligands, such as antibody *(53)*, folate *(95)*, and transferrin *(96)*. Both local and systemic administrations of PEI-based vectors have been applied to deliver gene to different organs, including brain, lung, and tumor. Like PLL, PEI also suffers from high cytotoxicity.

PAMAM dendrimer is characterized by its dendritic symmetric structure, whose size and surface charge are governed by generation numbers in the synthesis process. The abundance of amine groups and good biocompatibility of the PAMAM make it an attractive gene delivery polymer. Haensler and Szoka *(97)* first demonstrated in vitro transfection mediated by PAMAM dendrimers. PAMAM dendrimers are efficient gene transfection polymers and able to transfect primary human fibroblasts, which are generally difficult to transfect. The transfection efficiency of PAMAM dendrimers is related to the generation of the dendrimers and DNA to dendrimer charge ratio.

3.4.2. Other Polymers

Promising initial results from the first generation commercial polymers provided the inspiration for the design and synthesis of new polymers for gene delivery. Generally, these new polymers were designed to address specific barriers, such as endosomal escape, biodegradability, biocompatibility, and stability.

As discussed previously, endosomal escape is a critical step for efficient gene delivery. Hoffman and coworkers *(98,99)* have evaluated a series of membrane-disruptive polymers based on pH-sensitive α-alkyl acrylic acids, such as polyethylacrylic acid and polypropylacrylic acid. Moreover, they have further functionalized the membrane-disruptive polymers by incorporating grafts such as serum-stabilizing PEG through acid-degradable linkers *(100)*.

Biodegradable polymers were also designed to address cytotoxicity and gene-releasing barriers. A class of spontaneously biodegradable polymers was reported for DNA condensation and gene delivery based on cationic polyesters, including poly(α-[4-aminobutyl]-L-glycolic acid)] *(94)*, poly(4-hydroxyl-1-proline ester) *(101)*, hyperbranched poly(amino ester)] *(102)*. These polymers showed efficacy for delivery of DNA both in vitro and in vivo with minimal toxicity. Another application of degradable polymers is the crosslinked low molecular weight (LMW) PEI (<2000 Da). High molecular weight PEI is highly efficient for gene delivery, but very toxic to many cell lines; however, LMW PEI is essentially nontoxic but very ineffective for gene delivery. In a study, PEI (800 Da) was crosslinked with diacrylates to generate degradable PEI (14–30 kDa), which was more efficient and less toxic than nondegradable commercial PEI (25 kDa) *(103)*. Many biocompatible and FDA-approved pharmaceutical excipients were also investigated as gene delivery agents, such as cyclodextrins *(104)* and chitosan *(105)*.

3.5. Combined Lipid/Polymer-Based Vectors

Another unique nonviral vector consists of a polycation-condensed polyplex core coated with a lipid outlayer, which was named LPD (liposome/poly-lysine/DNA) *(54,106)*. Depending on the charge of the lipids, LPD can be characterized as LPDI and LPDII coated with positively and negatively charged lipids, respectively.

LPDI was first developed based on the different DNA condensing ability between monovalent cationic lipids and multivalent polymers. In Gao and Huang's study, DNA was first condensed with PLL and coated with various cationic lipids. The resulting LPDI demonstrated higher transfection efficiency than corresponding lipoplexes and had the ability to transfect some difficult cell lines *(106)*. The structure of LPDI appeared to be more compact and spherical with a size of 100 nm or less compared to the lipoplexes' spaghetti-meatball structure with large and heterogeneous size. LPDII has a condensed DNA core with moderate excess of positive charges, which is coated with anionic lipids resulting in an overall negatively charged particle. Lee and Huang designed folate-targeted LPDII composed of PLL-condensed DNA and pH-sensitive liposomes (DOPE/CHEMS/folate-PEG-DOPE). The LPDII showed 20–30 times greater gene transfer activity than DC-Chol-based lipoplexes in KB cells and had much less cytotoxicity *(54)*.

3.6. Summary

Nonviral vectors are receiving considerable interest mainly because of their safety advantages over viral vectors. Significant progress has been made in this area during the past decade with a variety of lipids and polymers being created to construct nonviral vectors. However, current nonviral vectors are much less efficient than viral vectors in delivering genes, limiting their use mainly to in vitro gene transfer studies. An ongoing trend of nonviral vector development is to design lipid and polymer molecules and build the multiple functional components into single nanometer-sized vehicles that are able to overcome a series of barriers for gene transfer.

4. Physical Methods for Gene Delivery

4.1. Introduction

Viral and nonviral vectors enhance gene uptake and expression by interacting with the cells and/or other biological entities passively. Alternatively, external physical energy or force can be used to actively facilitate the passage of the genes across various barriers for gene transfer. Such approaches are collectively called physical methods. Based on the type of the physical energy/force employed and how they are used for gene delivery, they are categorized into injection, projectile gene delivery, hydrodynamic gene delivery, electroporation, ultrasound-mediated delivery, laser irradiation, magnetic force-mediated delivery, and electrical field-induced molecular vibration.

4.2. Injection

Injection works by mechanically inserting a hollow needle into the target and pushing the agent of interest in through the needle. This technique has been used in both in vitro and in vivo gene delivery. To transfect cultured cells, injection must be performed on individual cells at a one-by-one mode. Because of the micrometer size of the cells and the scale of operation, this technique is termed as microinjection. Genes can be microinjected into either the cytoplasm or the nucleus of any cells with transfection efficiency up to 100% *(107)*. Microinjection also allows semiquantitative introduction of DNA into the cells and specifically monitoring individual cells for expression of the exogenous DNA, rendering it highly valuable for studying the basic mechanism of gene transfer *(108)*. However, microinjection is a fastidious and time-consuming operation. Typically, only a few hundred cells can be transfected per experiment by an experienced experimenter. It is therefore traditionally used to transfect embryonic cells for creating transgenic animals *(109)*. In an effort to overcome the drawbacks of this technique, a robot system has been developed to increase its throughput and make the operation easier *(110)*.

Although microinjection is not applicable to in vivo gene delivery, conventional parenteral injection is the most reliable, simple, and efficient way for the delivery of agents of high molecular weight and high hydrophilicity among all drug delivery approaches such as oral and transdermal administration. As a result, injection is used in the majority of gene transfer studies to deliver DNA in vivo, no matter whether they are naked or carried in viral/nonviral vectors. Furthermore, injection is a necessary procedure required by some other physical approaches for in vivo gene delivery.

4.3. Projectile Gene Delivery

Projectile gene delivery is based on the use of kinetic energy to enhance gene transfer by accelerating DNA with its carrier to a high speed and shooting them into the cells or tissues. Depending on either solid or liquid DNA carriers used, this technology is divided into gene gun and jet injection, respectively.

4.3.1. Gene Gun

Gene gun is also referred as particle-mediated or bioballistic gene delivery. The "bullets" are typically made of metal microparticles of 1–5 μm in diameter coated with plasmid DNA and propelled by high-pressure helium gas. Gold is the most commonly used carrier material because of its high density and chemical inertness. Gene gun was originally developed for creation of transgenic plants. Recently, this technique is applied to mammalian cells with high efficiency *(111–113)*. More importantly from a clinical point of view, gene gun has been used in vivo to transfect different organs and tissues including skin *(111)*, liver *(111,114)*, muscle *(115)*, brain *(116)*, heart *(117)*, bladder *(118)*, and tumors *(119)*. This technique is generally safe. The major drawback is the shallow penetration of the particles into tissues. Surgical exposure would thus be needed for transfecting internal tissues. Moreover, a relatively small amount of cells is transfected per treatment. As a result, genetic vaccination is currently the major clinical application of gene gun technology with the skin as the primary target because it contains a large number of resident antigen-presenting cells and is easy to access. Strong immune responses have been elicited in various animals against diseases such as hepatitis B, malaria, and HIV, leading to human clinical trials of this technology *(1,120)*. Encouraging results have also been obtained in the studies using gene gun to treat cancer *(119,121)*, bladder disease *(118)*, and promote wound healing *(122)*.

4.3.2. Jet Injection

Jet injection works by shooting high-speed liquid jets of DNA solution or suspension rather than solid particles into a target tissue. Compared to gene gun, jet injection can have deeper penetration of up to 2 cm under the skin *(123)*. Because preparation of the DNA/metal microparticles is not needed, jet injection is also simpler and of lower cost than gene gun. The effectiveness of this technique has been demonstrated via gene transfer of the skin *(124)*, mucosa *(125)*, muscle *(126)*, and tumors *(127–129)* in various animals including monkeys *(124)*. Similar to gene gun, jet injection is also limited by the relatively small amount of cells that can be transfected in a treatment.

4.4. Electroporation

Electroporation uses short, intense electrical pulses across cells to promote the entry of macromolecules into cells. Increased permeability is believed to result from the formation of transient and reversible pores in the cell membrane as an applied external field exceeds its capacity. Presence of DNA may assist the perforation process *(130)*. Pores of 20–100 nm diameter in red blood cells have been observed following electroporation by freeze-fracture electron microscopy *(131)*. Entry of DNA into the cells through the pores is driven by electrophoretic force and influenced by a number of factors including the pulse duration, number of pulses, and electric field strength, concentration of DNA, composition of membrane, and type of cells *(132)*. Electroporation has been widely used for in vitro gene delivery. A commercially available electroporation-based gene transfer technology, Nucleofector™ (Amaxa Biosystems, Köln, Germany), has been shown to be able to transfect a wide range of hard-to-transfect cells such as human primary cells and stem cells with high efficiency and high viability *(133–135)*.

In vivo electroporation-based gene transfer is advancing rapidly. Skeletal muscle is the major target tissue because of its large mass, high degree of vascularization, and ease of access. Moreover, skeletal muscle cells have a lengthy life span and do not undergo cellular division, potentially allowing for long-term transgene expression even though the transgene does not integrate into the host genome. Gene therapy based on electroporation on skeletal muscle is used for either correction of muscle disorders or using muscle as bioreactor for secretion of therapeutic proteins. Electroporation usually starts with intramuscular injection of plasmid DNA into the muscle and is followed by electrical discharge from either needle electrodes or plate electrodes, which can be placed externally. A typical pulse condition used for in vivo electroporation in skeletal muscle is several 1 Hz square-wave pulses at 200 V/cm voltage intensity and 20 ms duration *(132,136,137)*. A variety of proteins such as Factor VIII and IX, interleukins, growth factors, monoclonal antibodies, and erythropoietin for various pathologies have been successfully produced via in vivo electroporation gene transfer in the skeletal muscles *(137–140)*. Electroporation has been shown to increase the transgene expression by about 100 to 1000-fold over the simple intramuscular plasmid injection and can last for several months or even over a year *(132,137,140,141)*.

Delivery of large DNA with electroporation has also been demonstrated. A full-length 12.5 kb mouse dystrophin DNA, which is beyond the packaging capacity of onco-retroviral vector, first-generation adenoviral vector, and AAV vector, was successfully delivered and expressed in mice *(142,143)*. Besides skeletal muscles, electroporation has been used for transfecting other organs and tissues including the skin *(144)*, liver *(145)*, heart *(146)*, lung *(147)*, eye *(148)*, brain *(149)*, and solid tumor *(150)* in animals from mice and rats to pigs *(151)*. Whereas electroporation on the skeletal muscle can be performed with needle electrodes or external plate electrodes, this approach becomes more invasive as accessibility to the target tissue decreases. Surgical exposure is needed for accessing the internal organs for electroporation such as liver *(145)* and lung *(147)*. Furthermore, the current eletroporation methodology seems not suitable for transfecting tissues over a large region such as a whole limb of human because of the extremely high voltage required to achieve enough voltage intensity. In spite of these drawbacks, electroporation is a relatively simple, safe, efficient approach and has shown great potential for many therapeutic applications.

4.5. Ultrasound-Mediated Gene Transfer

Acoustic energy in the form of ultrasound at megahertz frequencies can enhance both in vitro and in vivo gene transfer. The mechanism is not well understood but cavitation, the dynamic formation and destruction of gas microbubbles caused by an acoustic field, is believed to play an important role *(152)*. Several processes involved in the cavitation can contribute to the enhanced cellular uptake of the genes. The most important one may be the formation of pores in the cell membrane induced by the oscillating gas microbubbles *(153)*. The entry of the genes into the cells via the pores may be facilitated by local microstreaming also caused by the oscillating bubbles *(154)*. Moreover, high-velocity fluid microjets towards the adjacent cell membrane *(155)*, local high temperature *(156)*, shear stress *(154)*, and shock waves *(157)* generated during the violent collapse of the bubbles may also assist the DNA to enter the cells. This cavitation mechanism is supported by the significant enhancement of gene transfer efficiency with the use of microbubble-based contrast enhancing agents, which can promote acoustic cavitation by acting as cavitation nuclei and lower the threshold of energy for cavitation *(158–162)*.

Enhanced gene transfer by ultrasound has been demonstrated with various cell types in vitro *(158,161,163)*. Furthermore, this technique has shown encouraging results in increasing the transfer of therapeutic genes in animal models. One study showed significantly increased capillary density, blood flow, and blood pressure ratio in the skeletal muscles of rabbits with limb ischemia following ultrasonic delivery of hepatocyte growth factor gene *(159)*. Similar results were also observed in rat models of acute myocardial infarction *(164)*. In another study, delivery of a tumor suppressor gene, p53, into the rat carotid artery after balloon injury has shown considerable antirestenosis effect *(160)*. The potential of ultrasonic gene transfer for the treatment of cancers was also demonstrated by the high-level tumor-specific expression of interleukin 2 (IL-2) gene following systemic administration *(165)*.

Ultrasonic gene delivery offers many favorable characteristics. First, ultrasound at megahertz frequencies is used for medical diagnostics and has excellent record of safety in clinical practice. Microbubbles have also been approved for clinical use as contrast agents by FDA. Second, this technique allows for transfection of internal organs noninvasively as an external transducer is used. Third, the transfection area can be controlled by either focusing ultrasonic field on a small region or scanning it over a large area. Finally, ultrasonic treatment is easy to perform, and imaging can be conducted in real time, using a single piece of equipment. Despite these attractive features and the proof-of-principle results, however, ultrasound-mediated gene delivery is at a rudimentary stage. Much more work such as proving its efficacy in larger animals would be needed to deliver its promise.

4.6. Magnetic Force-Mediated Gene Delivery

Magnetic force-mediated gene delivery is also named magnetofection *(166)*. DNA is first associated with magnetic particles to form DNA–magnetic particle complexes. Magnetic force can thus be exerted on the complexes to influence their interactions with cells and tissues for enhancing the gene transfer. To transfect cultured cells, a magnet is placed under the cells to generate a pulling force on DNA–magnetic particle complexes suspended in culture medium, resulting in rapid deposition of the complexes on the cell surface. With a relatively small amount of DNA, a high local DNA concentration at the vicinity of the cells can thus be quickly generated, leading to a fast and high level of uptake by nonspecific endocytosis *(167)*. The high transfection efficiency of magnetofection has been demonstrated with various cell types including hard-to-transfect cells such as primary human airway epithelial cells, as well as a dissected airway epithelium organ model *(168–170)*. However, an ex vivo test on a dissected lamb's heart showed insignificant increase of transgene expression mediated by magnetic force *(171)*.

High level and localized magnetofection has also been observed in vivo in the gastrointestinal tract of rats and blood vessels of pigs *(166)*. Another study demonstrated the successful use of vascular endothelial growth factor, a therapeutic gene, to promote angiogenesis and arteriogenesis in rabbit models of limb ischemia by magnetofection *(172)*. Encouraging results of magnetofection for treating feline cancer have also been obtained in a veterinary clinical trial but the detailed information was not revealed *(170)*. In this study, the DNA–magnetic particle complexes were injected into the tumors and magnetic force was used to retain the genes within tumor rather than cause sedimentation. In spite of these positive results, an in vivo study showed that the magnetic force did not improve transfection efficiency in airway epithelial cells *(173)*. Many more investigations are thus needed, especially with large animals, to fully exploit the potential of this technology.

4.7. Hydrodynamic Gene Delivery

Transfection efficiency of naked DNA delivered by conventional intravascular injection is low because of multiple biological and physical barriers. However, the efficiency can be significantly enhanced by controlling the hydrodynamic behavior of the blood, specifically, by temporarily increasing the blood pressure and/or blocking the blood flow *(174)*. The mechanism for enhanced DNA uptake is not clear. One hypothesis states that enlarged fenestrae in the liver sinusoids and the formation of transient pores in the membrane of hepatocytes caused by elevated blood pressure in the liver are responsible *(175–177)*. This mechanism is supported because a rapid injection of a large volume of DNA solution in occluded limb muscles is necessary for high-level gene expression *(178–181)*. However, efficient gene transfer was also observed without increasing blood pressure in both liver and diaphragm muscle, suggesting that other cellular mechanisms enhanced by prolonged, static presence of the plasmid in the blood may play a major role for gene uptake *(182–184)*.

Results of hydrodynamic delivery of therapeutic genes in the liver are highly encouraging. Following a rapid injection of a large volume of DNA solution in the tail vein of mice or rats, sustained production of human factor IX in hemophilia B mice for 1 yr and erythropoietin in rats for at least 12 wk was observed *(185,186)*. To be more relevant for human use, local DNA delivery and blood blocking have been demonstrated in the liver of rabbits with the use of a catheter and balloon cuff *(187)*. A therapeutic gene encoding full-length dystrophin has also been efficiently delivered and expressed in the muscles of both limbs *(180)* and diaphragm *(183)* in dystrophic mice following surgical blood occlusion and local intravascular gene delivery. For larger animals, the procedure can be even simpler and minimally invasive by the use of a tourniquet or blood pressure cuff to block blood flow and a catheter to deliver DNA solution into the blood. The effectiveness of this technique has been demonstrated in monkeys *(179,181)*. In summary, hydrodynamic delivery is unique among all gene transfer techniques for its ability to efficiently and safely transfect a whole organ or muscles over a large region with either transient or long-term gene expression. It thus holds considerable potential for treating human conditions such as Duchenne's muscular dystrophy, peripheral artery disease of distal limbs, liver diseases, and diseases requiring sustained secretion of proteins.

4.8. Laser Irradiation

Laser is a highly focused beam of light. When irradiated on a cell, it can perforate the cell membrane, allowing the entry of genes in the surrounding medium into the cell *(188,189)*. In one study, cultured cells were illuminated at a one-by-one mode by a femtosecond laser with each cell taking 10–15 s, resulting in a transfection efficiency of 100% *(190)*. Suspended cells can also be transfected by using a laser beam to optically trap a cell and a second beam to perforate it *(191)*. Enhanced gene uptake can also be obtained without direct laser irradiation on cells but by exposing the cells to shock waves induced by laser *(192)*. Moreover, laser-mediated transfection can be potentiated by elevated temperature and presence of light absorbents in the cell culture medium *(192–194)*. In vivo studies showed that expression of a reporter gene injected in the rat skin was enhanced by the shock waves induced by laser without major side effects *(195)*. In another study, illumination with infrared laser focused in the muscle 2 mm below the skin of mice has induced high and persistent expression of erythropoietin gene without tissue damage *(196)*. However, the mechanism for laser-enhanced in vivo gene transfer is not well understood.

4.9. Molecular Vibration

Molecular vibration is a recently developed technique for gene transfer based on the use of a vibrating ultrahigh-voltage electric field *(197)*. Different from electroporation, which also uses electrical energy, the electrodes used in molecular vibration are not in contact with suspension-containing cells and DNA and there is thus no current through the suspension during transfection. The vibrating ultrahigh-voltage electric field can drive all molecules in the suspension of cells and DNA to vibrate, converting the electrical energy to vibrational energy of the molecules. When the vibrational energy exceeds the hydrophobic bonding energy of phospholipids constituting the cell membrane, exogenous macromolecules including DNA can penetrate the membrane and enter the cells. Molecular vibration has demonstrated high transfection efficiency and cell viability with a variety of cell types, as well as tissue explants. However, whether this approach can be used for in vivo gene transfer remains unknown.

4.10. Summary

Compared to the vector-based gene delivery methods, physical methods offer unique advantages for gene transfer. Assisted by the external physical energy or force, enhanced gene transfer can be obtained with naked DNA, eliminating many problems associated with the preparation, toxicity, and immunogenicity of viral and nonviral vectors. The transfection efficiency of physical methods is also relatively less dependent on cell cycle and type and DNA size. Particularly valuable for in vivo gene transfer, physical methods have excellent targeting ability at the tissue and organ level. With these advantages, efficient gene transfer to various hard-to-transfect cells as well as a variety of animals has been observed using the physical methods.

Physical methods are relatively new in gene delivery technologies. The fundamental physics and biology involved in the transfection processes are generally not well understood. Side effects such as tissue damage are inherent to some methods because enhanced gene uptake is based on the perturbation of integrity of cells. Moreover, most physical techniques are at rudimentary stage of development and much work has to be done before they can reach human clinical trials. Nonetheless, physical methods for gene delivery have demonstrated encouraging results and hold great potential for further improvement.

5. Concluding Remarks

Gene therapy promises to revolutionize medicine. Realization of the promise must, however, be established on safe and efficient gene delivery technology. Viral vectors are dominant tools for therapeutic gene transfer at present. They are efficient in gene transfer but suffer from high toxicity and immunogenicity. Improving their safety profiles is one focus of current research on viral vectors. Progresses are continuously being made in this area and viral vectors are projected to remain as the major gene delivery tools in the foreseeable future. Nonviral vectors are receiving increasing attention in recent years. They are safer than viral vectors and offer great design flexibility. Although current nonviral vectors have unacceptably low transfection efficiency, advance in this direction may eventually lead to virus-like, multifunctional nanodevices capable of efficient and safe gene transfer. Physical gene delivery methods enhance gene transfer by employing external energy or force. This group of methods is at its early stage of development, but is likely to play a more important role in the future. Currently, none of the three classes of gene delivery techniques is satisfying for routine practical use but this area is under intense investigation and progressing rapidly. Development of clinically useful gene delivery technology is thus, we believe, challenging but highly promising.

Acknowledgment

This work was supported by National Science Foundation Grant DMI-0425626.

References

1. Gregersen, J. P. (2001) DNA vaccines. *Naturwissenschaften* **88,** 504–513.
2. Freeman, S. M., Abboud, C. N., Whartenby, K. A., et al. (1993) The "bystander effect": tumor regression when a fraction of the tumor mass is genetically modified. *Cancer Res.* **53,** 5274–5283.
3. Blaese, R. M., Culver, K. W., Miller, A. D., et al. (1995) T lymphocyte-directed gene therapy for ADA- SCID: initial trial results after 4 years. *Science* **270,** 475–480.
4. Schroder, A. R., Shinn, P., Chen, H., Berry, C., Ecker, J. R., and Bushman, F. (2002) HIV-1 integration in the human genome favors active genes and local hotspots. *Cell* **110,** 521–529.
5. Wu, X., Li, Y., Crise, B., and Burgess, S. M. (2003) Transcription start regions in the human genome are favored targets for MLV integration. *Science* **300,** 1749–1751.
6. Cavazzana-Calvo, M., Hacein-Bey, S., de Saint Basile, G., et al. (2000) Gene therapy of human severe combined immune deficiency (SCID)-X1 disease. *Science* **288,** 669–672.
7. Romano, G. (2005) Current development of lentiviral-mediated gene transfer. *Drug News Perspect.* **18,** 128–134.
8. Kumar, M., Keller, B., Makalou, N., and Sutton, R. E. (2001) Systematic determination of the packaging limit of lentiviral vectors. *Hum. Gene Ther.* **12,** 1893–1905.
9. De Palma, M., Venneri, M. A., and Naldini, L. (2003) In vivo targeting of tumor endothelial cells by systemic delivery of lentiviral vectors. *Hum. Gene Ther.* **14,** 1193–1206.
10. Totsugawa, T., Kobayashi, N., Maruyama, M., et al. (2003) Lentiviral vector: a useful tool for transduction of human liver endothelial cells. *ASAIO J.* **49,** 635–640.
11. Consiglio, A., Gritti, A., Dolcetta, D. (2004) Robust in vivo gene transfer into adult mammalian neural stem cells by lentiviral vectors. *Proc. Natl Acad. Sci. U. S. A.* **101,** 14,835–14,840.
12. Ma, Y., Ramezani, A., Lewis, R., Hawley, R. G., and Thomson, J. A. (2003) High-level sustained transgene expression in human embryonic stem cells using lentiviral vectors. *Stem Cells* **21,** 111–117.
13. Woods, N. B., Mikkola, H., Nilsson, E., Olsson, K., Trono, D., and Karlsson, S. (2001) Lentiviral-mediated gene transfer into haematopoietic stem cells. *J. Intern. Med.* **249,** 339–343.
14. Condiotti, R., Curran, M. A., Nolan, G. P., et al. (2004) Prolonged liver-specific transgene expression by a non-primate lentiviral vector. *Biochem. Biophys. Res. Commun.* **320,** 998–1006.
15. Stein, C. S., Kang, Y., Sauter, S. L., et al. (2001) In vivo treatment of hemophilia A and mucopolysaccharidosis type VII using nonprimate lentiviral vectors. *Mol. Ther.* **3,** 850–856.

16. Cefai, D., Simeoni, E., Ludunge, K. M., et al. (2005) Multiply attenuated, self-inactivating lentiviral vectors efficiently transduce human coronary artery cells in vitro and rat arteries in vivo. *J. Mol. Cell Cardiol.* **38,** 333–344.

17. Wong, L. F., Ralph, G. S., Walmsley, L. E., et al. (2005) Lentiviral-mediated delivery of Bcl-2 or GDNF protects against excitotoxicity in the rat hippocampus. *Mol. Ther.* **11,** 89–95.

18. Vigna, E. and Naldini, L. (2000) Lentiviral vectors: excellent tools for experimental gene transfer and promising candidates for gene therapy. *J. Gene Med.* **2,** 308–316.

19. Zufferey, R., Dull, T., Mandel, R. J., et al. (1998) Self-inactivating lentivirus vector for safe and efficient in vivo gene delivery. *J. Virol.* **72,** 9873–9880.

20. Harui, A., Suzuki, S., Kochanek, S., and Mitani, K. (1999) Frequency and stability of chromosomal integration of adenovirus vectors. *J. Virol.* **73,** 6141–6146.

21. Kochanek, S., Schiedner, G., and Volpers, C. (2001) High-capacity 'gutless' adenoviral vectors. *Curr. Opin. Mol. Ther.* **3,** 454–463.

22. Kim, I. H., Jozkowicz, A., Piedra, P. A., Oka, K., and Chan, L. (2001) Lifetime correction of genetic deficiency in mice with a single injection of helper-dependent adenoviral vector. *Proc. Natl Acad. Sci. U. S. A.* **98,** 13,282–13,287.

23. Dudley, R. W., Lu, Y., Gilbert, R., et al. (2004) Sustained improvement of muscle function one year after full-length dystrophin gene transfer into mdx mice by a gutted helper-dependent adenoviral vector. *Hum. Gene Ther.* **15,** 145–156.

24. Kotin, R. M., Siniscalco, M., Samulski, R. J., et al. (1990) Site-specific integration by adeno-associated virus. *Proc. Natl Acad. Sci. U. S. A.* **87,** 2211–2215.

25. Kearns, W. G., Afione, S. A., Fulmer, S. B., et al. (1996) Recombinant adeno-associated virus (AAV-CFTR) vectors do not integrate in a site-specific fashion in an immortalized epithelial cell line. *Gene Ther.* **3,** 748–755.

26. Nakai, H., Storm, T. A., and Kay, M. A. (2000) Increasing the size of rAAV-mediated expression cassettes in vivo by intermolecular joining of two complementary vectors. *Nat. Biotechnol.* **18,** 527–532.

27. Flotte, T. R. (2005) Recent developments in recombinant AAV-mediated gene therapy for lung diseases. *Curr. Gene Ther.* **5,** 361–366.

28. Wang, L. and Herzog, R. W. (2005) AAV-mediated gene transfer for treatment of hemophilia. *Curr. Gene Ther.* **5,** 349–360.

29. Athanasopoulos, T., Graham, I. R., Foster, H., and Dickson, G. (2004) Recombinant adeno-associated viral (rAAV) vectors as therapeutic tools for Duchenne muscular dystrophy (DMD). *Gene Ther.* **11**(Suppl 1), S109–S121.

30. Ruitenberg, M. J., Eggers, R., Boer, G. J., and Verhaagen, J. (2002) Adeno-associated viral vectors as agents for gene delivery: application in disorders and trauma of the central nervous system. *Methods* **28,** 182–194.

31. Martin, K. R., Klein, R. L., and Quigley, H. A. (2002) Gene delivery to the eye using adeno-associated viral vectors. *Methods* **28,** 267–275.

32. Cottard, V., Mulleman, D., Bouille, P., Mezzina, M., Boissier, M. C., and Bessis, N. (2000) Adeno-associated virus-mediated delivery of IL-4 prevents collagen-induced arthritis. *Gene Ther.* **7,** 1930–1939.

33. Latchman, D. S. (2001) Gene delivery and gene therapy with herpes simplex virus-based vectors. *Gene* **264,** 1–9.

34. Moriuchi, S., Wolfe, D., Tamura, M., et al. (2002) Double suicide gene therapy using a replication defective herpes simplex virus vector reveals reciprocal interference in a malignant glioma model. *Gene Ther.* **9,** 584–591.
35. Mata, M., Glorioso, J., and Fink, D. J. (2003) Development of HSV-mediated gene transfer for the treatment of chronic pain. *Exp. Neurol.* **184**(Suppl 1)**,** S25–S29.
36. Fink, D. J., Glorioso, J., and Mata, M. (2003) Therapeutic gene transfer with herpes-based vectors: studies in Parkinson's disease and motor nerve regeneration. *Exp. Neurol.* **184** (Suppl 1)**,** S19–S24.
37. Natsume, A., Mata, M., Wolfe, D., et al. (2002) Bcl-2 and GDNF delivered by HSV-mediated gene transfer after spinal root avulsion provide a synergistic effect. *J. Neurotrauma* **19,** 61–68.
38. Smith, G. L. and Moss, B. (1983) Infectious poxvirus vectors have capacity for at least 25,000 base pairs of foreign DNA. *Gene* **25,** 21–28.
39. Chen, B., Timiryasova, T. M., Andres, M. L., et al. (2000) Evaluation of combined vaccinia virus-mediated antitumor gene therapy with p53, IL-2, and IL-12 in a glioma model. *Cancer Gene Ther.* **7,** 1437–1447.
40. Kaufman, H. L., Flanagan, K., Lee, C. S., Perretta, D. J., and Horig, H. (2002) Insertion of interleukin-2 (IL-2) and interleukin-12 (IL-12) genes into vaccinia virus results in effective anti-tumor responses without toxicity. *Vaccine* **20,** 1862–1869.
41. Timiryasova, T. M., Gridley, D. S., Chen, B., et al. (2003) Radiation enhances the anti-tumor effects of vaccinia-p53 gene therapy in glioma. Technol *Cancer Res. Treat.* **2,** 223–235.
42. Gulley, J., Chen, A. P., Dahut, W., et al. (2002) Phase I study of a vaccine using recombinant vaccinia virus expressing PSA (rV-PSA) in patients with metastatic androgen-independent prostate cancer. *Prostate* **53,** 109–117.
43. Wagner, E., Cotten, M., Foisner, R., and Birnstiel, M. L. (1991) Transferrin–polycation–DNA complexes: the effect of polycations on the structure of the complex and DNA delivery to cells. *Proc. Natl Acad. Sci. U. S. A.* **88,** 4255–4259.
44. Abdelhady, H. G., Allen, S., Davies, M. C., Roberts, C. J., Tendler, S. J., and Williams, P. M. (2003) Direct real-time molecular scale visualisation of the degradation of condensed DNA complexes exposed to DNase I. *Nucleic Acids Res.* **31,** 4001–4005.
45. Sternberg, B., Sorgi, F. L., and Huang, L. (1994) New structures in complex formation between DNA and cationic liposomes visualized by freeze-fracture electron microscopy. *FEBS Lett.* **356,** 361–366.
46. Schaffer, D. V., Fidelman, N. A., Dan, N., and Lauffenburger, D. A. (2000) Vector unpacking as a potential barrier for receptor-mediated polyplex gene delivery. *Biotechnol. Bioeng.* **67,** 598–606.
47. Dash, P. R., Read, M. L., Barrett, L. B., Wolfert, M. A., and Seymour, L. W. (1999) Factors affecting blood clearance and in vivo distribution of polyelectrolyte complexes for gene delivery. *Gene Ther.* **6,** 643–650.
48. Hong, K., Zheng, W., Baker, A., and Papahadjopoulos, D. (1997) Stabilization of cationic liposome-plasmid DNA complexes by polyamines and poly(ethylene glycol)-phospholipid conjugates for efficient in vivo gene delivery. *FEBS Lett.* **400,** 233–237.

49. Toncheva, V., Wolfert, M. A., Dash, P. R., et al. (1998) Novel vectors for gene delivery formed by self-assembly of DNA with poly(L-lysine) grafted with hydrophilic polymers. *Biochim. Biophys. Acta* **1380,** 354–368.

50. Kircheis, R., Wightman, L., Schreiber, A., et al. (2001) Polyethylenimine/DNA complexes shielded by transferrin target gene expression to tumors after systemic application. *Gene Ther.* **8,** 28–40.

51. Mishra, S., Webster, P., and Davis, M. E. (2004) PEGylation significantly affects cellular uptake and intracellular trafficking of non-viral gene delivery particles. *Eur. J. Cell Biol.* **83,** 97–111.

52. Frederiksen, K. S., Abrahamsen, N., Cristiano, R. J., Damstrup, L., and Poulsen, H. S. (2000) Gene delivery by an epidermal growth factor/DNA polyplex to small cell lung cancer cell lines expressing low levels of epidermal growth factor receptor. *Cancer Gene Ther.* **7,** 262–268.

53. Chiu, S. J., Ueno, N. T., and Lee, R. J. (2004) Tumor-targeted gene delivery via anti-HER2 antibody (trastuzumab, Herceptin) conjugated polyethylenimine. *J. Control. Release* **97,** 357–369.

54. Lee, R. J. and Huang, L. (1996) Folate-targeted, anionic liposome-entrapped polylysine-condensed DNA for tumor cell-specific gene transfer. *J. Biol. Chem.* **271,** 8481–8487.

55. Labat-Moleur, F., Steffan, A. M., Brisson, C., et al. (1996) An electron microscopy study into the mechanism of gene transfer with lipopolyamines. *Gene Ther.* **3,** 1010–1017.

56. Tseng, W. C., Haselton, F. R., and Giorgio, T. D. (1997) Transfection by cationic liposomes using simultaneous single cell measurements of plasmid delivery and transgene expression. *J. Biol. Chem.* **272,** 25,641–25,647.

57. Hasegawa, S., Hirashima, N., and Nakanishi, M. (2001) Microtubule involvement in the intracellular dynamics for gene transfection mediated by cationic liposomes. *Gene Ther.* **8,** 1669–1673.

58. Xu, Y. and Szoka, F. C. Jr. (1996) Mechanism of DNA release from cationic liposome/DNA complexes used in cell transfection. *Biochemistry* **35,** 5616–5623.

59. Hui, S. W., Langner, M., Zhao, Y. L., Ross, P., Hurley, E., and Chan, K. (1996) The role of helper lipids in cationic liposome-mediated gene transfer. *Biophys. J.* **71,** 590–599.

60. Wagner, E., Plank, C., Zatloukal, K., Cotten, M., and Birnstiel, M. L. (1992) Influenza virus hemagglutinin HA-2 N-terminal fusogenic peptides augment gene transfer by transferrin–polylysine–DNA complexes: toward a synthetic virus-like gene-transfer vehicle. *Proc. Natl Acad. Sci. U. S. A.* **89,** 7934–7938.

61. Lee, H., Jeong, J. H., and Park, T. G. (2001) A new gene delivery formulation of polyethylenimine/DNA complexes coated with PEG conjugated fusogenic peptide. *J. Control. Release* **76,** 183–192.

62. Boussif, O., Lezoualc'h, F., Zanta, M. A., et al. (1995) A versatile vector for gene and oligonucleotide transfer into cells in culture and in vivo: polyethylenimine. *Proc. Natl Acad. Sci. U. S. A.* **92,** 7297–7301.

63. Pack, D. W., Hoffman, A. S., Pun, S., and Stayton, P. S. (2005) Design and development of polymers for gene delivery. *Nat. Rev. Drug Discov.* **4,** 581–593.

64. Legendre, J. Y. and Szoka, F. C. Jr. (1992) Delivery of plasmid DNA into mammalian cell lines using pH-sensitive liposomes: comparison with cationic liposomes. *Pharm. Res.* **9,** 1235–1242.

65. Lukacs, G. L., Haggie, P., Seksek, O., Lechardeur, D., Freedman, N., and Verkman, A. S. (2000) Size-dependent DNA mobility in cytoplasm and nucleus. *J. Biol. Chem.* **275,** 1625–1629.

66. Lechardeur, D., Sohn, K. J., Haardt, M., et al. (1999) Metabolic instability of plasmid DNA in the cytosol: a potential barrier to gene transfer. *Gene Ther.* **6,** 482–497.

67. Ryan, K. J. and Wente, S. R. (2000) The nuclear pore complex: a protein machine bridging the nucleus and cytoplasm. *Curr. Opin. Cell Biol.* **12,** 361–371.

68. Brunner, S., Sauer, T., Carotta, S., Cotten, M., Saltik, M., and Wagner, E. (2000) Cell cycle dependence of gene transfer by lipoplex, polyplex and recombinant adenovirus. *Gene Ther.* **7,** 401–407.

69. Zanta, M. A., Belguise-Valladier, P., and Behr, J. P. (1999) Gene delivery: a single nuclear localization signal peptide is sufficient to carry DNA to the cell nucleus. *Proc. Natl Acad. Sci. U. S. A.* **96,** 91–96.

70. Branden, L. J., Mohamed, A. J., and Smith, C. I. (1999) A peptide nucleic acid-nuclear localization signal fusion that mediates nuclear transport of DNA. *Nat. Biotechnol.* **17,** 784–787.

71. Chan, C. K. and Jans, D. A. (1999) Enhancement of polylysine-mediated transferrinfection by nuclear localization sequences: polylysine does not function as a nuclear localization sequence. *Hum. Gene Ther.* **10,** 1695–1702.

72. Chan, C. K., Senden, T., and Jans, D. A. (2000) Supramolecular structure and nuclear targeting efficiency determine the enhancement of transfection by modified polylysines. *Gene Ther.* **7,** 1690–1697.

73. Bremner, K. H., Seymour, L. W., Logan, A., and Read, M. L. (2004) Factors influencing the ability of nuclear localization sequence peptides to enhance nonviral gene delivery. *Bioconjug. Chem.* **15,** 152–161.

74. Tousignant, J. D., Gates, A. L., Ingram, L. A., et al. (2000) Comprehensive analysis of the acute toxicities induced by systemic administration of cationic lipid: plasmid DNA complexes in mice. *Hum. Gene Ther.* **11,** 2493–2513.

75. Freimark, B. D., Blezinger, H. P., Florack, V. J., et al. (1998) Cationic lipids enhance cytokine and cell influx levels in the lung following administration of plasmid: cationic lipid complexes. *J. Immunol.* **160,** 4580–4586.

76. Qin, L., Ding, Y., Pahud, D. R., Chang, E., Imperiale, M. J., and Bromberg, J. S. (1997) Promoter attenuation in gene therapy: interferon-gamma and tumor necrosis factor-alpha inhibit transgene expression. *Hum. Gene Ther.* **8,** 2019–2029.

77. Bird, A. P. (1986) CpG-rich islands and the function of DNA methylation. *Nature* **321,** 209–213.

78. Krieg, A. M. (1999) Direct immunologic activities of CpG DNA and implications for gene therapy. *J. Gene Med.* **1,** 56–63.

79. Felgner, P. L., Gadek, T. R., Holm, M., et al. (1987) Lipofection: a highly efficient, lipid-mediated DNA-transfection procedure. *Proc. Natl Acad. Sci. U. S. A.* **84,** 7413–7417.

80. Gao, X. and Huang, L. (1991) A novel cationic liposome reagent for efficient transfection of mammalian cells. *Biochem. Biophys. Res. Commun.* **179**, 280–285.

81. Lee, E. R., Marshall, J., Siegel, C. S., et al. (1996) Detailed analysis of structures and formulations of cationic lipids for efficient gene transfer to the lung. *Hum. Gene Ther.* **7**, 1701–1717.

82. Caplen, N. J., Alton, E. W., Middleton, P. G., et al. (1995) Liposome-mediated CFTR gene transfer to the nasal epithelium of patients with cystic fibrosis. *Nat. Med.* **1**, 39–46.

83. McLachlan, G., Ho, L. P., Davidson-Smith, H., et al. (1996) Laboratory and clinical studies in support of cystic fibrosis gene therapy using pCMV-CFTR-DOTAP. *Gene Ther.* **3**, 1113–1123.

84. Gill, D. R., Southern, K. W., Mofford, K. A., et al. (1997) A placebo-controlled study of liposome-mediated gene transfer to the nasal epithelium of patients with cystic fibrosis. *Gene Ther.* **4**, 199–209.

85. Laemmli, U. K. (1975) Characterization of DNA condensates induced by poly (ethylene oxide) and polylysine. *Proc. Natl Acad. Sci. U. S. A.* **72**, 4288–4292.

86. Akinc, A. and Langer, R. (2002) Measuring the pH environment of DNA delivered using nonviral vectors: implications for lysosomal trafficking. *Biotechnol. Bioeng.* **78**, 503–508.

87. Forrest, M. L. and Pack, D. W. (2002) On the kinetics of polyplex endocytic trafficking: implications for gene delivery vector design. *Mol. Ther.* **6**, 57–66.

88. Mislick, K. A., Baldeschwieler, J. D., Kayyem, J. F., and Meade, T. J. (1995) Transfection of folate-polylysine DNA complexes: evidence for lysosomal delivery. *Bioconjug. Chem.* **6**, 512–515.

89. Suh, W., Chung, J. K., Park, S. H., and Kim, S. W. (2001) Anti-JL1 antibody-conjugated poly (L-lysine) for targeted gene delivery to leukemia T cells. *J. Control. Release* **72**, 171–178.

90. Leamon, C. P., Weigl, D., and Hendren, R. W. (1999) Folate copolymer-mediated transfection of cultured cells. *Bioconjug. Chem.* **10**, 947–957.

91. Harbottle, R. P., Cooper, R. G., Hart, S. L., et al. (1998) An RGD-oligolysine peptide: a prototype construct for integrin-mediated gene delivery. *Hum. Gene Ther.* **9**, 1037–1047.

92. Zatloukal, K., Wagner, E., Cotten, M., et al. (1992) Transferrinfection: a highly efficient way to express gene constructs in eukaryotic cells. *Ann. N. Y. Acad. Sci.* **660**, 136–153.

93. Choi, Y. H., Liu, F., Kim, J. S., Choi, Y. K., Park, J. S., and Kim, S. W. (1998) Polyethylene glycol-grafted poly-L-lysine as polymeric gene carrier. *J. Control. Release* **54**, 39–48.

94. Lim, Y. B., Han, S. O., Kong, H. U., et al. (2000) Biodegradable polyester, poly[alpha-(4-aminobutyl)-L-glycolic acid], as a non-toxic gene carrier. *Pharm. Res.* **17**, 811–816.

95. Benns, J. M., Maheshwari, A., Furgeson, D. Y., Mahato, R. I., and Kim, S. W. (2001) Folate-PEG-folate-graft-polyethylenimine-based gene delivery. *J. Drug Target* **9**, 123–139.

96. Kircheis, R., Blessing, T., Brunner, S., Wightman, L., and Wagner, E. (2001) Tumor targeting with surface-shielded ligand–polycation DNA complexes. *J. Control. Release* **72,** 165–170.

97. Haensler, J. and Szoka, F. C. Jr. (1993) Polyamidoamine cascade polymers mediate efficient transfection of cells in culture. *Bioconjug. Chem.* **4,** 372–379.

98. Murthy, N., Robichaud, J. R., Tirrell, D. A., Stayton, P. S., and Hoffman, A. S. (1999) The design and synthesis of polymers for eukaryotic membrane disruption. *J. Control. Release* **61,** 137–143.

99. Kyriakides, T. R., Cheung, C. Y., Murthy, N., Bornstein, P., Stayton, P. S., and Hoffman, A. S. (2002) pH-sensitive polymers that enhance intracellular drug delivery in vivo. *J. Control. Release* **78,** 295–303.

100. Murthy, N., Campbell, J., Fausto, N., Hoffman, A. S., and Stayton, P. S. (2003) Design and synthesis of pH-responsive polymeric carriers that target uptake and enhance the intracellular delivery of oligonucleotides. *J. Control. Release* **89,** 365–374.

101. Lim, Y., Choi, Y. H., and Park, J. (1999) A Self-Destroying Polycationic Polymer: Biodegradable Poly(4-hydroxy-l-proline ester). *J. Am. Chem. Soc.* **121,** 5633–5639.

102. Lim, Y., Kim, S. M., Lee, Y., et al. (2001) Cationic hyperbranched poly(amino ester): a novel class of DNA condensing molecule with cationic surface, biodegradable three-dimensional structure, and tertiary amine groups in the interior. *J. Am. Chem. Soc.* **123,** 2460–2461.

103. Forrest, M. L., Koerber, J. T., and Pack, D. W. (2003) A degradable polyethylenimine derivative with low toxicity for highly efficient gene delivery. *Bioconjug. Chem.* **14,** 934–940.

104. Gonzalez, H., Hwang, S. J., and Davis, M. E. (1999) New class of polymers for the delivery of macromolecular therapeutics. *Bioconjug. Chem.* **10,** 1068–1074.

105. Borchard, G. (2001) Chitosans for gene delivery. *Adv. Drug Deliv. Rev.* **52,** 145–150.

106. Gao, X. and Huang, L. (1996) Potentiation of cationic liposome-mediated gene delivery by polycations. *Biochemistry* **35,** 1027–1036.

107. Graessmann, M. and Graessmann, A. (1983) Microinjection of tissue culture cells. *Methods Enzymol.* **101,** 482–492.

108. Masuda, T., Akita, H., and Harashima, H. (2005) Evaluation of nuclear transfer and trans-cription of plasmid DNA condensed with protamine by microinjection: the use of a nuclear transfer score. *FEBS Lett.* **579,** 2143–2148.

109. Hammer, R. E., Pursel, V. G., Rexroad, C. E. Jr., et al. (1985) Production of transgenic rabbits, sheep and pigs by microinjection. *Nature* **315,** 680–683.

110. Matsuoka, H., Komazaki, T., Mukai, Y., et al. (2005) High throughput easy microinjection with a single-cell manipulation supporting robot. *J. Biotechnol.* **116,** 185–194.

111. Yang, N. S., Burkholder, J., Roberts, B., Martinell, B., and McCabe, D. (1990) In vivo and in vitro gene transfer to mammalian somatic cells by particle bombardment. *Proc. Natl Acad. Sci. U. S. A.* **87,** 9568–9572.

112. O'Brien, J. and Lummis, S. C. (2004) Biolistic and diolistic transfection: using the gene gun to deliver DNA and lipophilic dyes into mammalian cells. *Methods* **33,** 121–125.

113. Murphy, R. C. and Messer, A. (2001) Gene transfer methods for CNS organotypic cultures: a comparison of three nonviral methods. *Mol. Ther.* **3,** 113–121.

114. Kuriyama, S., Mitoro, A., Tsujinoue, H., et al. (2000) Particle-mediated gene transfer into murine livers using a newly developed gene gun. *Gene Ther.* **7,** 1132–1136.

115. Lauritzen, H. P., Reynet, C., Schjerling, P., et al. (2002) Gene gun bombardment-mediated expression and translocation of EGFP-tagged GLUT4 in skeletal muscle fibres in vivo. *Pflugers Arch.* **444,** 710–721.

116. Sato, H., Hattori, S., Kawamoto, S., et al. (2000) In vivo gene gun-mediated DNA delivery into rodent brain tissue. *Biochem. Biophys. Res. Commun.* **270,** 163–170.

117. Matsuno, Y., Iwata, H., Umeda, Y., et al. (2003) Nonviral gene gun mediated transfer into the beating heart. *ASAIO J.* **49,** 641–644.

118. Chuang, Y. C., Yang, L. C., Chiang, P. H., et al. (2005) Gene gun particle encoding preproenkephalin cDNA produces analgesia against capsaicin-induced bladder pain in rats. *Urology* **65,** 804–810.

119. Lin, M. T., Pulkkinen, L., Uitto, J., and Yoon, K. (2000) The gene gun: current applications in cutaneous gene therapy. *Int. J. Dermatol.* **39,** 161–170.

120. Dean, H. J., Haynes, J., and Schmaljohn, C. (2005) The role of particle-mediated DNA vaccines in biodefense preparedness. *Adv. Drug Deliv. Rev.* **57,** 1315–1342.

121. Trimble, C., Lin, C. T., Hung, C. F., et al. (2000) Comparison of the CD8+ T cell responses and antitumor effects generated by DNA vaccine administered through gene gun, biojector, and syringe. *Vaccine* **21,** 4036–4042.

122. Davidson, J. M., Krieg, T., and Eming, S. A. (2000) Particle-mediated gene therapy of wounds. *Wound Repair Regen.* **8,** 452–459.

123. Furth, P. A., Shamay, A., and Hennighausen, L. (1995) Gene transfer into mammalian cells by jet injection. *Hybridoma* **14,** 149–152.

124. Haensler, J., Verdelet, C., Sanchez, V., et al. (1999) Intradermal DNA immunization by using jet-injectors in mice and monkeys. *Vaccine* **17,** 628–638.

125. Lundholm, P., Asakura, Y., Hinkula, J., Lucht, E., and Wahren, B. (1999) Induction of mucosal IgA by a novel jet delivery technique for HIV-1 DNA. *Vaccine* **17,** 2036–2042.

126. Horiki, M., Yamato, E., Ikegami, H., Ogihara, T., and Miyazaki, J. Needleless in vivo gene transfer into muscles by jet injection in combination with electroporation. *J. Gene Med.* **6,** 1134–1138.

127. Walther, W., Stein, U., Fichtner, I., Malcherek, L., Lemm, M., and Schlag, P. M. (2001) Nonviral in vivo gene delivery into tumors using a novel low volume jet-injection technology. *Gene Ther.* **8,** 173–180.

128. Walther, W., Stein, U., Fichtner, I., et al. (2002) Intratumoral low-volume jet-injection for efficient nonviral gene transfer. *Mol. Biotechnol.* **21,** 105–115.

129. Walther, W., Stein, U., Siegel, R., Fichtner, I., and Schlag, P. M. (2005) Use of the nuclease inhibitor aurintricarboxylic acid (ATA) for improved non-viral intra-tumoral in vivo gene transfer by jet-injection. *J. Gene Med.* **7,** 477–485.

130. Spassova, M., Tsoneva, I., Petrov, A. G., Petkova, J. I., and Neumann, E. (1994) Dip patch clamp currents suggest electrodiffusive transport of the polyelectrolyte DNA through lipid bilayers. *Biophys. Chem.* **52,** 267–274.

131. Chang, D. C. (1992) Structure and dynamics of electric field-induced membrane pores as revealed by rapid-freezing electron microscopy. In: *Guide to Electroporation and Electrofusion* (Chang, D. C., Chassy, B. M., Saunders, J. A., and Sowers, A. E., eds), Academic Press, San Diego, pp. 9–27.

132. Mir, L. M., Bureau, M. F., Gehl, J., et al. (1999) High-efficiency gene transfer into skeletal muscle mediated by electric pulses. *Proc. Natl Acad. Sci. U. S. A.* **96,** 4262–4267.

133. Leclere, P. G., Panjwani, A., Docherty, R., Berry, M., Pizzey, J., and Tonge, D. A. (2005) Effective gene delivery to adult neurons by a modified form of electroporation. *J. Neurosci. Methods* **142,** 137–143.

134. Hamm, A., Krott, N., Breibach, I., Blindt, R., and Bosserhoff, A. K. (2002) Efficient transfection method for primary cells. *Tissue Eng.* **8,** 235–245.

135. Lakshmipathy, U., Pelacho, B., Sudo, K., et al. (2004) Efficient transfection of embryonic and adult stem cells. *Stem Cells* **22,** 531–543.

136. Jeong, J. G., Kim, J. M., Ho, S. H., Hahn, W., Yu, S. S., and Kim, S. (2004) Electrotransfer of human IL-1Ra into skeletal muscles reduces the incidence of murine collagen-induced arthritis. *J. Gene Med.* **6,** 1125–1133.

137. Bettan, M., Emmanuel, F., Darteil, R., et al. (2000) High-level protein secretion into blood circulation after electric pulse-mediated gene transfer into skeletal muscle. *Mol. Ther.* **2,** 204–210.

138. Long, Y. C., Jaichandran, S., Ho, L. P., Tien, S. L., Tan, S. Y., and Kon, O. L. (2005) FVIII gene delivery by muscle electroporation corrects murine hemophilia A. *J. Gene Med.* **7,** 494–505.

139. Tjelle, T. E., Corthay, A., Lunde, E., et al. (2004) Monoclonal antibodies produced by muscle after plasmid injection and electroporation. *Mol. Ther.* **9,** 328–336.

140 Rizzuto, G., Cappelletti, M., Maione, D., et al. (1999) Efficient and regulated erythropoietin production by naked DNA injection and muscle electroporation. *Proc. Natl Acad. Sci. U. S. A.* **96,** 6417–6422.

141. Muramatsu, T., Arakawa, S., Fukazawa, K., et al. (2001) In vivo gene electroporation in skeletal muscle with special reference to the duration of gene expression. *Int. J. Mol. Med.* **7,** 37–42.

142. Murakami, T., Nishi, T., Kimura, E., et al. (2003) Full-length dystrophin cDNA transfer into skeletal muscle of adult mdx mice by electroporation. *Muscle Nerve* **27,** 237–241.

143. Ferrer, A., Foster, H., Wells, K. E., Dickson, G., and Wells, D. J. (2004) Long-term expression of full-length human dystrophin in transgenic mdx mice expressing internally deleted human dystrophins. *Gene Ther.* **11,** 884–893.

144. Medi, B. M., Hoselton, S., Marepalli, R. B., and Singh, J. (2005) Skin targeted DNA vaccine delivery using electroporation in rabbits. I: efficacy. *Int. J. Pharm.* **294,** 53–63.

145. Kobayashi, S., Dono, K., Takahara, S., et al. (2003) Electroporation-mediated ex vivo gene transfer into graft not requiring injection pressure in orthotopic liver transplantation. *J. Gene Med.* **5,** 510–517.

146. Harrison, R. L., Byrne, B. J., and Tung, L. (1998) Electroporation-mediated gene transfer in cardiac tissue. *FEBS Lett.* **435,** 1–5.

147. Dean, D. A., Machado-Aranda, D., Blair-Parks, K., Yeldandi, A. V., and Young, J. L. (2003) Electroporation as a method for high-level nonviral gene transfer to the lung. *Gene Ther.* **10,** 1608–1615.

148. Dezawa, M., Takano, M., Negishi, H., Mo, X., Oshitari, T., and Sawada, H. (2002) Gene transfer into retinal ganglion cells by in vivo electroporation: a new approach. *Micron* **33,** 1–6.

149. Wei, F., Xia, X. M., Tang, J., et al. (2003) Calmodulin regulates synaptic plasticity in the anterior cingulate cortex and behavioral responses: a microelectroporation study in adult rodents. *J. Neurosci.* **23,** 8402–8409.

150. Bettan, M., Ivanov, M. A., Mir, L. M., Boissiere, F., Delaere, P., and Scherman, D. (2000) Efficient DNA electrotransfer into tumors. *Bioelectrochemistry* **52,** 83–90.

151. Babiuk, S., Baca-Estrada, M. E., Foldvari, M., et al. (2004) Increased gene expression and inflammatory cell infiltration caused by electroporation are both important for improving the efficacy of DNA vaccines. *J. Biotechnol.* **110,** 1–10.

152. Bao, S., Thrall, B. D., and Miller, D. L. (1997) Transfection of a reporter plasmid into cultured cells by sonoporation in vitro. *Ultrasound Med. Biol.* **23,** 953–959.

153. Marmottant, P. and Hilgenfeldt, S. (2003) Controlled vesicle deformation and lysis by single oscillating bubbles. *Nature* **423,** 153–156.

154. Wu, J. (2002) Theoretical study on shear stress generated by microstreaming surrounding contrast agents attached to living cells. *Ultrasound Med. Biol.* **28,** 125–129.

155. Brujan, E. A. (2004) The role of cavitation microjets in the therapeutic applications of ultrasound. *Ultrasound Med. Biol.* **30,** 381–387.

156. Holt, R. G. and Roy, R. A. (2001) Measurements of bubble-enhanced heating from focused, MHz-frequency ultrasound in a tissue-mimicking material. *Ultrasound Med. Biol.* **27,** 1399–1412.

157. Zhong, P., Lin, H., Xi, X., Zhu, S., and Bhoghte, E. S. (1999) Shock wave-inertial microbubble interaction: methodology, physical characterization, and bioeffect study. *J. Acoust. Soc. Am.* **105,** 1997–2009.

158. Li, T., Tachibana, K., Kuroki, M., and Kuroki, M. (2003) Gene transfer with echo-enhanced contrast agents: comparison between Albunex, Optison, and Levovist in mice—initial results. *Radiology* **229,** 423–428.

159. Taniyama, Y., Tachibana, K., Hiraoka, K., et al. (2002) Development of safe and efficient novel nonviral gene transfer using ultrasound: enhancement of transfection efficiency of naked plasmid DNA in skeletal muscle. *Gene Ther.* **9,** 372–380.

160. Taniyama, Y., Tachibana, K., Hiraoka, K., et al. (2002) Local delivery of plasmid DNA into rat carotid artery using ultrasound. *Circulation* **105,** 1233–1239.

161. Frenkel, P. A., Chen, S., Thai, T., Shohet, R. V., and Grayburn, P. A. (2002) DNA-loaded albumin microbubbles enhance ultrasound-mediated transfection in vitro. *Ultrasound Med. Biol.* **28,** 817–822.

162. Lawrie, A., Brisken, A. F., Francis, S. E., Cumberland, D. C., Crossman, D. C., and Newman, C. M. (2000) Microbubble-enhanced ultrasound for vascular gene delivery. *Gene Ther.* **7,** 2023–2027.

163. Bao, S., Thrall, B. D., and Miller, D. L. (1997) Transfection of a reporter plasmid into cultured cells by sonoporation in vitro. *Ultrasound Med. Biol.* **23,** 953–959.

164. Kondo, I., Ohmori, K., Oshita, A., et al. (2004) Treatment of acute myocardial infarction by hepatocyte growth factor gene transfer: the first demonstration of myocardial transfer of a "functional" gene using ultrasonic microbubble destruction. *J. Am. Coll. Cardiol.* **44,** 644–653.

165. Anwer, K., Kao, G., Proctor, B., et al. (2000) Ultrasound enhancement of cationic lipid-mediated gene transfer to primary tumors following systemic administration. *Gene Ther.* **7,** 1833–1839.

166. Scherer, F., Anton, M., Schillinger, U., et al. (2002) Magnetofection: enhancing and targeting gene delivery by magnetic force in vitro and in vivo. *Gene Ther.* **9,** 102–109.

167. Huth, S., Lausier, J., Gersting, S. W., et al. (2004) Insights into the mechanism of magnetofection using PEI-based magnetofectins for gene transfer. *J. Gene Med.* **6,** 923–936.

168. Gersting, S. W., Schillinger, U., Lausier, J., et al. (2004) Gene delivery to respiratory epithelial cells by magnetofection. *J. Gene Med.* **6,** 913–922.

169. Krotz, F., Sohn, H. Y., Gloe, T., Plank, C., and Pohl, U. (2003) Magnetofection potentiates gene delivery to cultured endothelial cells. *J. Vasc. Res.* **40,** 425–434.

170. Schillinger, U., Brill, T., Rudolph, C., et al. (2005) Advances in magnetofection-magnetically guided nucleic acid delivery. *J. Magnetism Magnetic Mater.* **293,** 501–508.

171. Griesenbach, U., Dean, P., Marshall, N., et al. (2004) Magnetofection and Ultrasound To Increase "Naked" DNA Delivery to the Myocardium, *Mol. Ther.* **9**(Suppl. 1), 358.

172. Jiang, H., Zhang, T., and Sun, X. (2005) Vascular Endothelial Growth Factor Gene Delivery by Magnetic DNA Nanospheres Ameliorates Limb Ischemia in Rabbits. *J. Surg. Res.* **126,** 48–54.

173. Xenariou, S., Griesenbach, U., Ferrari, S., et al. (2004) Magnetofection to enhance airway gene transfer. *Mol. Ther.* **9**(Suppl. 1)**,** 180.

174. Herweijer, H. and Wolff, J. A. (2003) Progress and prospects: naked DNA gene transfer and therapy. *Gene Ther.* **10,** 453–458.

175. Andrianaivo, F., Lecocq, M., Wattiaux-De Coninck, S., Wattiaux, R., and Jadot, M. (2004) Hydrodynamics-based transfection of the liver: entrance into hepatocytes of DNA that causes expression takes place very early after injection. *J. Gene Med.* **6,** 877–883.

176. Kobayashi, N., Nishikawa, M., Hirata, K., and Takakura, Y. (2004) Hydrodynamics-based procedure involves transient hyperpermeability in the hepatic

cellular membrane: implication of a nonspecific process in efficient intracellular gene delivery. *J. Gene Med.* **6,** 584–592.

177. Zhang, G., Gao, X., Song, Y. K., et al. (2004) Hydroporation as the mechanism of hydrodynamic delivery. *Gene Ther.* **11,** 675–682.

178. Budker, V., Zhang, G., Danko, I., Williams, P., and Wolff, J. (1998) The efficient expression of intravascularly delivered DNA in rat muscle. *Gene Ther.* **5,** 272–276.

179. Zhang, G., Budker, V., Williams, P., Subbotin, V., and Wolff, J. A. (2001) Efficient expression of naked dna delivered intraarterially to limb muscles of nonhuman primates. *Hum. Gene Ther.* **12,** 427–438.

180. Liang, K. W., Nishikawa, M., Liu, F., Sun, B., Ye, Q., and Huang, L. (2004) Restoration of dystrophin expression in mdx mice by intravascular injection of naked DNA containing full-length dystrophin cDNA. *Gene Ther.* **11,** 901–908.

181. Hagstrom, J. E., Hegge, J., Zhang, G., et al. (2004) A facile nonviral method for delivering genes and siRNAs to skeletal muscle of mammalian limbs. *Mol. Ther.* **10,** 386–398.

182. Liu, F. and Huang, L. (2001) Improving plasmid DNA-mediated liver gene transfer by prolonging its retention in the hepatic vasculature. *J. Gene Med.* **3,** 569–576.

183. Liu, F., Nishikawa, M., Clemens, P. R., and Huang, L. (2001) Transfer of full-length Dmd to the diaphragm muscle of Dmd(mdx/mdx) mice through systemic administration of plasmid DNA. *Mol. Ther.* **4,** 45–51.

184. Budker, V., Budker, T., Zhang, G., Subbotin, V., Loomis, A., and Wolff, J. A. (2000) Hypothesis: naked plasmid DNA is taken up by cells in vivo by a receptor-mediated process. *J. Gene Med.* **2,** 76–88.

185. Miao, C. H., Thompson, A. R., Loeb, K., and Ye, X. (2001) Long-term and therapeutic-level hepatic gene expression of human factor IX after naked plasmid transfer in vivo. *Mol. Ther.* **3,** 947–957.

186. Maruyama, H., Higuchi, N., Nishikawa, Y., et al. (2002) High-level expression of naked DNA delivered to rat liver via tail vein injection. *J. Gene Med.* **4,** 333–341.

187. Eastman, S. J., Baskin, K. M., Hodges, B. L., et al. (2002) Development of catheter-based procedures for transducing the isolated rabbit liver with plasmid DNA. *Hum. Gene Ther.* **13,** 2065–2077.

188. Kurata, S., Tsukakoshi, M., Kasuya, T., and Ikawa, Y. (1986) The laser method for efficient introduction of foreign DNA into cultured cells. *Exp. Cell Res.* **162,** 372–378.

189. Tao, W., Wilkinson, J., Stanbridge, E. J., and Berns, M. W. (1987) Direct gene transfer into human cultured cells facilitated by laser micropuncture of the cell membrane. *Proc. Natl Acad. Sci. U. S. A.* **84,** 4180–4184.

190. Tirlapur, U. K. and Konig, K. (2002) Targeted transfection by femtosecond laser. *Nature* **418,** 290–291.

191. Shirahata, Y., Ohkohchi, N., Itagak, H., and Satomi, S. (2001) New technique for gene transfection using laser irradiation. *J. Investig. Med.* **49,** 184–190.

192. Terakawa, M., Ogura, M., Sato, S., et al. (2004) Gene transfer into mammalian cells by use of a nanosecond pulsed laser-induced stress wave. *Opt. Lett.* **29,** 1227–1229.

193. Umebayashi, Y., Miyamoto, Y., Wakita, M., Kobayashi, A., and Nishisaka, T. (2003) Elevation of plasma membrane permeability on laser irradiation of extracellular latex particles. *J. Biochem. (Tokyo)* **134,** 219–224.

194. Palumbo, G., Caruso, M., Crescenzi, E., Tecce, M. F., Roberti, G., and Colasanti, A. (1996) Targeted gene transfer in eucaryotic cells by dye-assisted laser optoporation. *J. Photochem. Photobiol. B* **36,** 41–46.

195. Ogura, M., Sato, S., Nakanishi, K., et al. (2004) In vivo targeted gene transfer in skin by the use of laser-induced stress waves. *Lasers Surg. Med.* **34,** 242–248.

196. Zeira, E., Manevitch, A., Khatchatouriants, A., et al. (2003) Femtosecond infrared laser-an efficient and safe in vivo gene delivery system for prolonged expression. *Mol. Ther.* **8,** 342–350.

197. Song, L., Chau, L., Sakamoto, Y., Nakashima, J., Koide, M., and Tuan, R. S. (2004) Electric field-induced molecular vibration for noninvasive, high-efficiency DNA transfection.

9

Stem Cell Technology and Drug Development

Helen L. Fillmore and Susanna Wu-Pong

Abstract

Stem cell technology holds the potential to offer new therapeutic options for poorly treated and debilitating diseases either by cell replacement therapies or by identification of drugs using high-throughput stem cell cultures differentiated into specific mature cell types. This promising technology is based on isolation of pluripotent cells that can be expanded in culture and differentiated into a mature phenotype. In addition to the social and political considerations, many methodological issues, such as choice of cell type, culture and differentiation methods, stability of cells in culture, and transplantation techniques, remain. This chapter reviews these issues and presents some novel potential applications.

Key Words: Stem cells; drug development; embryonic stem cells; toxicology; nuclear reprogramming; culture; differentiation; adult stem cells.

1. Introduction

Stem cells obtained from embryos and adult tissue hold great promise for drug development, cell replacement therapies, and basic science discoveries in areas such as developmental biology and cancer. The use of human embryonic stem (ES) cells for research has generated nationwide debate in scientific, political, and consumer groups for several reasons. Currently the technology requires the destruction of human embryos to obtain ES cells. This has led to a ban on federally funded research monies on new ES cell research in the United States. However, in 2001 President Bush allowed federal funding for continued research involving 21 established stem cell lines (Human Embryonic Stem Cell Registry: http://stemcells.nih.gov/research/registry/). Recently, the United Kingdom has also created the UK Stem Cell Bank, a British resource for stem cell lines (http://www.ukstemcellbank.org.uk/).

From: *Biopharmaceutical Drug Design and Development*
Edited by: S. Wu-Pong and Y. Rojanasakul © Humana Press Inc., Totowa, NJ

Stem cells can also be obtained from adult tissue, and although lagging behind in research compared to the ES cells, these cells are being aggressively studied for their potential applications. This chapter will examine the scientific and therapeutic applications of both cell types, including the potential for individualized cell therapy, then review predictions for promising new applications for this powerful technology.

2. Background

The power of stem cell technology resides in their ability, unlike mature cells, to differentiate into a variety of cell types. Stem cells can be derived from embryonic, fetal, or adult tissues. Stem cell differentiation is critical to their potential as therapeutic entities, with ES cells currently believed to hold great potential for medical advancement and therapies. Stem cells exhibit unlimited or prolonged self-renewal capacity, have the ability to differentiate into one or more cell types, and are regulated in vivo by extrinsic and intrinsic factors.

Clearly these unique properties require that stem cells have important biological functions. The ability of stem cells to differentiate is critical to the correct morphologic development of the fetus. ES cells are considered pluripotent in that they are self-replicating and can differentiate into cell types that comprise the three germ layers of an embryo. In contrast, totipotent stem cells give rise to pluripotent cells and cells of extraembryonic tissue (i.e., placental cells).

In addition to fetal development, stem cell differentiation enables maintenance of organs and tissues in the adult. Somatic or adult stem cells (ASCs), considered multipotent, are present in low numbers within mature tissues throughout the body and usually reside in specialized microenvironments or 'niches'. ASCs have been identified in several adult tissues, for example, in skin, bone marrow stroma, and brain (reviewed in ref. *[1]*). The earliest identified and most studied ASC is the bone marrow hematopoietic stem cell which can differentiate into the many types of blood cells. Asymmetrical division of ASCs leads to pools of transient amplifying cells that differentiate into mature cells via both extrinsic and intrinsic regulatory control mechanisms *(2)*. This stem pool in adult tissues provides a continuous source of cell precursors required for tissue renewal and replenishment.

In terms of drug discovery and cell replacement strategies, both ASC and ES cell technology hold great potential. Both cell types have relative advantages and disadvantages, and key differences exist between the two. Compared to ASCs, the major advantage of ES cells is pluripotency and the ability to differentiate into all mature cell types. ES cells are also relatively easy to grow in culture and, because of significant technical advances, can be grown without the use of nonhuman feeder cell layers *(3)*. A major disadvantage of using ES cells is the potential of rejection by the host immune system as discussed below. In terms of drug discovery, however, the main concern is the long-term maintenance in culture and risk of genomic instability.

In contrast, the major advantage of ASCs is the potential for using a patient's own cells for transplantation thereby avoiding potential of rejection by the patients' immune system. However, the primary disadvantage of ASCs is that currently the methods for isolation, expansion, and controlled differentiation lag behind that of ES cell technology.

Whereas the mouse has proven to be a useful model in studying the complex cellular mechanisms involved in the control of pluripotency and differentiation, hES (human ES) cells will be used for human cell therapy. Progress with hES cells has lagged behind mouse models, although significant success has been achieved in differentiating the cells into mature somatic cells, such as cardio-myocytes, hematopoietic, and endothelial cells (Table 1). However, for use in cell replacement therapies, hES cells must be developed with minimal foreign additives and cells to reduce the possibility of immune response *(2,25)*.

3. Stem Cell Methodology

3.1. Isolation and Propagation

3.1.1. ES Cells

ES cells are obtained from the inner cell mass of a blastocyst which is a tissue structure formed 4–5 d after fertilization. Standardized procedures are used to create ES cell cultures. The outer trophectoderm layer of the blastocyst is removed, leaving the inner mass of 30–34 cells. The cells are typically plated in serum-containing media onto irradiated mouse embryonic fibroblast cells. Mouse ES cells then grow in aggregates which can be removed, mechanically dissociated, and replated. Homogenous cells are then selectively replated, expanded, and passaged (reviewed in ref. *[4]*).

Mouse ES cells are relatively simple to grow in culture and will expand readily in the presence of leukemia inhibitor factor (LIF) and serum. hES cells generally require cell feeder layers, such as embryonic fibroblasts, to grow in serum-free culture supplemented with bFGF. The presence of feeder layers complicates the use of ES cells in humans because of the potential risk of contamination by feeder cell proteins. Human feeder cells have also been tested; the primary draw-back is the risk of contamination of human feeder cells with infectious agents and possible transfer of these agents to the ES cells (*see* ref. *[5]*).

Because of the drawbacks of using feeder cells, feeder-free systems are being developed for ES cell culture *(3)*. These systems still require bFGF and possi-bly other elements like TGFβ or conditioned media. Despite efforts to maintain cells in their undifferentiated state, spontaneous differentiation can occur, espe-cially with feeder-free systems.

Table 1
Examples of Therapeutic Use of Stem Cells

	Differentiation–induction	Markers of differentiation	Applications
Neural	Removal of LIF	Neurotransmitter production	Neuro-degenerative disorders
	Forced expression of genes such as Sox1, Nurr1	Electrophysiologic properties	Brain/spinal cord trauma
	Stromal feeder cells	Response to neurotransmitters	Multiple sclerosis (oligodendro-cytes)
	Growth factors (FGF2, EGF, PDGF, NGF, etc.)	Dopamine, GABA, serotonin receptors	
	Low MW agents	Neuron cell adhesion molecule	
		Production of myelin (oligodendrocytes)	
Mesenchymal	Removal of LIF	Electrical properties	Heart disease
	Ascorbic acid	Cardiac alpha myosin	(cardiomyocytes)
	Cardiogenol	Cardiac troponin 1/T	Bone fractures
	5-Aza-deoxycytidine	Atrial natiuretic factor	and
	Retinoic acid		osteoarthritis
	Vitamin D		(osteoblasts,
	Vitamin C		chondrocytes, osteoclasts)
Epidermal	BMP4	Keratins	Skin substitutes
	Ascorbate	Involucrin	
Hematopoietic	Hematopoietic cytokines		Hemophilias, anemias, etc.
	Bone morphogenic proteins		
	Plating on methylcellulose		
Hepatocyte	Sodium butyrate	Albumin	In vitro drug
	Activin A	P450 activity	metabolism
	Acidic fibroblast growth factor	Glycogen storage	assays
	Hepatocyte growth factor	Tyrosine aminotransferase	Hepatitis, cirrho-sis, hepatic
		Glucose 6 phosphatase	injury, etc.

(Continued)

Table 1 (*Continued*)

	Differentiation–induction	Markers of differentiation	Applications
Endothelial and vascular smooth muscle		LDL uptake Angiogenesis-like activity	
Pancreatic	Force Pax4 expression PI3-K inhibitors Activin Exendin-4 Nicotinamide Random differentiation	c-Peptide Insulin expression or secretion Glucagon Somatastatin Pdx1 Tyrosine aminotransferase Glucose 6 phosphatase	Diabetes

Data from refs. *(5,15,26)*

As mentioned earlier, ES cells may also exhibit genomic instability in culture, particularly in feeder-free systems, and will therefore require periodic monitoring of genomic integrity *(6)*. The observed genomic instability has implications for the preservation of the ES cells in the NIH registry in terms of their utility for stem cell development. The NIH Human Embryonic Stem Cell Registry lists stem cell lines that are eligible for federal research funding, and contact information on how to obtain the cell lines. However, the lines available in the past have reported genetic mutations, and to date the majority of the lines listed on the site are no longer available for shipping *(7)*.

Another challenge for ES cultures is retention of pluripotency in vitro. ES cell pluripotency relies on both intracellular and extracellular signaling pathways and appears to include proteins involved in gene silencing. One such group of proteins is the polycomb family members which are involved in inhibiting differentiation pathways by silencing critical transcription factor genes. These proteins transcriptionally repress developmental genes in concert with pluripotency proteins (for review, *see* ref. *[8]*).

3.1.2. Adult Stem Cells

ASC isolation or enrichment is obviously more difficult compared to ES cells because tissues tend to be comprised of several cell types. Many methods are used for obtaining and preparing cells; the methodology varies from one study to another. In general, the tissue is either mechanically or enzymatically dissociated to obtain single cells. Depending on the tissue, unique properties of resident stem cells can be utilized to attempt separation from the more mature cell types. For example, somatic stem cells from specific tissues express different proteins than their more mature counterparts including cell surface proteins which scientists have taken advantage of to isolate the stem cells. Antibodies to differentially expressed cell surface proteins can be used to sort stem cells from the other cell types by using technologies such as fluorescence-activate cell sorting (FACS), magnetic cell sorting, or buoyancy gradient centrifugation (i.e., Percoll). Although these are powerful tools for stem cell isolation, these methods may poorly affect cell survival *(9)*.

Other methods have also been successfully used to improve cell yield. For example, the isolation of mesenchymal stem cells from bone marrow can be accomplished using adhesion to culture plates followed by a series of expansion and preferential passaging to reduce contaminating cells. In the case of neural stem cells, there are several methods used to isolate adult neural stem/progenitor cells including antiadhesive reagents that may be used to aid in the separation of the attached cells from the 'neurospheres' which contain neural 'stem' cells (reviewed in ref. *[10]*). Recently, isolation of an attached population of adult neuronal stem cells was described by Walton et al. *(11)* (described below in Section 2.4).

For ASCs to be useful clinically, a stable and sufficient supply of cells must be available. Like any other primary cell line, the life span of any cell in culture is limited because of senescence-associated growth arrest. For example, human mesenchymal stem cells (hMSCs) isolated from young donors can normally undergo 24–40 population doublings. Kassem et al. (reviewed in ref. *[12]*) has studied telomerase-dependent senescence to determine whether cultured hMSCs life span could be extended in culture. The authors forced expression of human telomerase reverse transcriptase to restore telomerase activity and, thus, prevent telomere shortening. The modified cells not only showed extended life span, but also exhibited genetic instability and cell transformation. In general, senescence could be a limitation for cultivation of ASCs and progenitor cells derived from either ES or ASCs.

3.2. Differentiation

The ability to selectively and efficiently differentiate ES cells into specific lineages and tissue specific cells will determine their usefulness as research and therapeutic tools. Mouse ES cells have been successfully used to study differentiation, gene function, and molecular pathways (Table 1). One of the primary challenges facing ES cell development involves obtaining purified colonies of mature cells following in vitro differentiation. Several techniques are available

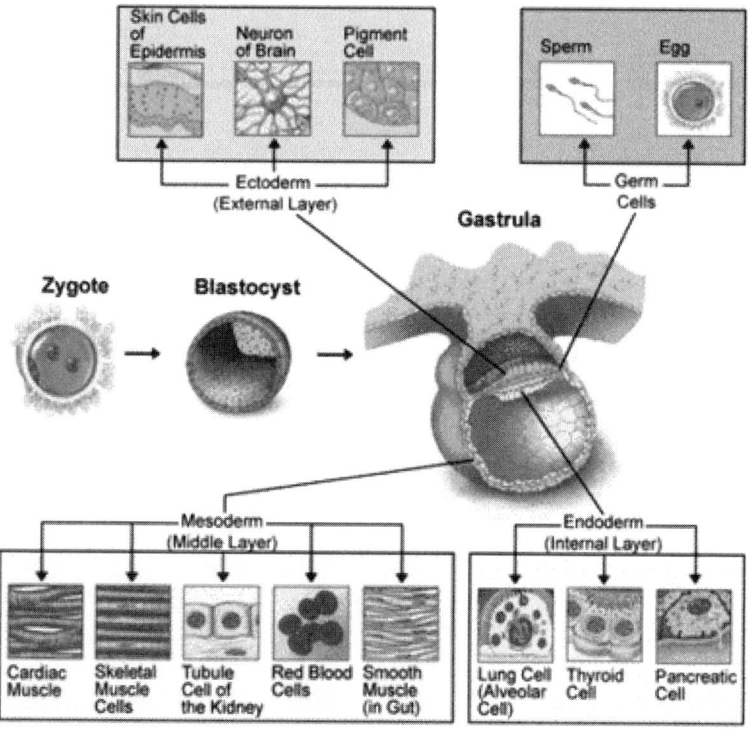

Fig. 1. Differentiation of human tissues. (Source: http://www.ncbi.nlm.nih.gov/About/primer/genetics_cell.html.).

for population enrichment and can include targeted introduction of stem cell specific genes fused with a reporter gene that allow easy detection and/or isolation of specific cells. Cell selection using cell surface markers and FACS can also be used to purify differentiated stem cells (*9*).

ASC differentiation in culture has also been studied but to a lesser extent compared to ES cells (reviewed in refs. *[8,12]*). The focus of many papers has been to determine whether ASCs can be induced to differentiate into the desired cell type. Studies have shown that ASCs are indeed pluripotent: ASCs derived from multipotent adult progenitor cells and unrestricted somatic stem cells from human umbilical cord blood can differentiate into cell types consistent with the three germ layers of a blastocyst. In addition, hMSCs can be induced to differentiate into osteoblasts in culture using hormones, growth factors, and extracellular matrix proteins.

The mechanisms of hMSC cell differentiation in culture were also examined (*12*). The authors discovered that expression of Dlk1/Pref-1, an EGF-like (epidermal growth factor) protein, maintains the undifferentiated phenotype of the cell in culture. The authors also used proteomic methods to compare the proteins involved in EGF- vs platelet-derived growth factor (PDGF)-induced differentiation of hMSCs. The proteins involved in EGF- and PGDF-induced differentiation are 90% similar and differ only by the PGDF-induced activation of the PI3K pathway. Identification of such control points can conceivably be used to finely control differentiation pathways.

3.3. Transdifferentiation

Transdifferentiation is defined as the differentiation of a pluripotent cell type from one organ to a specific cell type of another organ. The field of trans-differentiation is still in its infancy and well-accepted protocols are yet to be developed. Some contend that observed reports of transdifferentiation in the literature are actually the result of contamination or ASC fusion with a mature cell.

A recent example of transdifferentiation has been described by Harris et al. *(13)*, who report that bone marrow-derived cells (BMDC) appear to transdiffer-entiate into epithelial cells. The authors demonstrate that epithelial cells develop from mature BMDC in the absence of cell fusion, using a Z/EG Cre-reporter mouse. Male β actin Cre mice were used to collect BMDC donor cells which were then transplanted into irradiated females who ubiquitously express Cre recombinase. In this mouse strain, if fusion between male donor and female acceptor cells occurs, then Cre recombinase (from the female) induces recombi-nation in the male genome, and EGFP expression is permitted. Thus the presence of the Y chromosome concurrent with the absence of EGFP expression in cells of the recipient mouse was used to demonstrate the absence of fusion. The authors suggest that transdifferentiation occurs in their system because of minimal tissue injury or immunodepletion prior to transplant.

3.4. Transplant

Individualized cell therapy is dependent on the ability of hES cells to be successfully transplanted into the recipient and results in functioning cells that will integrate into the desired tissue. Transplants of differentiated mouse stem cells into a mouse model have demonstrated successful integration into host tissue (for reviews *see* refs. *[14,15]*). Experimental efforts to transplant hES cells into a mouse recipient have produced some success. Although some report very poor reconstitution in the target tissue, others report distribution through-out the host tissue and differentiation in a region-specific manner.

A recent report by Walton et al. *(11)* discusses the identification and selection of adult human neural progenitor (AHNP) cells that are stably grown in culture and generate glial and neuronal cell types both in vitro and in vivo. The cells are isolated from surgical samples by plating onto uncoated plastic dishes and propagating in the presence of bovine pituitary extract, fetal calf serum, epidermal and basic fibroblast growth factors. The cultured cells are stable in morphology and size in culture for over 60 population doublings. The cells remain growth factor dependent for growth and show no signs of indefinite growth or gross cytogenetic abnormalities. Cells transplanted into the right ventricle of C56/B6 mice were detectable after 30 d, adopted neuronal morphologies, but mainly concentrated along the injection site. Therefore, Walton et al. demonstrate successful transplant of human ASCs into a mouse model.

4. Scientific Applications

Stem cells can be used as an in vitro model to facilitate drug development or for the creation of more sophisticated transgenic models to study normal function and disease. Stem cell and transgenic models can be used in many phases of drug discovery. One of the major advantages for using stem cells for drug development and testing, in terms of their pharmacological or toxicological effects, is the ability of using differentiated cells of specific organs without the use of transforming factors. In addition, stem cells are invaluable tools that are used to pursue a more detailed understanding of complex processes such as differentiation and gene function. In summary, these models can be used for the identification of new drug targets and therefore lead to novel drug compounds.

4.1. Nuclear Reprogramming and Genetic Manipulation

Genetic manipulation in an in vitro or in vivo model may be necessary to improve the predictive ability or relevance of the model. For example, an investigator may wish to control or add expression of genes that are critical to the phenotype or hypothesis in question. Isolated ES cells can be genetically manipulated like any other cultivated cell (*see* Chapter 7). New genes can be introduced into the ES cell genome, existing genes can be silenced, or some combination can be used. Such approaches are useful for studying gene function and expression, particularly in differentiation, or to improve cell properties for cell transplantation therapy.

Nuclear transfer is a process in which an entire nucleus from a somatic cell can be transferred into an enucleated oocyte. The resulting ES cells contain the genome of the donor, but by some mechanism reprogram the cell back to an undifferentiated state. This technique, called nuclear reprogramming, has been effectively used to study the genetic changes in differentiation and their role in totipotency and cancer (for review *see [8]*). Nuclear reprogramming could also provide a new source of cells for cell replacement therapy that involves harvesting of adult somatic cells and oocytes for the creation of a pluripotent, undifferentiated cell.

Nuclear reprogrammed cells may be developed for use in cell transplant. Reprogrammed cells derived from the patient would be syngeneic and compatible with the patient's immune system. Besides the usual ethical questions that arise from nuclear or gene manipulation for human therapeutics, this approach is labor intensive and expensive and therefore may not be conducive to widespread use. Other problems are also associated with the methodology used: nuclear transfer into an oocyte can result in abnormalities during development; fusion with ES cells generates tetraploid cells; oocyte or ES cell extracts exposed to permeabilized somatic cells result in no functional reprogramming; explantation of germ cells (which become reprogrammed to the new host tissue) requires, of course, the use of germline cells (8).

4.2. Pharmacokinetics, Pharmacodynamics, and Toxicology

Currently, drug metabolism studies employ some combination of in vivo or in vitro models to determine the likely metabolic pathways of drugs. Animal models are suitable for preclinical studies, but drugs must eventually be tested in humans during clinical trials. Pharmaceutical companies prefer to obtain as much preclinical information as possible to minimize the risk of severe adverse reactions in Phase I clinical studies. Therefore, when available, human liver tissue is used to assess drug metabolism and toxicity. ES cell sources of hepatocytes could produce a more readily available and reproducible source of cells for metabolism studies. Selective differentiation of ES cells into hepatocyte-like cells has been demonstrated (Table 1).

Similarly, absorption and transport studies initially are executed using cell or animal models. Whereas intestinal absorption models using a human cancer cell line, Caco-2 colonic adenocarcinoma cells, is considered adequate, human cell transport models for other epithelial or endothelial cell barriers are of variable quality and availability (16,17). Such models, if optimized and made available using ES technology, can be useful tools for elaborating the mechanisms of drug transport and metabolism at the cellular level. Optimization of drug transport across these biological barriers can also be a useful application of such in vitro models.

Though ES-based model for pharmacokinetics is still in its infancy, ES cells are currently in use for toxicology testing. Embryotoxicity of compounds is tested by measuring the effect of the agent on ES cell differentiation. Single cell models can be used. For example, Buesen et al. (18) measure the impact of compounds on ES cell differentiation into cardiomyocytes as a toxicological end point for determining teratogenicity.

Multiple end points can also be used for greater predictive ability of toxicity and mutagenicity tests, such as the validated embryonic stem cell test (EST). EST uses an embryoid body (aggregates of ES cell) rather than the ES cell culture model (19,20). In the EST, mouse ES cell and 3T3 fibroblast cells are used to test compounds' effect on differentiation as a predictor for embryotoxicity.

The three measures are: (1) influence on ES cell differentiation into myocardium cells, (2) ES cell cytotoxicity, and (3) 3T3 cell cytotoxicity. The combined use of these assays improves the predictability of in vivo embryotoxicity when compared to the use of a single test alone. The primary limitation of the EST is its labor-intensive nature and the limited number of cell lineages represented in the test.

In addition to embryotoxicity, tests for cytotoxicity and mutagenicity are also in development using ES cells, embryonic carcinoma, and embryonic germ cells. Nonhuman ES cell models have also been tested for embryotoxicity, including Xenopus or chicken embryos, whole embryos, and other vertebrate and invertebrate cell lines (for review *see* ref. *[21]*).

ASCs can also be used to develop in vitro models for disease. For example, the ASC from diseased tissue may retain physiologic dysfunction when induced to differentiate into the respective somatic cell. The cell can then be used to study disease processes or treatment in vitro, or further developed into a transgenic animal model. The success of using stem cells for in vitro disease models has been mixed. However, important insights into hematopoiesis, vascular and angiogenesis, as well as cardiovascular, renal, pancreatic systems have been obtained. Thus, ASCs from diseased tissue and their differentiated somatic cells could provide a renewable source of cells to examine the biochemistry and treatment of disease (*see* refs. *[14,22]*).

The availability of an unlimited pool of virtually any cell type can also be an invaluable research tool. Within a single individual, hundreds of cell types coexist harmoniously, several even within a single tissue. Each of these cell types has unique properties despite a common genome. Therefore, using ASCs to create separate arrays of cell types from an individual patient could be used to improve our understanding of the fundamental concepts of biological variability and differentiation.

A variation on the theme of ASCs obtained from individual patients, should the technology one day become routine and cost effective, is individualized drug optimization studies. Each human DNA sequence is unique; thus each person is also unique in how they respond to drugs. Therefore, advanced knowledge of a patient's response to a drug, whether the response is toxicity, drug efficacy, or no effect, would allow dosing regimen optimization before actually starting the treatment regimen. Furthermore, because tissues differ in their expression of transport and receptor molecules, metabolic capabilities, blood flow, intracellular protein expression, one could hypothesize that different cell types even from the same patient could require different drug concentrations to produce either a therapeutic or toxic effect. Individualized cell arrays could conceivably be used to screen for ideal drug and concentration combinations for an individual patient. This approach may be especially useful when certain cells are particularly vulnerable to toxicity, such as dividing cells' susceptibility to chemotherapeutic agents. An emerging example of this concept is illustrated by Torrance et al. *(23)* who altered oncogene expression in cancer cell line, then compared suscepti-bility of the different cells to antitumor agents. Another example would be the isolation of "cancer" stem cells from individual patient tumors which could then be used to identify individualized drug treatment regimens.

Arrays of the same cells from different individuals could also be an invaluable research tool. For example, arrays of hepatocytes that express different cyto-chrome enzyme polymorphisms could be used to rapidly determine the enzymes responsible for drug metabolism. The advantage of using a hepatic cell array instead of purified enzyme arrays may be the provision of a more relevant biomatrix for metabolic assessment.

5. Therapeutic Applications – Cell Replacement Therapy

Because differentiation into neuronal cells appears to be the default pathway for ES cell on removal of LIF, neurological applications are the logical starting point for ES cell therapeutic development (*see* ref. *[5]*). The process of neuronal differentiation into specific subtypes is facilitated by the addition of growth factors and plating with feeder cells. If successful, neuronal cells can be used to treat degenerative disorders such as Parkinson's or Alzheimer's or brain and spinal cord injuries (Table 1).

Although cell replacement therapy remains a potential treatment option for Parkinson's disease, several obstacles must be overcome in order to achieve successful clinical outcomes. These obstacles include reliable methods of cultivation and expansion of stem cells as mentioned earlier and optimization of methods for cell delivery. For example, the selection of cell delivery catheter will be critical to the successful transplant of cells. Commercially available catheters produce either unacceptable levels of tissue damage, poor cell deli-very, or unintended flow paths of delivered cells. An improved coaxial tube cell delivery device was developed that allows for subsequent infusions of growth factors and infusions at multiple sites along the insertion paths. Once a suitable nondopaminergic region of the brain has been identified for cell delivery and a suitable treatment planning system designed for optimal flow and delivery of the proper cell concentration, such catheters could be used for targeted delivery to the brain (for review, *see* ref. *[10]*).

The use of fetal stem cells for the treatment of Parkinson's started as early as 1990. These early studies demonstrated some clinical benefit for the patient. For example, in 1996 Defer et al. *(24)* showed that –fetal mesencephalic cell implantation resulted in bilateral improvements in motor function, fluorodopa uptake, good implant survival, and dopaminergic reinnervation following uni-lateral implantation. Other studies have shown graft survival for up to 10 yr, normalization of dopamine release, reversal of impairment of cortical activation, and even resulted in significant clinical improvement for some patients. Despite promising early studies, development of clinical therapies is still in very early stages. Patient response to these therapies is uneven, and dyskinesias have been reported as a result of the intervention. Other concerns listed above, such as purity of culture and genomic instability, as well as the potential role of genetic manipulation of stem cells, are issues that continue to be relevant for clinical trials (for review *see* ref. *[25]*).

In addition to treating Parkinson's, neural cell replacement therapy may be used to treat spinal cord injury. hES-derived neural cell transplants have resulted in detection of human neurons, astrocytes, oligodendrocytes, and glia in recipient mice brains. ES cell-derived oligodendrocytes in some cases have even been found to form myelin sheaths. When injected near the injury into a spinal cord, partial functional recovery was observed in spinal cord-injured mice, possibly because of ES cell differentiation into oligodendrocyte myelination. However, the success of the transplant continues to be limited by the differentiation efficiency of the hES cells prior to transplant (*see* refs. *[5,15]*).

Stem cells are also in development for the creation of pancreatic islet cells for the treatment of diabetes. Islet-like cells have been created in vitro by treatment with selective growth factors, differentiating agents, or genetic manipulation (Table 1). Specifically, such cells (induction of β-geo or the transcription factor Pax4 or inhibition of phosphoinositide 3 kinase) improve hyperglycemia when implanted into diabetic mouse models. Removal of cells differentiated with the Pax4 or PI3K inhibitor also resulted in reversal of hypoglycemic effects (*see* ref. *[15]*).

A major limitation of the use of allogeneic stem cells for individualized therapy is similar to any other cell transplant methods: immune rejection of the transplanted tissue. However, autologous stem cells can be created using nuclear transfer, used similarly in cloning. Nuclear transfer involves the removal of the nucleus of a stem cell (or the nucleus of a fertilized egg in traditional cloning), then replacement with the nucleus of the patient's cell. These reprogrammed cells once expanded in culture can either be directly transplanted into damaged tissue for tissue repair or genetically modified to add or remove a gene prior to transplant. Theoretically, nuclear reprogramming and genetic manipulation of a patient's somatic cell could allow customizable cell replacement therapy. This technique is also obviously labor intensive and expensive, which ultimately may limit its viability as a therapeutic intervention.

Other limitations of allogeneic stem cell transplant include risk of transmission of infectious diseases or latent genetic diseases such as a predisposition to cancer. Extensive testing of the donor cells could minimize many of these potential risks. Another serious concern involves the risk of ES cells forming teratomas in vivo. Including a suicide gene into the cells could allow for external control of transplant cell death if such an event occurs.

The in vivo issues are only part of the spectrum of technical problems that must be resolved for a cell-based therapy to be feasible for human use. Other development issues to consider include expanding the manufacturing capacity of the product without compromising the quality of the product. Successful pharmaceutical applications of any type of stem cells rely on the availability of an unlimited source of cells and the standardization of protocols for differentiation. Cultures should also be able to withstand freeze-thawing, as pharmaceutical companies require preservation of master cell lines to maintain the integrity of the product as originally designed, tested, and validated. Testing protocols for the cells must be developed and validated to ensure that the product meets standards of purity throughout the manufacturing process in terms of cell population and a proven absence or acceptable reduction in levels of contaminants. Development of good manufacturing practices (GMPs) for stem cell therapeutics will also become necessary. As ES cell or ASC therapies get closer to the market, these and other unforeseen issues will require the expertise of scientists from many disciplines to enable the approval of these therapeutic agents.

6. Stem Cells and Biotechnology

The development of stem cells for human therapeutics is reminiscent of the evolution of biotechnology for the design and development of biopharmaceuticals. Both may involve the use of complex technology that involves foreign cellular components, xenogeneic cells, or molecules. The use of foreign molecules created appropriate concern about the safety of recombinant products in humans. The result has been the creation of manufacturing and quality control protocols that established acceptable limits for endotoxin and other contaminants that result from the manufacturing process. Today, biopharmaceutical products are made routinely when only two decades ago the field was fraught with uncertainty about the feasibility of the endeavor.

The next phase of pharmaceutical biotechnology will involve the more complex technologies such as gene and stem cell therapies. In retrospect we may view the development of protein pharmaceuticals nostalgically as the era when rapid developments in biotechnology were achieved. Though protein drugs often have 3D structure and a molecular weight that can be orders of magnitude greater than a traditional drug, still the product is often only one active molecule formulated into a relatively simple aqueous delivery system. One can theorize that the addition of a single component to a recombinant molecule can exponentially complicate the manufacturing and quality control efforts. To use a simplistic example, the use of a lipid carrier to a protein or DNA drug adds not only manufacturing concerns for the second molecule but also necessitates characterization and reproducible production of the correct spatial and physicochemical relationship with the first molecule (i.e., the liposome). One must only peruse the liposome literature to observe that considerable effort is spent solely on characterization and reproducibility of the liposome/drug complex, which is in fact a simple system compared to a viral vector or a cell.

The biopharmaceutical industry must also routinely ensure the stability of the cells used to create the recombinant product. The whole system of Master Cell Banks (the original clone using which the product was created) and Working Cell Banks (the cells used in manufacturing that were derived from some Master Cell Bank cells) has been created to preserve the integrity and drug-producing capacity of the original cells. Stem cell technology may be able to rely on such a system to preserve the master cells; however, stem cells and their counterparts that exist along the differentiation spectrum may not be as amenable to freezing and long-term storage as many of the more robust bacterial and mammalian cells that are used for biopharmaceutical manufacture. Cell viability, clonogenicity, genetic stability, and extent and nature of differentiation are all variables that will require attention during process development (reviewed in ref. *[10]*).

If the challenge of creating therapeutic stem cells appears overwhelming and impossible, one must consider the analytical and genetic tools that are currently available that were nonexistent in the early days of biopharmaceutical drug development. Advances in cell cloning, cell culture, genomics, gene expression analysis (microarrays), and bioanalytical methods will enable the development of this technology when it would have been impossible 20 yr ago. Undoubtedly developmental, political, and ethical challenges continue to exist for ES cells but the opportunities to provide meaningful treatments for conditions such as neurodegenerative diseases demand a full exploration of the potential of this technology.

References

1. Preston, S. L., Alison, M. R., Forbes, S. J., Direkze, N. C., Pousom, R., and Wright, N. A. (2003) The new stem cell biology: something for everyone. *J. Clin. Pathol: Mol. Pathol.* **56,** 86–96.
2. Watt, F. M. and Hogan, B. L. (2000) Out of Eden: stem cells and their niches. *Science* **287,** 1427–1430.
3. Stacey, G. N., Cobo, F., Nieto, A., Talavera, P., Healy, L., and Concha, A. (2006) The development of 'feeder' cells for the preparation of clinical grade hES cell lines: challenges and solutions. *J. Biotechnol.* **125,** 583–588.
4. Regenerative Medicine 2006. (2006) Department of Health and Human Services. August, http://stemcells.nih.gov/info/scireport/2006report.htm
5. Pouton, C. W. and Hayes, J. M. (2005) Pharmaceutical applications of embryonic stem cells. *Adv. Drug Deliv. Rev.* **57,** 1918–1934.

6. Maitra, A., Arking, D. E., Shivapurkar, N., et al. (2005) Genomic alterations in cultured human embryonic stem cells. *Nat. Genet.* **37,** 1099–1103.
7. NIH Human Embryonic Stem Cell Registry (continually updated), National Institutes of Health, http: //stemcells.nih.gov/research/registry/defaultpage.asp.
8. Hochedlinger, K. and Jaenisch, R. (2006) Nuclear reprogramming and pluripotency. *Nature* **441,** 1061–1067.
9. Park, P. C., Selvarajah, S., Bayani, J., Zielenska, M., and Squire, J. A. (2006) Stem cell enrichment approaches. *Semin. Cancer Biol.* Apr 29 (epub ahead of print).
10. Fillmore, H. L., Holloway, K. L., and Gillies, G. T. (2005) Cell replacement efforts to repair neuronal injury: a potential paradigm for the treatment of Parkinson's disease. *Neurorehabilitation* **20,** 233–242.
11. Walton, N. M., Sutter, B. M., Chen, H. X., et al. (2006) Derivation and large-scale expansion of multipotent astroglial neural progenitors from adult human brain. *Development* **133**(18), 3671–3681.
12. Kassem, M. (2006) Stem Cells. Potential therapy for age-related diseases. *Ann. N. Y. Acad. Sci.* **1067,** 436–442.
13. Harris, R. G., Herzog, E. L., Ruscia, E. M. B., Grove, J. E., Van Arname, J. S., and Krause, D. S. (2004) Lack of a fusion requirement for development of bone marrow-derived epithelia. *Science* **305,** 90–94.
14. Hook, L., O'Brien, C., and Allsopp, T. (2005) ES cell technology: An introduction to genetic manipulation, differentiation, and therapeutic cloning. *Adv. Drug Deliv. Rev.* **57,** 1904–1917.
15. Keller, G. (2005) Embryonic stem cell differentiation: emergence of a new era in biology and medicine. *Genes Dev.* **19,** 1129–1155.
16. De Angelis, E., Moss, S. H., and Pouton, C. W. (1996) Endothelial cell biology and culture methods for drug transport studies. *Adv. Drug Deliv. Rev.* **18,** 193–218.
17. Tuma, P. L. and Hubbard, A. L. (2003) Transcytosis: crossing Cellular Barriers. *Phys. Rev.* **83,** 871–932.
18. Buesen, R., Visan, A., Genschow, E., Slawik, B., Spielmann, H., and Seiler, A. (2004) Trends in improving the embryonic stem cell test (EST): an overview. *ALTEX* **21,** 15–22.
19. Scholz, G., Genschow, E., Pohl, I., et al. (1999) Prevalidation of the embryonic stem cell test (EST)—a new in vitro embryotoxicity test. *Toxicol. In Vitro* **13,** 675–681.
20. Spielmann, H., Pohl, I., Döring, B., Liebsch, M., and Moldenhauer, F. (1997) The embryonic stem cell test (EST), an in vitro embryotoxicity test using two permanent mouse cell lines: 3T3 fibroblasts and embryonic stem cells. *In Vitro Toxicol.* **10,** 119–127.
21. Rohwedel, J., Guan, K., Hegert, C. and Wobus, A. M. (2001) Embryonic stem cells as an in vitro model for mutagenicity, cytotoxicity and embryotoxicity studies: present state and future prospects. *Toxicol. In Vitro* **15,** 741–753.
22. Wobus, A. M. and Boheler, K. (2005) Embryonic stem cells: Prospects for developmental biology and cell therapy. *Physiol. Rev.* **85,** 635–678.
23. Torrance, C. J., Agrawal, V., Vogelstein, B., and Kinzler, K. W. (2001) Use of isogenic human cancer cells for high-throughput screening and drug discovery. *Nat. Biotechnol.* **19,** 940–945.

24. Defer, G. L., Geny, C., Ricolfi, F., et al. (1996) Long-term outcome of unilaterally transplanted Parkinsonian patients: I. clinical approach. *Brain* **119,** 41–50.

25. Lindvall, O., Kokaia, Z., and Martinez-Serrano, A. (2004) Stem cell therapy for human neurodegenerative disorders—how to make it work. *Nat. Med.* **10,** S42–S50.

26. Menendez, P., Want, L., and Bhatia, M. (2005) Genetic manipulation of human embryonic stem cells: a system to study early human development and potential therapeutic applications. *Curr. Gene Ther.* **5,** 375–385.

10

Small Nucleic Acid-Based Drugs: Successes and Pitfalls

A. Yin Zhang and Susanna Wu-Pong

Abstract

Since the discovery of the application of small nucleic acid molecules as inhibitors of gene expression, literally thousands of new research papers have used this technique for functional genomics or to test these molecules as potential therapeutic agents. During this time, the field has evolved from simple phosphodiester oligonucleotide (ODN) binding to complementary mRNA in vitro, to the development of hundreds of chemically modified ODNs with improved properties and mechanisms of action. RNA-based gene inhibitors have also powerfully merged onto the scene with first ribozymes, and now siRNA, providing new approaches to gene inhibition. This chapter will review the area of small nucleic acid drugs in terms of their design, delivery, and application.

Key Words: Antisense; oligonucleotide; siRNA; delivery; pharmacokinetics; drug.

1. Overview

Since the first report of the potential use of oligonucleotides (ODNs) as antisense agents that inhibit viral replication in cell culture *(1)*, antisense technology has become an increasingly powerful tool for gene functional analysis. Along with the completion of the Human Genome Project, antisense technology allows the direct utilization of sequence information and translation into broad applications, such as functional genomics, target validation, and therapeutics. In fact, the first Food and Drug Administration (FDA)-approved antisense (AS) ODN drug, Vitravene™ (fomivirsen, Isis Pharmaceuticals), is used in clinical setting for the treatment of cytomegalovirus (CMV) retinitis in acquired immune deficiency syndrome (AIDS) patients. The AS ODN therapeutic class is expected to experience considerable expansion with numerous drug candidates (e.g., antisense to Bcl-2, ribonucleotide reductase, c-Raf, etc.) currently in various phases of clinical trials for oncological and immunological conditions. Another emerging technology is RNA interference (RNAi) or gene

From: *Biopharmaceutical Drug Design and Development*
Edited by: S. Wu-Pong and Y. Rojanasakul © Humana Press Inc., Totowa, NJ

silencing. Although the technology was only discovered a short time ago, its high efficacy and specificity has attracted a great deal of research. Numerous in vivo studies are currently underway, and RNAi-based therapies for antiviral, anticancer, and neurological diseases hold great promise.

Antisense and RNAi technologies represent a "new pharmacology." In contrast to the traditional small molecule drugs that target proteins, the products of gene expression, antisense and RNAi nucleic acids target specific genes and thus inhibit protein synthesis. The drug–receptor interaction, occurring between antisense and RNAi nucleic acid molecules (the "drug") and cellular mRNA (the "receptor"), relies on sequence-specific Watson–Crick hybridization. These nucleic acid macromolecules are typically 15–25 nucleotides long and have molecular weights exceeding 5 kDa. Their hydrophilic polyanionic charge presents another challenge to traverse the hydrophobic plasma membrane. As a result, a major pharmaceutical concern is the effective delivery of these macromolecules both in vitro and in vivo in order to realize the full potential of the new technologies. Many chemical, physical, and molecular strategies have been developed to improve nucleic acid cellular delivery. This chapter will present an overview of nucleic acid-based drug development and therapeutic potential, focusing on cellular pharmacology, structural considerations in drug design, a brief discussion of the advances in nucleic acid delivery strategies, clinical applications, as well as future directions of nucleic acids as emerging therapeutic entities.

2. Types of Nucleic Acid Drugs

2.1. Antisense Oligonucleotides

Three nucleic acid-based strategies have been extensively studied for regulating gene expression at the posttranscriptional level (Fig. 1). These molecules either destabilize mRNA or prevent its translation, thereby inhibiting the synthesis of the target protein. The earliest generation of nucleic acid drugs is based on antisense technology. Single-stranded ODNs, typically 15–20 nucleotides long, are designed to hybridize to complementary sequences of target mRNA via Watson–Crick basepairing. The formation of an ODN/mRNA duplex recruits the activation of endogenous ribonuclease H (RNase H) enzyme, which degrades the hybridized mRNA strands *(2)*. The exact recognition elements for RNase H activation are not well understood. However, in vitro studies have demonstrated RNase H-mediated mRNA cleavage using AS ODNs containing as few as five phosphodiester (PD)-linked nucleotides *(3)*. Such a property leads to the design of chimeric ODNs exploiting the RNase H-activating ability of PD ODNs combined with enhanced nuclease stability and target affinity afforded by newer ODN analogs.

In addition to RNase H activation, AS ODNs can inhibit protein expression via multiple mechanisms during the protein biosynthetic pathway. Starting with

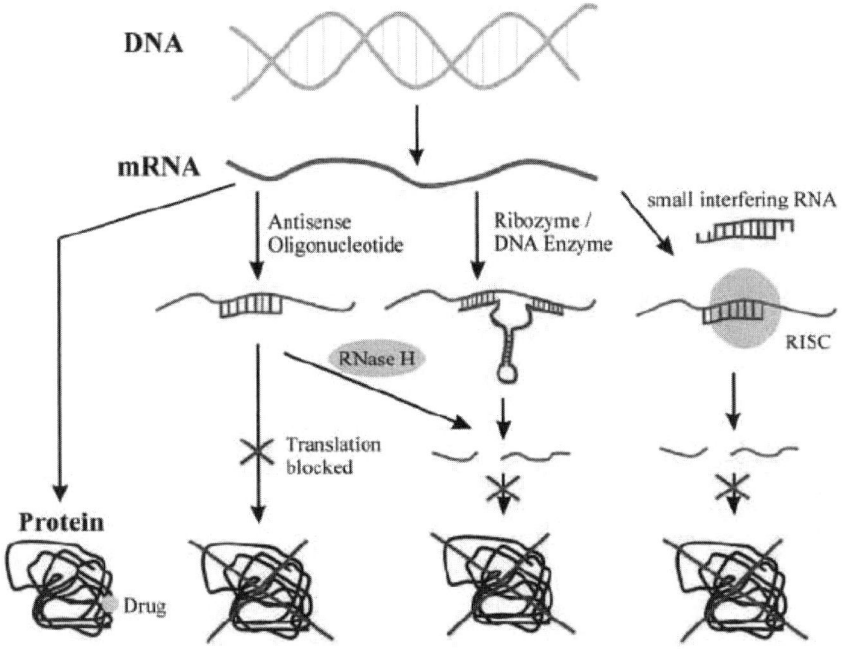

Fig. 1. Comparison of different strategies to inhibit protein synthesis at the posttranscriptional level. In contrast to the traditional small synthetic drug targeting posttranslationally, antisense oligonucleotides bind to mRNA and prevent the synthesis of target protein. They either block protein translation via steric hindrance or induce mRNA degradation via activation of RNase H enzyme. Ribozymes and DNAzymes can cleave target mRNA through their catalytic activity. RNA interference is achieved with the incorporation of siRNAs into the RISC protein complex and the subsequent induction of mRNA degradation. *(Figure adapted from ref. 85).*

transcription in the nucleus, AS ODNs may theoretically hybridize to double-stranded DNA through Hoogsteen basepairing to form an ODN/DNA triplex, thus inhibiting transcription. Alternatively, AS ODNs may interfere with the processing of pre-mRNA into the mature form by inhibiting 5′-cap formation, intron–exon splicing, or polyadenylation. Once the mature mRNA is transported from the nucleus to the cytoplasm, protein translational arrest can occur because of steric hindrance of ribosomal binding to mRNA.

2.2. RNA Interference

Much attraction and excitement in the nucleic acid research field has come from the discovery of an endogenous gene-silencing phenomenon called RNA interference. Originally described as a natural defense mechanism against viral pathogens or transposon mobilization in plants and fungi, the seminal discovery of the RNAi phenomenon in nematode worm *Caenorhabditis elegans* by Fire and coworkers *(4)* suggests that RNAi is an evolutionary conserved pathway for gene regulation. Since then, the RNAi field has been advancing with tremendous momentum, with the demonstration of the RNAi mechanism in mammalian cells and animal models *(5,6)*.

The antiviral RNAi process comprises at least four sequential steps. First, RNAi is initiated by the cleavage of foreign long double-stranded RNA by the endogenous Dicer enzyme into fragments of 21–23 nucleotides short-interfering RNAs (siRNAs) with two nucleotide overhangs. Next, siRNAs are incorporated into an inactive 360-kDa ribonucleoprotein complex. In the third step, unwinding of the siRNA duplex occurs in an ATP-dependent fashion which generates an active RNA-induced silencing complex (RISC). Finally, RISC uses the antisense strand of the siRNAs to guide Watson–Crick mRNA basepairing and cleavage, thus promoting target mRNA degradation of similar foreign RNA sequences *(5,7)*.

This viral defense mechanism can be used for therapeutic purposes. For example, siRNAs can be introduced into cells as synthetic 21-basepair siRNAs or via in vivo generation of siRNAs and short hairpin RNAs from plasmid DNA constructs *(8)* to target degradation of complementary sequences. In addition, naturally occurring, endogenous short RNA molecules have been discovered to be involved in gene regulation in mammals. These micro-RNAs (miRNAs) are processed in a similar fashion as the synthetic long double-stranded RNA by the Dicer enzyme to yield effector siRNAs. Approximately 1650 of these distinct siRNAs have been identified, which play critical regulatory roles in gene expression controlling a variety of cellular processes *(9,10)*.

2.3. Ribozymes

In addition to complementary sequence binding, ODNs can be engineered to possess catalytic activity, as in the case of ribozymes and DNAzymes. These catalytic ODNs bind to target mRNA in an antisense fashion and cleave or edit RNA molecules. Several natural ribozyme motifs have been identified, including hammerhead, hairpin, group I and group II intron, ribonuclease P, and hepatitis delta virus ribozymes. Hammerhead ribozyme, the smallest among the group (approximately 30 nucleotides long), is the most widely studied because of its simplicity and the ability to be incorporated into a variety of flanking sequence motifs without altering site-specific cleavage properties. The catalytic domain of hammerhead ribozyme recognizes the "XUY" sequence motif on the substrate mRNA, with X being any base, U being uridine, and Y being any base except guanine *(11)*. Many studies have demonstrated the effectiveness of ribozymes in vitro and in vivo, but short duration of activity is a major drawback because of nuclease susceptibility in the biological system *(12,13)*. Chemical modification of ribozymes is not as straightforward as AS ODNs or siRNAs because the introduction of modified nucleotides may lead to conformational changes that abolish the catalytic activity.

2.4. Transcription Factor Decoys

Instead of inhibiting protein synthesis at the posttranscriptional level, ODNs can also be designed to act at the transcriptional level. Decoy ODNs are double-stranded ODNs containing a *cis*-transcription recognition sequence which binds to transcription factors. These molecules compete with the respective DNA promoter region to bind and sequester transcription factors, resulting in decreased availability of transcription factors and reduced rate of gene transcriptions *(14)*. Targeting transcription factors is an attractive therapeutic approach because a single transcription factor can regulate a number of genes in coordinated signaling pathways. By the same token, potential unwanted effects can be problematic if the targeted transcription factor performs multiple biological functions. The recent disappointing results from the PREVENT IV trial on the use of E2F decoys to prevent vein graft failure suggest that the technology still needs "fine-tuning" *(15)*.

3. Nucleic Acid Medicinal Chemistry

3.1. Antisense Oligonucleotides

One of the major successes in bringing nucleic acids one-step closer to clinical applications is the ability to modify the native nucleic acid chemical structure. Such modification not only retains drug activity, but more importantly enhances serum stability, tissue distribution, potency, and safety profile. Extensive research using AS ODNs as the prototype has concentrated on three types of modifications, including the phosphate backbone, ribose sugar (mainly the $2'$ position), and nucleic acid base (Fig. 2 and Table 1). Phosphorothioate (PS) ODNs representing the earliest modification, commonly referred to as "first generation" AS ODNs, remain the most widely used class of antisense compounds currently being tested in clinical trials. By substituting one of the nonbridging oxygen atoms in the PD bond with sulfur, nucleotide stability in biological fluids against endo- and exonucleases is remarkably improved with a serum half-life of 9–10 hr compared to less than 1 hr for unmodified PD ODNs *(16)*. In addition, PS ODNs display relatively high affinity for target mRNA, RNase H activation, and high aqueous solubility. Numerous in vitro studies have demonstrated successful PS AS ODN-induced target knockdown of cellular proteins, leading to alterations in their associated cellular processes, such as proliferation, apoptosis, drug uptake, and drug sensitivity.

The major drawback of PS modification relates to the polyanionic nature of the ODN backbone, rendering them prone to bind to certain proteins in a

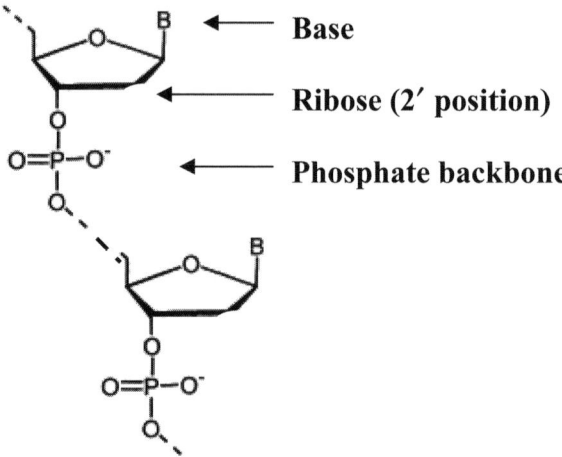

Fig. 2. Common modification sites for oligonucleotides. Chemical modifications can occur with unnatural bases (denoted as *B*), modified sugars (especially at the 2′ position), or altered phosphate backbone.

sequence-independent manner, thus causing nonspecific effects and cellular toxicity. Interactions with heparin-binding proteins, such as basic fibroblast growth factor, as well as cell surface proteins, including epidermal growth factor receptor and vascular endothelial growth factor receptor (VEGFR), have been reported *(17,18)*. G3139 (Genasense, Genta), an 18-mer PS ODN targeting the initiation codon of Bcl-2 mRNA, has demonstrated chemosensitizing effects in patients with chronic lymphocytic leukemia and malignant melanoma. The drug is currently under both FDA and European Medicine Agency (EMEA) review for marketing approval (http://www.genta.com). In spite of impressive clinical data, in vitro studies suggest that G3139 exerts its cytotoxicity predominantly by nonsequence-specific mechanisms. G3139 produced cytotoxicity to the same extent in Bcl-2 overexpressing melanoma cells compared to mock-transfected cells despite greater than 1000-fold differences in Bcl-2 protein expression between the two clones. Furthermore, whereas downregulation using ODN achieved more than 90% reduction in Bcl-2 protein expression, no cytotoxicity or chemosensitization was observed *(19)*.

Although PD ODNs possess polyanionic charges as well, they show much less pronounced nonspecific protein binding than PS ODNs of the same sequence, suggesting that the sulfur substitution is responsible for these interactions. Introduction of sulfur atoms creates chiral centers at phosphorus atoms, leading to a mixture of Rp- and Sp-diastereomers from ODN chemical synthesis. To determine whether the nonspecific protein interactions of PS ODNs are

Table 1
Chemical Modifications to Optimize Pharmaceutical Properties

	Stability/ nuclease resistance	RNase H activity	Potency/ RNA affinity	Toxicity	Tissue distribution
First generation Phosphorothioate (PS)	Stable	Yes	Slightly ↓ vs unmodified	Nonsequence-specific toxicity, because of binding to cellular proteins	Extensive, mainly kidney and liver
Second generation 2′-*O*-methyl (OMe) 2′-*O*-methoxyethyl (MOE)	↑ vs PS	No	↑ vs PS	↓ vs PS	Similar to PS
Third generation Locked nucleic acid (LNA) Morpholino (PMO)	↑ vs PS	No	↑ vs PS	↓ vs PS	Improved tissue distri-bution?

related to stereoregularity, Benimetskaya and coworkers compared competitive binding affinities of Rp-, Sp-isomers, and racemic mixtures of PS ODNs to various cellular proteins using an alkylating PD ODN probe. Surprisingly, binding affinities of each PS species were almost identical, suggesting that nonspecific binding to studied proteins was independent of chirality *(20)*.

In order to reduce nonspecific effects associated with PS ODNs, end-modified mixed backbone ODNs have been synthesized where stretches of PS linkages are placed on the 5′- or 3′-ends of unmodified PD ODNs. Although such design affords reasonable resistance against serum nucleases in vitro, rapid degradation, most notably by endonucleases, has been observed after intravenous administration.

In search for improved ODN chemistries obliterating undesirable nonsequence-specific effects associated with PS modification, "second generation" ODNs bring promising results. Modification at the 2′ position of the ribose moiety with an electronegative substituent not only leads to fewer nonspecific effects but also shows enhanced nuclease resistance and binding affinity to target mRNA than PS ODNs. 2′-*O*-methyl (OME) and 2′-*O*-methoxyethyl (MOE) are the most important members of the class. The modification produces a RNA-like C3′-endo conformation to the ODN, which greatly increases hybridization affinity *(21)*.

The favorable properties of second generation ODNs are, however, counter-balanced by the abrogation of the ability to activate RNase H, which is a crucial aspect of the antisense mechanism. Nevertheless, potent inhibition of target protein expression has been achieved in an RNase H-independent mechanism. Baker and coworkers investigated the antisense mechanism of 2′-*O*-MOE ODNs that target the 5′ cap region of the human intercellular adhesion molecule 1 (ICAM-1) transcript. They observed a 20-fold decrease in IC_{50} (2.1 vs 41 nM) for inhibition of ICAM-1 expression with 2′-*O*-MOE in comparison to PS modification. Polysome profile analysis further showed that the enhanced activity was because of steric translational arrest by interfering with the formation of the 80S translation initiation complex *(22)*.

For most antisense applications, RNase H-mediated activity is preferred to ensure high potency. The dilemma of the second generation ODNs has been addressed by adopting a mixed backbone approach. By incorporating 2′-modified nucleotides on both 3′- and 5′-ends of the molecule while maintaining PS modifications at the center portion, favorable physiochemical, biological, and pharmacokinetic properties of both chemistries are preserved. Studies comparing 2′-*O*-MOE/PS gapmers and PS ODNs have consistently shown the superiority of the gapmers in enhancing potency, nuclease resistance, and tissue half-life *(23)*. More importantly, the increased in vivo stability enables a longer duration of action and less frequent dosing regimen *(24)*. Currently, most second generation AS ODNs undergoing clinical trials are based on the gapmer design.

Newer improved modifications continue to emerge. Locked nucleic acid (LNA) is a third generation AS ODN that possesses superior stability against nucleolytic degradation, high target affinity, potent biological activity, and low in vivo toxicity. LNA modifications are unique in that these are ribonucleotides containing a methylene bridge between the 2′-oxygen and 4′-carbon furanose ring. The constraint on the sugar moiety results in a restricted 3′-endo conformation, thus structurally mimicking RNA molecules, leading to a remarkably high binding affinity to target mRNA *(25)*. Nuclear magnetic resonance (NMR) data reveal increases of 9.6°C in the melting temperature of LNA/RNA duplex per LNA monomer introduced *(26)*. Similar to the second generation 2′-ribose-modified ODNs, LNAs inevitably suffer the same problem of the inability to activate RNase H. Therefore, a mixed backbone gapmer design is typically used to alleviate such problems. Kurreck and coworkers conducted comprehensive studies in characterizing the optimal design of LNAs to confer activity and stability. They found that a stretch of 7–8 DNA PD monomers in the center of a chimeric DNA/LNA ODN is required for the activation of RNase H. Further, at least three residues of LNA monomers at each end of a chimeric ODN are required to protect from nuclease degradation in human serum *(25)*. LNAs also exhibit better safety profile than PS ODN when injected into rat brain parenchyma *(27)*, suggesting the feasibility of central nervous system (CNS) applications such as treatment for Alzheimer's disease and brain tumors.

Another promising third generation ODN modification is the morpholino oligonucleotide (PMO). Distinct from the first and second generation ODNs, PMO has a novel nonionic chemistry wherein the deoxyribose moiety of DNA is replaced with a six-membered morpholine ring, and the charged PD internucleoside linkage is replaced with phosphorodiamidate linkages. PMO affords excellent resistance to nucleases and serum enzymes both in vitro and in vivo. Intratumoral injection of a 20-mer PMO targeted against c-myc has produced significant intracellular concentrations and distribution of intact PMO in the tumor tissues *(28)*. Furthermore, the neutral charge of PMO avoids the problem of "stickiness" of PS backbones, thus greatly minimizing unwanted nonspecific protein binding. Likewise, one might expect improved cellular uptake of PMO because of its lack of ionic charges. Ghosh and coworkers *(29)* have reported more efficient in vitro delivery of PMOs using physical methods rather than the more commonly used cationic lipid strategy.

PMOs inhibit protein synthesis either through steric blockade of ribosomal assembly preventing translation or through interference with intron–exon splicing of pre-mRNA preventing appropriate translation of targeted mRNA *(29)*. Potent target protein inhibition and functional efficacy of PMO-based agents have been demonstrated in several preclinical cancer models *(30,31)*.

Whether these favorable pharmaceutical and biological properties of PMO can be translated into a viable therapeutic entity awaits further evaluation in the clinical setting.

3.2. siRNAs

Though single-stranded RNA is notoriously susceptible to degradation, unmodified double-stranded siRNAs are inherently more resistant to nuclease degradation than their single-stranded counterparts *(32)*. However, chemical modifications are still warranted to increase desirable in vivo properties. Much of the same chemistries developed for AS ODNs can be applied to siRNAs. Modifications with PS linkages, 2'-*O*-methyl, 2'-fluoro substitutions have been shown to retain gene-silencing activity while improving serum stability *(33–35)*. Certain rules regarding compatible positions for chemical modifications have also emerged. For example, modifications at the 5'-terminus of the antisense strand or the 3'- or 5'-terminus of the sense strand are well tolerated. However, modification at the 3'-terminus of the antisense strand will completely abolish RNAi activity *(36,37)*.

Morrissey and coworkers reported the first demonstration of in vivo activity with systemically delivered, chemically modified siRNAs in a hepatitis B virus (HBV) mouse model. Substitutions with a combination of 2'-fluoro, 2'-*O*-methyl, 2'-deoxy sugars, PS linkages, and terminal capping not only enhanced siRNA in vitro serum stability by three orders of magnitude, from 5 min to greater than 15 hr, but HBV replication was also reduced by 10-fold, whereas the unmodified siRNA did not show any antiviral effect. Although the results were encouraging, less than 1% of the injected siRNA reached the liver target organ despite the use of extremely high doses *(38)*. Much research is still needed to find modification strategies that will improve siRNA tissue targeting and cellular uptake.

4. Challenges for the Successful Development of Nucleic Acid Drugs
4.1. Efficient Delivery

For any molecular or chemical entity to be considered as a drug candidate, certain pharmaceutical attributes have to be met. These include: (1) stability in the biological system against metabolism and degradation; (2) ability to reach the target site and achieve effective therapeutic concentration at the site of action; (3) minimal nonspecific binding to serum proteins and other biomolecules; and (4) absence of immune response with the exception of immunotherapy. Since the introduction of antisense technology more than 25 yr ago, much progress has been made towards improving the pharmaceutical properties of ODNs. Most notably, advances in nucleotide chemistry have greatly expanded the nucleic acid analog family, characterized by enhanced resistance against nucleolytic degradation in biological fluid and higher affinity towards target mRNA. In addition, increased understanding of sequence-related nonspecific effects has allowed more careful design of candidate molecules, such as avoiding immuno-stimulatory CpG motifs.

Despite the encouraging prospects of nucleotide chemistry and sequence design, efficient cellular delivery remains a major hurdle in the successful development of nucleic acid therapeutics. Although delivery methods such as cationic liposomes can greatly enhance cellular uptake of ODNs across the plasma membrane, the delivery of free ODNs to the site of action (i.e., cyto-plasm and nucleus) is still a rather inefficient process. The delivery challenge is even more pronounced in the case of targeted delivery to solid tumors because the majority of ODN candidates are intended for anticancer therapy. ODNs must first avoid interactions with components in blood, including opsonizing proteins and lipoproteins. Then ODNs must extravasate into the tumor inter-stitium and migrate through the tumor bed to enter the target neoplastic cells. Extensive neovascularization and irregular endothelial fenestration may initially aid the selective delivery of ODNs into the tumor tissue. However, poor distri-bution of ODNs throughout the tumor bed is a major problem because of the heterogeneous and underdeveloped vascular network of the tumor. Another major hurdle is the transport of ODNs across the tumor interstitial space. Because most tumors have reduced lymphatic drainage, this leads to fluid accumulation and buildup of positive pressure inside the tumor core, thus extravasation of fluid and molecules into the tumor tissue is greatly hindered. Lastly, ODNs must withstand the abundant release of nucleases and cellular enzymes from necrotic cells as well as avoid being "trapped" in the necrotic regions of the tumor. In sum, current research efforts focusing on the development of delivery systems that mediate efficient cellular uptake, accurate in vivo targeting, and sustained release of drugs are priorities for the successful application of nucleic acid in the clinical setting.

4.2. Target Nonspecificity

The need to combine high target specificity with high biological potency is one of the most important challenges facing both traditional AS ODNs and siRNAs. Although the concept of nucleic acid drugs is based on the specific Watson–Crick basepairing between the drug molecule and target mRNA, numerous nonspecific "off-target" effects are frequently observed. As discussed earlier, the PS backbone is highly prone to nonspecific binding to heparin-binding proteins present in the serum, leading to unpredictable biological consequences. In addition, degradation products of ODNs may be another contributing factor for nonspecificity. Using microarray analysis, Bilanges and coworkers observed that the extent of "off-target" signature of both AS ODN and siRNA increased in a temporal fashion. By 72 hr of incubation, gene expression profiles were no longer discernable between AS ODN and its mismatched control ODN *(39)*.

A separate study was carried out to assess the specificity of siRNA using a similar microarray approach. Surprisingly, all 16 siRNAs targeting the coding region of insulin-like growth factor receptor and 8 siRNAs targeting mitogen-activated protein kinase 1 (MAPK14) consistently produced unique transcription expression profiles that were siRNA treatment-specific rather than target specific. In fact, only a few common genes were altered by the different siRNAs targeting the same transcript. Furthermore, as few as 11 contiguous nucleotides of sequence identity were found to be sufficient to direct silencing of nontargeted genes *(40)*. Therefore, until more advanced siRNA and AS ODN design becomes available with minimal target nonspecificity, stringent experimental controls and multiple target sequences should be employed to ensure that the observed phenotype is because of specific target knockdown.

5. Delivery of Nucleic Acids
5.1. Cellular Uptake
5.1.1. Barriers to Efficient Target Delivery

During the evolutionary process, eukaryotic cells have developed cellular barriers to protect from disadvantageous nonself gene uptake and to ensure genetic integrity. In order to be biologically active, ODNs must circumvent all barriers to reach their target site of action at sufficiently high concentrations. Following the passage of extracellular barriers, such as extreme pH, nucleases, immune defense, and scavenger systems, ODNs encounter multiple levels of cellular barriers. The first barrier is obviously the plasma membrane at the point of entry. Owing to ODNs' hydrophilic and macromolecular (>5 kDa) nature, internalization across the hydrophobic membrane via passive diffusion is extremely inefficient. Although small amounts of ODNs do gain entry across the plasma membrane via endocytosis, these molecules have to escape from the endosomal membrane as the second barrier. In fact, release from the endosomal compartment must be rapid, otherwise ODNs' degradation by lysosomal enzymes inevitably occurs. Thirdly, ODNs must traffic across the cytoplasm to the nucleus if the site of action is within the nucleus. High mobility and resistance against intracellular nucleases are important factors for the successful transit of ODNs.

On the other hand, studies employing delivery strategies bypassing the endocytic pathway, such as microinjection and electroporation, do suggest that once ODNs are free in the cytosol, rapid transport to the nucleus occurs within minutes *(41)*. And finally, entry of ODNs across the nuclear membrane presents the last barrier. The nuclear envelope is richly embedded with integral proteins called nuclear pore complexes (NPCs). Improved transport across the NPCs can be achieved when ODNs are tagged with a nuclear localization sequence (NLS). Preferential nuclear uptake of ODNs also occurs during mitosis because of the disassembly of the nuclear envelope *(42)*. Such an uptake property could be exploited for the selective uptake of ODNs by rapidly dividing cells as in the case of tumor cells. Hence, thorough understanding of the various cellular barriers and development of strategies to overcome these barriers will be an integral part of the optimal delivery and efficacy of ODN molecules.

5.1.2. Mechanisms of Cellular Uptake

Owing to the presence of cellular barriers to efficient cellular uptake of ODNs as discussed earlier, much research has focused on elucidating the mechanisms of ODN cellular uptake in order to design a more rational delivery strategy. The exact mechanism of ODN cellular internalization is still not well defined. The best accepted theory entails adsorptive endocytosis, pinocytosis, or a combination of both mechanisms *(43,44)*. In the case of adsorptive endocytosis, adsorption of ODNs on the cell surface leads to endocytic internalization, whereas pinocytosis refers to the process of cell engulfing water and dissolved solutes from the fluid phase. The relative contribution of each process depends on ODN concentration, ODN chemistry, and cell type. Once inside the cell, ODNs are sequestered in the endosomal compartment as evidenced by the punctate cytoplasmic distribution of fluorescently labeled ODNs within the cell and intracellular release of ODNs after treatment of lysosomatropic agents *(45)*. Rapid endosomal escape is a critical factor in the pharmacological actions of ODNs because entrapped ODNs face either lysosomal degradation or recycling back to the cell surface. The exact mechanism of how ODNs escape the endosomal vesicles still remains unclear. As mentioned previously, after endosomal exit, the highly migratory ODNs accumulate in the nucleus within minutes. Numerous putative ODN cell surface binding proteins, ranging from 22 to 143 kDa, have been identified to facilitate ODNs cellular internalization *(46–48)*. In addition, uptake mediated through a membrane pore or porin-like transporter has been suggested as nonendocytic mechanisms *(49–51)*.

Surprisingly, in vivo delivery of ODNs appears to be a more efficient process than in vitro delivery. Several studies have described the successful attainment of antisense effects with "naked" ODNs. In addition, local application, such as intracerebral or intrathecal injection of ODNs, also appears to be an effective method to achieve extensive ODN nuclear localization and antisense activity *(52)*. The mechanisms for in vivo cellular uptake of ODNs are thought to occur via large membrane disruption as a result of mechanical force during injection or transient membrane pores created by the injection procedure *(53)*. Membrane pore formation in the targeted site, such as hepatocytes, is a consequence of elevated pressure in the hepatic circulation generated from large volume and high injection speed, leading to enlargement of liver fenestrae and hepatocyte membrane permeabilization. The process favors a physical mechanism because effective delivery does not exhibit any specificity towards DNA structures or types of macromolecules.

5.2. Chemical Delivery Methods

Because the topic of nucleic acid drug delivery is discussed elsewhere in this book, we will only present a brief overview of the subject in terms of how the subject has been applied to small nucleic acid drug delivery. Since the first report of using cationic lipid, N-[1-(2,3-dioleyloxi)propyl]-N,N,N-trimethyl-ammonium chloride (DOTMA), as a vehicle for the transfer of DNA into eukaryotic cells, many new cationic lipids have been synthesized and are currently the most widely used nonviral carrier system for in vitro and in vivo nucleic acid delivery *(54)*. Cationic liposomes are vesicles formulated from phospholipids and sometimes sterols with an average size from 100 to 200 nm. DNA readily associates with cationic liposomes through electrostatic interactions between the negative charges on the nucleic acids and positive charges on the liposomal head groups, thereby forming DNA/lipid complexes known as lipoplexes. The net positive charge of the lipoplexes ensures high affinity to the cell membrane, and cellular entry occurs via an endocytic mechanism *(55)*. Similar to trafficking of "naked" ODNs, endosomal escape of complexed ODNs is a rate-limiting step in the delivery process. Numerous studies have demonstrated the punctate vesicular localization of fluorescently labeled ODNs following cell transfection. One hypothesis proposes a "flip-flop" model for the release of ODNs from the endosome. Anionic lipids from the cytoplasmic-facing endosomal membrane are proposed to "flip-flip" to the luminal side on membrane destabilization by the cationic liposomes. The anionic lipids then diffuse laterally into the liposome, neutralize the positive charges of the cationic lipids, thus leading to the release of complexed ODNs into the cytoplasm *(56)*. A common strategy to facilitate ODNs' endosomal release is the inclusion of a neutral helper lipid such as dioleoylphosphatidylethanolamine (DOPE) to the liposomal formulation. DOPE is known to destabilize endosomal membranes to promote membrane fusion between liposomal and endosomal membranes, thereby facilitating the release of ODNs to the site of action.

In addition to enhanced cellular uptake of ODNs, cationic liposomes also protect ODNs from serum nucleases. High storage stability and versatile in vivo administration routes further contribute to its widespread applications. An inherent drawback of liposomal delivery is cytotoxicity, which is closely associated with the charge ratio between the cationic lipid species in the formulation and ODNs. Higher charge ratios are generally more cytotoxic and appear to be cell type specific *(57)*. Omidi and coworkers examined the toxic effects of commercial lipid formulations via a microarray approach, where they observed more than twofold alterations in gene expression in 6–17% of genes tested. Such lipid-induced gene expression changes also led to an increase in apoptosis *(58)*. Hence, careful optimization of cationic liposomes used and thorough evaluation of liposome-induced nonspecific effects are imperative to minimize cytotoxicity and to avoid misinterpretation of experimental data.

Successful application of liposome-mediated ODN delivery in in vivo settings faces additional challenges. In the presence of serum proteins, lipoplexes can form aggregates and get trapped in "first-pass" organs, such as the liver, spleen, and lung. Such problems can be overcome by local administration to the target site. Bertrand and coworkers demonstrated knockdown of green fluorescent protein (GFP) expression in nude mice bearing GFP-expressing HeLa tumor xenograft following intratumoral administration of liposome-formulated siRNA *(59)*. Furthermore, in vivo studies have shown that systemic administration of lipoplexes often results in clearance by the reticuloendothelial (RES) system. Incorporation of the hydrophilic polymer, polyethylene glycol (PEG), into the liposome formulation has been shown to reduce opsonization and phagocytosis by cells of the RES, thus resulting in sustained circulation of lipoplexes. The addition of PEG effectively shields the charge of lipoplexes and prevents nonspecific binding to serum proteins and cells. Based on these advantages, PEG-containing formulation called stabilized antisense lipid particles has been developed and has exhibited significantly prolonged plasma circulation time in comparison to free nucleic acids or plain lipoplexes. The art of PEGylation is currently an area of intensive research in the pharmaceutical industry as more biotechnology-derived products (i.e., nucleic acids, monoclonal antibodies) are moving through the drug development phase. The first PEGylated nucleic acid drug, Macugen® (Pegaptanib, Eyetech and Pfizer) as an ODN aptamer targeting the vascular endothelial growth factor (VEGF) protein, has received FDA approval for the treatment of age-related macular degeneration (AMD).

ODNs have also been successfully delivered with cationic polymers, including dendrimers and polyethylenimine (PEI) polymers, as well as biodegradable nanoparticles and microspheres. Khan and coworkers showed that local injection of microsphere-formulated ODNs into the rat brain not only prolonged ODN delivery but also improved subcellular distribution of ODNs. Instead of the punctate vesicular, presumably endosomal localization of fluorescently labeled ODNs in rat neuronal cells when delivered "naked," the microsphere formulation was characterized by a more diffuse cytosolic and nuclear fluorescence, possibly indicating absence of endosomal sequestration of ODNs and thus more bioavailable active molecules *(60)*.

5.3. Molecular Delivery Methods

In addition to chemical formulation methods, more targeted delivery of ODNs has been achieved by attaching receptor-specific carrier molecules or signaling peptides. Receptor-mediated endocytosis exploits the exclusive presence or the overexpression of receptors on certain tissues or cell types. Receptor ligands can be directly coupled to ODNs or conjugated to particle-complexed ODNs. Commonly used ligands include transferrin, folate, low-density lipoprotein (LDL), and epidermal growth factor (EGF). Use of ligands is a particularly attractive strategy for anticancer therapy because tumor cells often express tumor-specific markers or overexpress signaling receptors on their cell surface because of increased requirement for growth factors.

Transferrin receptor mediates the uptake of transferrin–iron complexes, so overexpression is found in rapidly proliferating cells. Transferrin is also highly expressed in cerebral endothelial cells, thus making it an attractive target for CNS drug delivery. Transferrin has been incorporated into polyplexes, not only providing selective targeting to tumor cells but also enabling effective shielding of surface charges, thus extending the complex circulation time in vivo and its delivery to the intended target site *(61)*. Using a similar principle, Hudson et al. *(62)* has shown the improved cellular uptake of transferrin receptor antibody-conjugated ribozymes.

Steroid and lipid conjugation to nucleic acids is another approach for selective targeting to liver cells via a receptor-mediated endocytic mechanism coupled with increased membrane permeability. Lorenz and coworkers systemically examined the cellular uptake of siRNAs conjugated with derivatives of cholesterol, lithocholic acid, or lauric acid in human liver cells. Out of 30 modified siRNA candidates, cholesterol and lauric acid modifications exhibited highest activity (i.e., downregulation of β-galactosidase expression). Furthermore, conjugation to the sense strand was superior to that of the antisense strand or both strands *(63)*. Therapeutic efficacy of intravenous delivery of cholesterol-conjugated siRNA has also been reported. Soutschek and coworkers synthesized cholesterol-conjugated siRNA targeting apolipoprotein B (apoB) in mice. Conjugation substantially enhanced the pharmacokinetic profile of siRNA with a 15-fold increase in plasma half-life, a 15-fold decrease in total clearance, and broad tissue distribution including liver, jejunum, heart, kidney, lung, and fat tissue. More importantly, conjugated siRNA efficiently reduced apoB mRNA in liver and jejunum, decreased plasma level of apoB protein, and translated to a clinically relevant reduction in total blood cholesterol level *(64)*. Although the exact cellular uptake mechanism is unclear, cholesterol-conjugated ODNs have been shown to bind plasma lipoproteins and then become internalized by cells via LDL and high-density lipoprotein (HDL) surface receptors.

Despite tremendous improvement in the enhancement of the cellular uptake of nucleic acids, many of the delivery strategies rely on the endocytic mechanism which still present a certain degree of challenge for ODNs to escape the endosomal vesicle. Therefore, much interest has generated to develop nonendocytic delivery methods. Peptides, either with fusogenic or nuclear localization properties, have been successfully used. These small peptides contain protein transduction domains, typically 10–16 amino acids in length, which show the ability to cross biological membranes efficiently and independently of transporters or specific receptors *(65,66)*.

The best known membrane-fusion peptide motifs are derived from viruses, such as Tat protein from human immunodeficient virus type 1(HIV-1) virus, SV40 from Simian virus, and Antennapedia homeodomain-derived peptide Ant from *Drosophila*. These peptides contain NLS which enables the recognition by cytoplasmic transport receptors and efficient nuclear uptake through the NPCs. Rapid and efficient nuclear delivery of peptide-bound ODNs, even in the presence of serum, has been observed in cell culture studies *(67,68)*.

6. Nucleic Acid Drugs In Vivo

6.1. Pharmacokinetics and Toxicology

The FDA approval of two ODN drugs is testimonial to the improvements in in vivo stability and activity that have been made in the design of these molecules. However, both Macugen and Vitravene are administered intravitreously indicating that systemic administration is still a difficult option for ODNs. The drugs distribute throughout the vitreous fluid, retina, and aqueous fluid and are slowly absorbed systemically after intravitreous injection. Drug and its metabolites are excreted via the urine (http://www.macugen.com). In contrast, Vitravene is primarily metabolized in the eye with little drug absorbed into the plasma (Vitravene package insert).

Because unmodified ODNs are rapidly cleared from the plasma, modified ODNs are necessary for in vivo use. Kinetics and tissue distribution is generally independent of sequence but varies somewhat between the types of modification. PS ODNs have a half-life of 0.5–60 hr, depending on the experimental conditions and assay, and are well distributed through most tissues, but primarily to the liver and kidney. The drugs are mostly renally eliminated with nuclease activity contributing to metabolism. Kinetics appears to be dose dependent with many studies demonstrating saturation in distribution (for review *see [69,70]*).

As discussed earlier cellular uptake is also widely observed, with investigators speculating on the mechanism of transport in vivo occurring via a receptor such as the scavenger receptor. However, Butler et al. *(71)* showed in scavenger receptor knockout mice that tissue and cellular distribution was unaltered between wild-type and knockout mice.

The improved half-life of PS ODNs over PD ODNs is in part because of greater resistance to nuclease degradation as well as higher protein binding and decreased renal excretion *(74)*. PS ODN plasma protein binding is significant and ODNs containing exclusively PS backbones are considered "sticky". Such ODN backbones most notably bind complement and can cause dose-dependent intravascular coagulation in monkeys after rapid infusion that produces plasma concentrations in excess of 40–50 μg/mL. Other adverse reactions include thrombocytopenia, hypotension, and hypoglycemia.

To further define the role of complement activation in the acute and transient toxicities of PS ODNs in monkeys, Henry and coworkers successfully prevented all hemodynamic and clinical symptoms following a 10-min infusion of 20 mg/kg 20-mer PS ODNs by pretreatment with a complement inhibitor. Furthermore, the addition of the inhibitor did not alter the plasma profile of ODNs, suggesting that the protection was mediated via the inhibition of complement activation rather than pharmacokinetic interaction *(72)*.

Such outcomes reinforce the mixed benefit that results from ODN sequence-independent protein interactions. In case of Macugen and other aptamers, these interactions can be therapeutically beneficial, but are potentially lethal in the case of complement binding. Therefore, mixed backbone chemistries are used to improve the toxicity profile of PS ODNs. Another unintended but possible therapeutic benefit of protein binding is the immune activation induced by sequences containing CpG motifs. This seemingly negative property of CpG-containing ODNs has been exploited in the clinical development of an ODN-based toll-like receptor 9 (TLR9) agonist for the treatment of cancer.

Because the clinically relevant doses in humans are considerably lower than those used in preclinical toxicology studies, PS ODNs are generally well tolerated. Less severe and dose-limiting toxicities include fever, fatigue, transient thrombocytopenia, and reversible hepatotoxicity *(73,74)*.

Other ODN modifications have similar properties to PS ODN. PMO are reported to have similar kinetic properties, but improved safety and efficacy compared to phosphorothioates *(75)*. 2-Methoxyester ODNs and end-modified mixed backbone ODNs are also reported to have improved in vivo stability and possibly oral bioavailability compared to PS ODNs *(76,77)*.

Because siRNA is still a relatively new field, in vivo data are limited though examples of successful in vivo applications are discussed above. RNA molecules have an even shorter in vivo half-life than PD ODNs, so chemical modification is also necessary when using single-stranded RNA. Because siRNA is administered as double-stranded molecules, the stability is improved but still requires chemical modification for in vivo use. In addition, maintenance of the double-stranded structure in vivo is required for improved stability and pharma-cologic activity, but may reduce cellular uptake. In vivo activity has been demonstrated in mice liver, and distribution is also primarily to the liver and kidney (for review *see [78]*).

Senn et al. compared intracerebroventricular (i.c.v.) injection of AS ODNs and siRNA for distribution and stability. Fluorescently labeled siRNA was virtually undetectable in the brain after administration, even when used with detergent, whereas AS ODNs demonstrated good tissue distribution *(79)*.

When siRNAs targeting apoB were encapsulated in stable nucleic acid lipid particles (SNALP) and administered by intravenous injection to Cynomolgus monkeys at doses of 1 or 2.5 mg/kg, they exhibited an attractive safety profile. Unlike AS ODNs, siRNAs did not induce complement activation, impaired coagulation, proinflammatory cytokine production, or changes in hematological parameters. The only notable toxicity was a transient and reversible increase in liver enzymes which became normalized by day 6 *(80)*. In contrast, other studies have shown that liposome-formulated siRNAs were capable of producing dose-dependent inflammatory cytokine release both in vivo in mice and in vitro in human peripheral blood mononuclear cells (PBMC). However, such immune stimulation was absent using naked siRNA *(81)*.

The safety and clinical superiority of siRNA over AS ODNs still awaits compilation of extensive preclinical and clinical data. Based on lessons and experience gained from AS ODNs, in vivo delivery of duplex siRNA to target tissues will be a major challenge and unfortunately most delivery vehicles are not inert. Therefore, studies addressing optimal delivery routes/formulations to minimize siRNA toxicity will be a critical component along the therapeutic development pathway.

6.2. Clinical Trials

Since the FDA approval of the first AS ODN-based drug, Vitravene®, in 1998, numerous ODN drug candidates have entered the clinical development stages for the treatment of cancer, viral, inflammatory, and metabolic diseases (Table 2). In particular, the field of oncology has experienced the fastest grow-ing pipeline as increasing numbers of target genes and molecular pathways that associate with tumorigenesis, disease progression, and cell death are being elu-cidated. Protein members that become upregulated during or causally related to cancer progression and chemotherapy resistance are especially attractive ODN targets. In addition, because of the multigenic defective nature of tumor cells, combination treatments with ODNs and conventional chemo- or radiation ther-apies are promising strategies to restore functional molecular signaling and to promote treatment sensitization. Such combination strategies not only offer

Table 2
Nucleic Acids in Clinical Trials

Target	Compound	Size/chemistry	Company	Disease	Development phase	Combination treatment?
AS ODN						
Bcl-2	Genasense (G3139, Oblimersen)	18-mer/PS	Genta	Advanced or refractory CLL	NDA submitted to FDA	Yes, fludarabine plus cyclophosph-amide
	Genasense	Same as above	Same as above	Advanced malignant melanoma	MAA submitted for EU approval	Yes, dacarbazine
	Genasense	Same as above	Same as above	AML, HRPC, SCLC, NSCLC	II–III	Yes, cytarabine plus daunoru-bicin, docetaxel, carboplatin plus etoposide
Clusterin	OGX-011	21-mer/2'-O-MOE/PS gapmer	OncoGenex	Prostate, breast cancer, NSCLC	II	Yes, docetaxel
RNR R2 subunit	GTI-2040	20-mer/PS	Lorus therapeutics	Kidney, colon, lung, breast, prostate cancer	II	Yes, capecitabine
RNR R1 subunit	GTI-2501	20-mer/PS	Lours Therapeutics	HRPC	II	Yes, docetaxel
ICAM-1	Alicaforsen (ISIS 2302)	20-mer/PS	ISIS	Ulcerative colitis	II	No

212

Target	Name	Chemistry	Company	Indication	Phase	Approved
PTP-1B	ISIS 113715	20-mer/2′-O-MOE/PS gapmer	ISIS	Type 2 diabetes	II	No
TLR9	IMO-2055	Not disclosed	Idera	RCC	II	No
DNMT1	MG98	20-mer/second generation PS	Methylgene	RCC	II	Yes, interferon alpha
c-MYB	G4460	24-mer/PS	Genta	CML	I	No
c-MYC	AVI-4126	20-mer/PMO	AVI BioPharma	Prostate, breast cancer	I	No
SiRNA						
RSV nucleo-capsid "N" gene	ALN-RSV01	Not disclosed	Alynlam	RSV infection	I	No
VEGFR-1	Sirna-027	Not disclosed	Sirna	AMD	I	No
Ribozyme						
VEGFR-1	Angiozyme	Not disclosed	Ribozyme Pharma-ceuticals	Solid tumors	I	No

AMD, age-related macular degeneration; AML, acute myeloid leukemia; CLL, chronic lymphocytic leukemia; CML, chronic myelogenous leukemia; DNMT1, DNA methyltransferase 1; EU, European Union; HRPC, hormone-refractory prostate cancer; ICAM-1, intercellular adhesion molecule 1; MAA, marketing authorization application; NSCLC, non-small-cell lung cancer; NDA, new drug application; PTP-1B, protein tyrosine phosphatase 1B; RCC, renal cell carcinoma; RNR, ribonucleotide reductase; RSV, respiratory syncytial virus; SCLC, small-cell lung cancer; TLR9, toll-like receptor 9; VEGFR-1, vascular endothelial growth factor receptor-1.

potential additive or synergistic anticancer activities but also allow lowering of the effective chemotherapy dosing, thus minimizing patients' exposure to notorious toxicities of chemotherapy.

At the forefront of the development pipeline is a PS-modified anti-Bcl-2 ODN, Genasense (G3139, Oblimersen, Genta, Berkeley Heights, NJ), for the treatment of various types of hematologic cancers and solid tumors. Genta has recently filed new drug applications (NDAs) for FDA and EMEA approval of Genasense plus chemotherapy in chronic lymphocytic leukemia (CLL) and advanced melanoma, respectively (http://www.genta.com).

Bcl-2 is a critical antiapoptotic protein acting on the mitochondrial level, thus playing a key role in promoting cell survival. Overexpression of Bcl-2 is found in many tumor types and associates with resistance to conventional anticancer treatment. Therefore, targeting Bcl-2 serves as an attractive strategy to enhance chemotherapy-induced apoptotic cell death. A pivotal phase III randomized trial was designed to test whether Genasense in combination with fludarabine and cyclophosphamide (Flu/Cy) was superior to Flu/Cy alone in patients with advanced CLL. The primary end point was complete response (CR) or partial nodular response (nPR). The addition of Genasense significantly increased the proportion of patients who met primary end point from 7% in the chemotherapy-only arm to 17% in the combination arm (P=0.025). Furthermore, the median duration of CR/nPR was 22 mo for the Flu/Cy group, whereas the median duration has not yet been reached in the Genasense group (P=0.03). Although the clinical efficacy data are quite impressive, the practicality of dosing regimen remains less desirable. Genasense was administered as a continuous intravenous infusion for 7 d for each 21-d chemotherapy cycle *(82)*. Safety data compiled from earlier phase I and II trials indicate that dosing is limited by the development of a serious "cytokine release syndrome", characterized by fever, hypotension, back pain, and decreasing leukocyte counts *(83)*. Perhaps the incorporation of more advanced second or third generation ODN chemistries will afford improved tissue half-life in vivo, decreased toxicity, and more patient-friendly dosing regimen.

Seeking to improve the "drugability" of AS ODNs, a number of AS ODNs containing second and third generation backbone have entered clinical trials. For example, AVI-4126 is a PMO targeted against c-MYC proto-oncogene, to treat various types of solid tumors. A phase I single dose study in patients with prostate and breast cancers demonstrated that ODN not only was well tolerated but also was bioavailable in tumor tissues, suggesting the feasibility of using PMO in the clinical setting *(28)*. OGX-011, another promising compound, is a second generation PS ODN targeting the translation site of clusterin mRNA. The 21-mer ODN has a gapmer structure with four 2′-*O*-MOE modifications on either end of the PS backbone. Preclinical studies in mice have confirmed that second generation ODN is more effective than conventional PS ODN in decreasing clusterin mRNA and protein expression, coupled with longer in vivo half-life *(23)*. On the basis of preclinical results, phase I clinical trial in prostate cancer patients receiving OGX-011 and androgen ablation therapy prior to prostatectomy was designed to determine both pharmacokinetic and safety profiles, as well as OGX-011 tissue concentrations and clusterin protein inhibition in the target tissue. The study concluded that once weekly intravenous infusion could achieve sufficiently high drug concentrations in target prostate tissues. Furthermore, OGX-011 exhibited dose-dependent inhibition in clusterin expression in prostate cells and lymph nodes, up to 91% protein reduction with the highest dose tested *(84)*. The impressive phase I data helped to establish optimal therapeutic dose and the drug is currently in phase II trials in prostate, breast, and non-small-cell lung cancers.

Going beyond AS ODNs, the first IND for a therapeutic siRNA was filed in 2004. Sirna-027, targeting VEGFR-1, is the first siRNA to be tested in a human clinical trial for the wet form of AMD, a leading cause of adult blindness. Sirna-027 decreases the production of VEGF, which is the main culprit leading to pathologic angiogenesis in AMD. Phase I data collected thus far indicated that sirna-027 was safe and well tolerated with 100% of patients showing visual acuity stabilization and 23% of patients experiencing clinically significant improvement in visual acuity after 8 wk following a single injection. Because the siRNA was delivered locally by intravitreal injection, problems associated with systemic delivery, such as stability in blood and target organ uptake, were avoided.

Numerous nucleic acid-based molecules are being tested in various phases of clinical trials as summarized in Table 2. This list is likely to expand substantially as additional protein targets currently being validated in preclinical studies successfully advance into clinical development. Some promising candidates include Mcl-1 (ISIS 20408, Isis), insulin-like growth factor binding proteins (IGFBPs, OGX-225, OncoGenex), and eukaryotic initiation factor-4E (LY2275796, Eli Lilly), all of which target malignant tumor growth.

7. Conclusion

In the span of less than 30 yr, the field of small nucleic acid drug therapy has emerged with two FDA-approved drugs and dozens more in clinical trials. The last three decades have yielded many new surprising insights into this field, including the discovery of catalytic properties, nucleic acid conformations considered "unusual" at the time, and a vast array of new tools for gene downregulation. ODNs were also for a time considered to be the next "magic bullet", though as with most "magic bullet" designations, detailed investigation uncovers the limitations of an almost endlessly promising new field. The limitations such as stability, specificity, and in vivo properties are being addressed individually and as a result will continue to allow the emergence of more small nucleic acid drug molecules as therapeutic agents.

References

1. Zamecnik, P. C. and Stephenson, M. L. (1978) Inhibition of Rous sarcoma virus replication and cell transformation by a specific oligodeoxynucleotide. *Proc. Natl Acad Sci U. S. A.* **75**(1), 280–284.
2. Crook, S. T. (1998) Molecular mechanisms of antisense drugs: RNase H. *Antisense Nucleic Acid Drug Dev.* **8**(2), 133–134.
3. Monia, B. P., Lesnik, E. A., Gonzalez, C., et al. (1993) Evaluation of 2'-modified oligonucleotides containing 2'-deoxy gaps as antisense inhibitors of gene expression. *J. Biol. Chem.* **268**(19), 14,514–14,522.
4. Fire, A., Xu, X., Montgomery, M. K., Kostas, S. A., Driver, S. E., and Mello, C. C. (1998) Potent and specific genetic interference by double-stranded RNA in *Caenorhabditis elegans*. *Nature* **391**, 806–811.
5. Elbashir, S. M., Harborth, J., Lendeckel, W., Yalcin, A., Weber, K., and Tuschl, T. (2001) Duplexes of 21-nucleotide RNAs mediate RNA interference in cultured mammalian cells. *Nature* **411**(6836), 494–498.
6. Soutschek, J., Akinc, A., Bramlage, B., et al. (2004) Therapeutic silencing of an endogenous gene by systemic administration of modified siRNAs. *Nature* **432**(7014), 173–178.
7. Zamore, P. D., Tuschl, T., Sharp, P. A., and Bartel, D. P. (2000) RNAi: double-stranded RNA directs the ATP-dependent cleavage of mRNA at 21 to 23 nucleotide intervals. *Cell* **101**(1), 25–33.
8. McManus, M. T., Haines, B. B., Dillon, C. P., et al. (2002) Small interfering RNA-mediated gene silencing in T lymphocytes. *J. Immunol.* **169**(10), 5754–5760.
9. Ambros, V. (2004) The functions of animal microRNAs. *Nature* **431**(7006), 350–355.
10. Uprichard, S. L., Boyd, B., Althage, A., and Chisari, F. V. (2005) Clearance of hepatitis B virus from the liver of transgenic mice by short hairpin RNAs. *Proc. Natl Acad. Sci. U. S. A.* **102**(3), 773–778.
11. Phylactou, L. A., Kilpatrick, M. W., and Wood, M. J. (1998) Ribozymes as therapeutic tools for genetic disease. *Hum. Mol. Genet.* **7**(10), 1649–1653.
12. Bramlage, B., Alefelder, S., Marschall, P., and Eckstein, F. (1999) Inhibition of luciferase expression by synthetic hammerhead ribozymes and their cellular uptake. *Nucleic Acids Res.* **27**(15), 3159–3167.
13. Parry, T. J., Cushman, C., Gallegos, A. M., et al. (1999) Bioactivity of anti-angiogenic ribozymes targeting Flt-1 and KDR mRNA.*Nucleic Acids Res.* **27**(13), 2569–2577.
14. Morishita, R., Higaki, J., Tomita, N., and Ogihara, T. (1998) Application of transcription factor "decoy" strategy as means of gene therapy and study of gene expression in cardiovascular disease. *Circ. Res.* **82**(10), 1023–1028.
15. Alexander, J. H., Hafley, G., Harrington, R. A., et al. (2005) Efficacy and safety of edifoligide, an E2F transcription factor decoy, for prevention of vein graft failure

following coronary artery bypass graft surgery: PREVENT IV: a randomized controlled trial. *JAMA* **294**(19), 2446–2454.

16. Campbell, J. M., Bacon, T. A., and Wickstrom, E. (1990) Oligodeoxynucleoside phosphorothioate stability in subcellular extracts, culture media, sera and cerebrospinal fluid. *J. Biochem. Biophys. Methods.* **20**(3), 259–267.

17. Guvakova, M. A., Yakubov, L. A., Vlodavsky, I., Tonkinson, J. L., and Stein, C. A. (1995) Phosphorothioate oligodeoxynucleotides bind to basic fibroblast growth factor, inhibit its binding to cell surface receptors, and remove it from low affinity binding sites on extracellular matrix. *J. Biol. Chem.* **270**(6), 2620–2627.

18. Levin, A. A. (1999) A review of the issues in the pharmacokinetics and toxicology of phosphorothioate antisense oligonucleotides. *Biochim. Biophys. Acta.* **1489**(1), 69–84.

19. Benimetskaya, L., Wittenberger, T., Stein, C. A., et al. (2004) Changes in gene expression induced by phosphorothioate oligodeoxynucleotides (including G3139) in PC3 prostate carcinoma cells are recapitulated at least in part by treatment with interferon-beta and -gamma. *Clin. Cancer Res.* **10**(11), 3678–3688.

20. Benimetskaya, L., Tonkinson, J. L., Koziolkiewicz, M., et al. (1995) Binding of phosphorothioate oligodeoxynucleotides to basic fibroblast growth factor, recombinant soluble CD4, laminin and fibronectin is P-chirality independent. *Nucleic Acids Res.* **23**(21), 4239–4245.

21. Freier, S. M. and Altmann, K. H. (1997) The ups and downs of nucleic acid duplex stability: structure-stability studies on chemically-modified DNA:RNA duplexes. *Nucleic Acids Res.* **25**(22), 4429–4443.

22. Baker, B. F., Lot, S. S., Condon, T. P., et al. (1997) 2′-O-(2-Methoxy)ethyl-modified anti-intercellular adhesion molecule 1 (ICAM-1) oligonucleotides selectively increase the ICAM-1 mRNA level and inhibit formation of the ICAM-1 translation initiation complex in human umbilical vein endothelial cells. *J Biol Chem.* **272**(18), 11,994–12,000.

23. Zellweger, T., Miyake, H., Cooper, S., et al. (2001) Antitumor activity of antisense clusterin oligonucleotides is improved in vitro and in vivo by incorporation of 2′-O-(2-methoxy)ethyl chemistry. *J. Pharmacol. Exp. Ther.* **298**(3), 934–940.

24. Chi, K. N., Eisenhauer, E., Fazli, L., et al. (2005) A phase I pharmacokinetic and pharmacodynamic study of OGX-011, a 2′-methoxyethyl antisense oligonucleotide to clusterin, in patients with localized prostate cancer. *J. Natl Cancer Inst.* **97**(17), 1287–1296.

25. Kurreck, J., Wyszko, E., Gillen, C., and Erdmann, V. A. (2002) Design of antisense oligonucleotides stabilized by locked nucleic acids. *Nucleic Acids Res.* **30**(9), 1911–1918.

26. Bondensgaard, K., Petersen, M., Singh, S. K., et al. (2000) Structural studies of LNA:RNA duplexes by NMR: conformations and implications for RNase H activity. *Chemistry* **6**(15), 2687–2695.

27. Wahlestedt, C., Salmi, P., Good, L., et al. (2000) Potent and nontoxic antisense oligonucleotides containing locked nucleic acids. *Proc. Natl Acad. Sci. U. S. A.* **97**(10), 5633–5638.

28. Devi, G. R., Beer, T. M., Corless, C. L., Arora, V., Weller, D. L., and Iversen, P. L. (2005) In vivo bioavailability and pharmacokinetics of a c-MYC antisense phosphorodiamidate morpholino oligomer, AVI-4126, in solid tumors. *Clin. Cancer Res.* **11**(10), 3930–3938.

29. Ghosh, C. and Iversen, P. L. (2000) Intracellular delivery strategies for antisense phosphorodiamidate morpholino oligomers. *Antisense Nucleic Acid Drug Dev.* **10**(4), 263–274.

30. London, C. A., Sekhon, H. S., Arora, V., Stein, D. A., Iversen, P. L., and Devi, G. R. (2003) A novel antisense inhibitor of MMP-9 attenuates angiogenesis, human prostate cancer cell invasion and tumorigenicity. *Cancer Gene Ther.* **10**(11), 823–832.

31. Iversen, P. L., Arora, V., Acker, A. J., Mason, D. H., and Devi, G. R. (2003) Efficacy of antisense morpholino oligomer targeted to c-myc in prostate cancer xenograft murine model and a Phase I safety study in humans. *Clin. Cancer Res.* **9**(7), 2510–2519.

32. Bertrand, J. R., Pottier, M., Vekris, A., Opolon, P., Maksimenko, A., and Malvy, C. (2002) Comparison of antisense oligonucleotides and siRNAs in cell culture and in vivo. *Biochem. Biophys. Res. Commun.* **296**(4), 1000–1004.

33. Braasch, D. A., Jensen, S., Liu, Y., et al. (2003) RNA interference in mammalian cells by chemically-modified RNA. *Biochemistry.* **42**(26), 7967–7975.

34. Braasch, D. A., Paroo, Z., Constantinescu, A., et al. (2004) Biodistribution of phosphodiester and phosphorothioate siRNA. *Bioorg. Med. Chem. Lett.* **14**(5), 1139–1143.

35. Czauderna, F., Fechtner, M., Dames, S., et al. (2003) Structural variations and stabilising modifications of synthetic siRNAs in mammalian cells. *Nucleic Acids Res.* **31**(11), 2705–2716.

36. Hamada, M., Ohtsuka, T., Kawaida, R., et al. (2002) Effects on RNA interference in gene expression (RNAi) in cultured mammalian cells of mismatches and the introduction of chemical modifications at the 3′-ends of siRNAs. *Antisense Nucleic Acid Drug Dev.* **12**(5), 301–309.

37. Harborth, J., Elbashir, S. M., Vandenburgh, K., et al. (2003) Sequence, chemical, and structural variation of small interfering RNAs and short hairpin RNAs and the effect on mammalian gene silencing. *Antisense Nucleic Acid Drug Dev.* **13**(2), 83–105.

38. Morrissey, D. V., Blanchard, K., Shaw, L., et al. (2005) Activity of stabilized short interfering RNA in a mouse model of hepatitis B virus replication. *Hepatology.* **41**(6), 1349–1356.

39. Bilanges, B. and Stokoe, D. (2005) Direct comparison of the specificity of gene silencing using antisense oligonucleotides and RNAi. *Biochem. J.* **388**(Pt 2), 573–583.

40. Jackson, A. L., Bartz, S. R., Schelter, J., et al. (2003) Expression profiling reveals off-target gene regulation by RNAi. *Nat. Biotechnol.* **21**(6), 635–637. Epub 2003 May 18.

41. Chin, D. J., Green, G. A., Zon, G., Szoka, F. C. Jr., and Straubinger, R. M. (1990) Rapid nuclear accumulation of injected oligodeoxyribonucleotides. *New Biol.* **2**(12), 1091–1100.

42. Krieg, A. M., Gmelig-Meyling, F., Gourley, M. F., Kisch, W. J., Chrisey, L. A., and Steinberg, A. D. (1991) Uptake of oligodeoxyribonucleotides by lymphoid cells is heterogeneous and inducible. *Antisense Res. Dev.* **1**(2), 161–171.
43. Yakubov, L. A., Deeva, E. A., Zarytova, V. F., et al. (1989) Mechanism of oligonucleotide uptake by cells: involvement of specific receptors? *Proc. Natl Acad. Sci. U. S. A.* **86**(17), 6454–6458.
44. Stein, C. A., Tonkinson, J. L., Zhang, L. M., et al. (1993) Dynamics of the internalization of phosphodiester oligodeoxynucleotides in HL60 cells. *Biochemistry.* **32**(18), 4855–4861.
45. Loke, S. L., Stein, C. A., Zhang, X. H., et al. (1989) Characterization of oligonucleotide transport into living cells. *Proc. Natl Acad. Sci. U. S. A.* **86**(10), 3474–3478.
46. Beltinger, C., Saragovi, H. U., Smith, R. M., et al. (1995) Binding, uptake, and intracellular trafficking of phosphorothioate-modified oligodeoxynucleotides. *J. Clin. Invest.* **95**(4), 1814–1823.
47. Laktionov, P., Dazard, J. E., Piette, J., et al. (1999) Uptake of oligonucleotides by keratinocytes. *Nucleosides Nucleotides* **18**(6–7), 1697–1699.
48. Yao, G. Q., Corrias, S., and Cheng, Y. C. (1996) Identification of two oligodeoxyribonucleotide binding proteins on plasma membranes of human cell lines. *Biochem. Pharmacol.* **51**(4), 431–436.
49. Wu-Pong, S., Weiss, T. L., and Hunt, C. A. (1992) Antisense c-myc oligodeoxyribonucleotide cellular uptake. *Pharm. Res.* **9**(8), 1010–1017.
50. Wu-Pong, S. (2000) Alternative interpretations of the oligonucleotide transport literature: insights from nature. *Adv. Drug Deliv. Rev.* **44**(1), 59–70.
51. Hanss, B., Leal-Pinto, E., Bruggeman, L. A., Copeland, T. D., and Klotman, P. E. (1998) Identification and characterization of a cell membrane nucleic acid channel. *Proc. Natl Acad. Sci. U. S. A.* **95**(4), 1921–1926.
52. Grzanna, R., Dubin, J. R., Dent, G. W., et al. (1998) Intrastriatal and intraventricular injections of oligodeoxynucleotides in the rat brain: tissue penetration, intracellular distribution and c-fos antisense effects. *Brain Res. Mol. Brain Res.* **63**(1), 35–52.
53. Budker, V., Budker, T., Zhang, G., Subbotin, V., Loomis, A., and Wolff, J. A. (2000) Hypothesis: naked plasmid DNA is taken up by cells in vivo by a receptor-mediated process. *J. Gene Med.* **2**(2), 76–88.
54. Felgner, J. H., Kumar, R., Sridhar, C. N., et al. (1994) Enhanced gene delivery and mechanism studies with a novel series of cationic lipid formulations. *J. Biol. Chem.* **269**(4), 2550–2561.
55. Zabner, J., Fasbender, A. J., Moninger, T., Poellinger, K. A., and Welsh, M. J. (1995) Cellular and molecular barriers to gene transfer by a cationic lipid. *J. Biol. Chem.* **270**(32), 18,997–19,007.
56. Zelphati, O., Uyechi, L. S., Barron, L. G., and Szoka, F. C. Jr. (1998) Effect of serum components on the physico-chemical properties of cationic lipid/oligonucleotide complexes and on their interactions with cells. *Biochim. Biophys. Acta.* **1390**(2), 119–133.
57. Dass, C. R. (2002) Cytotoxicity issues pertinent to lipoplex-mediated gene therapy in-vivo. *J. Pharm. Pharmacol.* **54**(5), 593–601.

58. Omidi, Y., Hollins, A. J., Benboubetra, M., Drayton, R., Benter, I. F., and Akhtar, S. (2003) Toxicogenomics of non-viral vectors for gene therapy: a microarray study of lipofectin- and oligofectamine-induced gene expression changes in human epithelial cells. *J. Drug Target.* **11**(6), 311–323.

59. Bertrand, J. R., Pottier, M., Vekris, A., Opolon, P., Maksimenko, A., and Malvy, C. (2002) Comparison of antisense oligonucleotides and siRNAs in cell culture and in vivo. *Biochem. Biophys. Res. Commun.* **296**(4), 1000–1004.

60. Khan, A., Sommer, W., Fuxe, K., and Akhtar, S. (2000) Site-specific administration of antisense oligonucleotides using biodegradable polymer microspheres provides sustained delivery and improved subcellular biodistribution in the neostriatum of the rat brain. *J. Drug Target.* **8**(5), 319–334.

61. Kircheis, R., Wightman, L., Schreiber, A., et al. (2001) Polyethylenimine/DNA complexes shielded by transferrin target gene expression to tumors after systemic application. *Gene Ther.* **8**(1), 28–40.

62. Hudson, A. J., Normand, N., Ackroyd, J., and Akhtar, S. (1999) Cellular delivery of hammerhead ribozymes conjugated to a transferrin receptor antibody. *Int. J. Pharm.* **182**(1), 49–58.

63. Lorenz, C., Hadwiger, P., John, M., Vornlocher, H. P., and Unverzagt, C. (2004) Steroid and lipid conjugates of siRNAs to enhance cellular uptake and gene silencing in liver cells. *Bioorg. Med. Chem. Lett.* **14**(19), 4975–4977.

64. Soutschek, J., Akinc, A., Bramlage, B., et al. (2004) Therapeutic silencing of an endogenous gene by systemic administration of modified siRNAs. *Nature* **432**(7014), 173–178.

65. Oehlke, J., Beyermann, M., Wiesner, B., et al. (1997) Evidence for extensive and non-specific translocation of oligopeptides across plasma membranes of mammalian cells. *Biochim. Biophys. Acta.* **1330**(1), 50–60.

66. Niidome, T., Wakamatsu, M., Wada, A., Hirayama, T., and Aoyagi, H. (2000) Required structure of cationic peptide for oligonucleotide-binding and -delivering into cells. *J. Pept. Sci.* **6**(6), 271–279.

67. Morris, M. C., Vidal, P., Chaloin, L., Heitz, F., and Divita, G. (1997) A new peptide vector for efficient delivery of oligonucleotides into mammalian cells. *Nucleic Acids Res.* **25**(14), 2730–2736.

68. Astriab-Fisher, A., Sergueev, D. S., Fisher, M., Shaw, B. R., and Juliano, R. L. (2000) Antisense inhibition of P-glycoprotein expression using peptide-oligonucleotide conjugates. *Biochem. Pharmacol.* **60**(1), 83–90.

69. Dvorchik, B. H. (2000) The disposition (ADME) of antisense oligonucleotides. *Curr. Opin. Mol. Ther.* **2**(3), 253–257.

70. Levin, A. A. (1999) A review of the issues in the pharmacokinetics and toxicology of phosphorothioate antisense oligonucleotides. *Biochim. Biophys. Acta.* **1489**(1), 69–84.

71. Butler, M., Crooke, R. M., Graham, M. J., et al. (2000) Phosphorothioate oligodeoxynucleotides distribute similarly in class A scavenger receptor knockout and wild-type mice. *J. Pharmacol. Exp. Ther.* **292**(2), 489–496.

72. Henry, S., Beattie, G., Yeh, G., et al. (2002) Complement activation is responsible for acute toxicities in rhesus monkeys treated with a phosphorothioate oligodeoxynucleotide. *Int. Immunopharmacol.* **2,** 1657–1666.

73. Stevenson, J. P., Yao, K.-S., Gallagher, M., et al. (1999) Phase I clinical/pharmacokinetic and pharmacodynamic trial of the c-raf-1 antisense oligonucleotide ISIS 5132 (CGP 69846A). *J. Clin. Oncol.* **17**(7), 2227–2236.

74. Yuen, A. R., Halsey, J., Fisher, G. A., et al. (1999) Phase I study of an antisense oligonucleotide to protein kinase C-α (ISIS 3521/CGP 64128A) in patients with cancer. *Clin. Cancer Res.* **5**(11), 3357–3363.

75. Amantana, A. and Iversen, P. L. (2005) Pharmacokinetics and biodistribution of phosphorodiamidate morpholino antisense oligomers. *Curr. Opin. Pharmacol.* **5**(5), 550–555.

76. Yu, R. Z., Geary, R. S., Monteith, D. K., et al. (2004) Tissue disposition of 2′-O-(2-methoxy) ethyl modified antisense oligonucleotides in monkeys. *J. Pharm. Sci.* **93**(1), 48–59.

77. Geary, R. S., Khatsenko, O., Bunker, K., et al. (2001) Absolute bioavailability of 2′-O-(2-methoxyethyl)-modified antisense oligonucleotides following intraduodenal instillation in rats. *J. Pharmacol. Exp. Ther.* **296**(3), 898–904.

78. Paroo, Z. and Corey, D. R. (2004) Challenges for RNAi in vivo. *Trends Biotechnol.* **22**(8), 390–394.

79. Senn, C., Hangartner, C., Moes, S., Guerini, D., and Hofbauer, K. G. (2005) Central administration of small interfering RNAs in rats: a comparison with antisense oligonucleotides. *Eur. J. Pharmacol.* **522**(1–3), 30–37.

80. Zimmermann, T., Lee, A., Akinc, A., et al. (2006) RNAi-mediated gene silencing in non-human primates. *Nature* Advanced online publication.

81. Judge, A. D., Scood, V., Shaw, J. R., Fang, D., McClintock, K., and MacLachlan, I. (2005) Sequence-dependent stimulation of the mammalian innate immune response by synthetic siRNA. *Nat. Biotechnol.* **23**, 457–462.

82. Rai, K., Moore, J., Boyd, T., et al. (2004). Phase 3 randomized trial of fludarabine/cyclophosphamide chemotherapy with or without Oblimersen sodium (Bcl-2 Antisense; Genasense; G3139) for patients with relapsed or refractory chronic lymphocytic leukemia (CLL). *Blood* **104,** abstract #338.

83. O'Brien, S. M., Cunningham, C. C., Golenkov, A. K., Turkina, A. G., Novick, S. C., and Rai, K. R. (2005) Phase I to II multicenter study of oblimersen sodium, a Bcl-2 antisense oligonucleotide, in patients with advanced chronic lymphocytic leukemia. *J. Clin. Oncol.* **23**(30), 7697–7702.

84. Chi, K.N., Eisenhauer, E., Fazli, L., et al. (2005) A phase I pharmacokinetic and pharmacodynamic study of OGX-011, a 2′-methoxyethyl antisense oligonucleotide to clusterin, in patients with localized prostate cancer. *J. Natl Cancer Inst.* **97**(17), 1287–1296.

85. Kurreck, J. (2003) Antisense technologies. Improvement through novel chemical modifications. *Eur. J. Biochem.* **270**, 1628–1644.

11

Therapeutic Strategies Targeting the Innate Antiviral Immune Response

Robert C. Tam, Zhi Hong, Miriana Moran, Andrei Varnavski, and Sung-Kwon Kim

Abstract

Sensing the presence of pathogens and responding quickly is one of the most important tasks during the early or innate immune response. This evolutionarily conserved immune system orchestrates the elimination of pathogens of viral, bacterial, and parasitic nature. The host immune response can destroy the pathogen at the site of infection or act to avoid the spread of infection until elimination and protective immunity are achieved via the adaptive immune response. In addition innate immune mechanisms participate in shaping and regulating the effector mechanisms of the pathogen-specific adaptive immune response. Of course the immune response is not always sufficient to clear the infection. Viruses, for example, have developed many immune evasion strategies to promote their continued survival in the host; any advantage the pathogen gains in this fight against the host immune system can lead to persistent infection, subsequent disease, and even death of the host. To overcome this it has been possible to therapeutically manipulate the immune response to facilitate eradication of virus using, for example, vaccines, adjuvants, or cytokines. Interferons (IFNs) are one such family of cytokines that have been used successfully to treat viral infections. This chapter specifically discusses the role of IFN in the innate immune response and how it has been used successfully to treat chronic viral hepatitis. More importantly it addresses the shortcomings of IFN treatment and how new therapeutic strategies that target endogenous IFN induction may provide the next wave of antiviral therapies. The seminal discovery of the pathogen recognition receptors called toll-like receptors (TLR) and their critical role in transducing pathogen recognition into innate immunity against the pathogen has led to significant interrogation of this pathway as therapeutic targets. The identification of synthetic ligands that act as selective TLR agonists in a similar manner to pathogens such as viruses has been the most exciting development, and the path of these potential new drugs to the clinic will be discussed.

Key Words: Innate immunity; host antiviral response; toll-like receptor agonists; interferon induction.

From: *Biopharmaceutical Drug Design and Development*
Edited by: S. Wu-Pong and Y. Rojanasakul © Humana Press Inc., Totowa, NJ

1. Innate Immunity and Viral Infection

The pioneering vision of Janeway *(1)* that cells of the innate immune system express germline-encoded receptors that recognize conserved motifs of invading pathogens has only recently been realized with the identification of toll-like receptors (TLRs) *(2)*. TLRs have been demonstrated to be the principal pathogen recognition receptors (PRRs) involved in detecting the presence of and initiating the first line of defense against many pathogens including viruses *(3)*. This family of receptors is able to induce signals in response to ligands from a broad range of viruses, bacteria, and parasites and initiate the proinflammatory response *(4)*. Recognition of a virus, or its molecular patterns, by selective TLRs such as TLR3, TLR7, and TLR9 on dendritic cells (DC) induces maturation and expression of high levels of type I interferons (IFNs) by these cells. These events and the subsequent expression of proinflammatory cytokines and costimulatory molecules on these cells lead to further recruitment of immune cells such as natural killer (NK) cells and macrophages and the transition to an adaptive antiviral immunity *(5)*. IFNs are transcriptionally regulated cytokines and are key players in early antiviral immunity. The action of IFNs on virus-infected cells and surrounding tissues elicits an antiviral state that can be characterized by the upregulation of IFN-stimulated genes (ISGs). Conversely, viruses have developed diverse strategies to counterattack host defenses in order to generate their progenies. Many viruses expend significant portions of their limited coding capacity in inhibiting IFN induction or signaling *(6)*.

1.1. Role of Type I IFN in Antiviral Response

IFNs were first discovered by Isaacs and Lindemann in 1957 *(7)* as a family of cytokines which act early in the innate immune response. In addition to their antiviral activity, the IFN cytokines have a role in regulating the ensuing immune responses. The IFN family is classified into type I and type II IFNs. The type I IFNs consist of at least 13 α genes encoding for 12 IFNα subtypes, one β gene encoding a single IFNβ, and a single ω gene encoding for IFNω. On viral infection, IFNα can be induced from cells of lymphoid origin whereas IFNβ can be induced from most cell types including fibroblasts *(8)*. Type II IFN consists of a single gene that encodes for IFNγ that binds uniquely to the IFNγ receptor. IFNγ is produced by T cells, neutrophils, and NK cells and is primarily involved in immune regulation *(8)*. Recently the cytokines IL-28A, IL-28B, and IL-29 also called IFNλ 1–3 have been shown to exert antiviral activity *(9,10)*. These new IFNs show little homology to IFNα and do not signal via IFN$\alpha\beta$- or IFNγ-receptors but like IFNα and IFNβ, IFNλs activate the STAT2 pathway *(10)*.

IFNs mediate their effects through interactions with type-specific receptors, which are different and nonredundant for the type I and type II IFNs. IFN$\alpha\beta$-receptor knockout mice (as well as IFNγ-receptor knockouts) cannot mount effective antiviral responses. The IFN receptors do not have enzymatic activity, but they set in motion complex signaling pathways that ultimately result in the transcription of hundreds of ISGs *(11)*. Once IFN is induced it is secreted from the cell and acts in an autocrine or paracrine loop to stimulate its receptor on the infected cell or to stimulate uninfected cells to acquire an antiviral state. This protective effect results from the expression of delayed type I IFN genes and the antiviral actions of diverse ISGs such as 2′,5′-oligo adenylate synthetase (2′,5′-OAS), RNAse L, double-stranded RNA (dsRNA)-dependent protein kinase (PKR), and Mx proteins, among others *(8)*. Both 2′,5′-OAS and PKR enzymes require dsRNA as a cofactor to become activated. The dsRNA is generated as an intermediate of virus replication during virus infection. Thus the antiviral state triggered by the IFNs is only initiated and becomes effective when the cells are infected. On activation, both the 2′,5′-OAS and PKR contribute to antiviral activity by interrupting protein synthesis, and arrest of virus replication, thereby preventing viral spreading *(12)*.

The secretion of IFNs is not only essential to mount an innate antiviral immune response, but equally crucial to trigger an adaptive immune response *(13)*. DCs orchestrate this transition by first sensing the virus through the TLR and then by secreting immunomodulatory cytokines (IL-12 and TNFα), as well as processing and presenting the viral antigens via MHC II to T cells. Cumulatively these events help to initiate the adaptive immune response to the virus.

1.2. Mechanisms of Viral Evasion of IFN System

Thus, the IFN system plays an important role in limiting virus spread at an early stage of infection. As such it has become a necessity for the survival of viruses to evolve multiple strategies to escape the IFN system. These include viral interference with specific components of the IFN induction pathway or with IFN effector functions (Table 1). For example, hepatitis C virus (HCV) infection fails to produce a robust, sustained antiviral response and subsequently leads to persistent chronic infection. This is likely because of the evolution by viruses of very effective innate immune evasion mechanisms. This can be demonstrated by the ability of the virus to rapidly reach high serum titers within 1 wk of infection despite the dramatically elevated expression levels of many ISGs *(14,15)*. Several anti-IFN mechanisms have been proposed that may blunt the host innate IFN response to HCV *(15,16)* HCV serine protease NS3-4A has been shown to block IRF3-mediated induction of type I IFN in vitro *(17)*, possibly through the disruption of RIG-I (DexD/H box-containing RNA helicase encoded by retinoid acid-inducible gene I) signaling *(18)*. In addition, NS5A, a HCV nonstructural protein linked to IFN resistance, can directly target the IFN signaling pathway by disrupting

Table 1
Viral Immune Evasion Strategies

Strategy	Viral protein	Virus
Interference with IFN signaling		
Recognition of dsRNA	NS1	Influenza
Activation of transcription factors		
NFκB, AP1	E3L	Poxviruses
	NS1	Influenza
IRF3—phosphorylation	VP35	Ebola
	cNS3/NS4A	[a]HCV
	NS1/NS2	RSV
IRF3—dimerization	ML	Thogoto
IRF3—activation	NSP-1	Rotavirus
	E6	HPV16
		Sendai
IRF3—nuclear transport	ICP0	HSV-1
		BVDV
IRF3—viral dominant negative analogs		HHV8
STAT	VP	HPIV
JAK1	CP, VP	Measles
Induction of SOCS-3	Core	HCV
		HSV-1
TYK2	E6	HPV18
Viral proteins that bind dsRNA	Tat	HIV
	γ34.5	HSV-1
Induce degradation of effector molecules		Poliovirus
[b]Binding DC-SIGN (sequesters virus in DC)		HIV-1

dsRNA, double-stranded RNA; RSV, respiratory syncytial virus; HPV16, human papilloma virus-16; HSV-1, herpes simplex virus-1; BVDV, bovine diarrhea virus; HHV8, human herpes virus-8; HPIV, human parainfluenza virus; DC-SIGN, DC-specific intercellular adhesion molecule grabbing nonintegrin.

Source from Refs. *(6,16)*, specific references for [a]*(17)* and [b]*(136)*.

the crosstalk between MAPK and JAK-STAT pathways *(19)*. Finally, specific HCV proteins may interfere with the function of innate effector cells such as NK cells *(14)*. For example it has been demonstrated that the ability of NK cells to activate the IFN-producing DCs is impaired in HCV-infected patients, owing to overproduction of the inhibitory NK receptor CD94-NKG2A and the immunoregulatory cytokines, TGFβ and IL-10 *(20)*. In contrast to HCV, hepatitis B virus (HBV) appears to act as a stealth virus evading the host innate immune response *(21)*. In support of this notion HBV does not induce any gene expression during the early phase of infection *(22)*. Masking detection of the virus by the IFN signaling pathway appears to be its main evasion tactic as HBV is readily suppressed by TLR ligands that activate the innate immune response in HBV transgenic mice *(23)*. However, the mechanisms by which HCV and HBV evade IFN can be circumvented by therapeutic administration of type I IFN.

2. IFN as Therapy for Chronic Viral Infection

IFNα has been shown to elicit strong antiviral responses to both HBV and HCV in vitro and in vivo *(24)*. Recombinant forms of IFNα are approved for treating chronic HCV *(25)* and HBV infections *(26)*. As a component of antiviral therapy for HCV, IFNα treatment has been successful in generating sustained HCV clearance, preventing disease progression such as cirrhosis and hepatocellular carcinoma, and perhaps more importantly, improving patient quality of life *(27)*. In the treatment of chronic HBV infection, IFNα has led to the suppression of HBV DNA and liver inflammation *(28)*. The timeline of key events in the history of the development of IFNα as an antiviral therapeutic is outlined in Table 2. Recent improvements through pegylation (PEG) of IFN has improved the pharmacokinetic profiles and provided more effective use of IFN *(29)*. The standard of care for treatment of chronic HCV infection as of 2002 is weekly subcutaneous injection of PEG-IFN given in combination with oral ribavirin *(30)*. For patients infected with HCV genotype 2/3, this combination treatment induced a sustained virologic response (SVR—undetectable HCV RNA 24 wk after treatment completion) in a staggering 76–82% of patients treated. Although this was a tremendous achievement, when the results from HCV genotypes 1 and 4 were factored in, the mean SVR was 46–52% *(27)*.

As with most biotherapeutics, the parenteral administration of IFNα is associated with numerous side effects that can lead to dose reduction or therapy cessation. These include flu-like symptoms, neuropyschiatric and sexual dysfunction side effects, neutropenia and thrombocytopenia, injection site reactions, exacerbation of underlying autoimmune disease, thyroid disease, alopecia, skin rashes, and headaches *(26,27,31–33)*. It is noteworthy that the gravity of these side effects can be reduced through dose modification without compromising efficacy. In addition patients can develop antibodies to IFNα that can potentially neutralize the activity of this cytokine, an effect that can result in treatment relapse *(25)*.

It is important to note that IFNα therapy has been successful and the treatment regimen has been significantly improved over the last 20 yr to generate a higher cure rate by optimizing its use (sustained blood levels and weekly dosing, tailoring dose regimens to reduce side effects instead of discontinuation, body weight adjustment for dosing) *(25,27)*. However, the primary problem with

Table 2
Development History of IFNα

Indication	Approval date	Ref.
First indication		
Hairy cell leukemia	1986	*(137)*
First antiviral indication		
Human papillomavirus infection (genital warts)	1988	*(138)*
Subsequent antiviral indications		
Chronic hepatitis C virus infection		
IFN monotherapy 24 wk (SVR 6–8%)	1991	*(139)*
IFN monotherapy 48 wk (SVR 12–16%)	1996	*(140)*
IFN + ribavirin (SVR 38–43%)	1998	*(141,142)*
PEG-IFN + ribavirin (SVR 54–56%)	2002	*(143,144)*
Chronic hepatitis B virus infection	1992	*(145)*

SVR, sustained virologic response defined as percentage of patients with undetectable HCV RNA 24 wk after treatment completion.

exogenous IFNα treatment is that the effective and physiologically optimal innate antiviral response varies from patient to patient. These problems may be overcome by using an equally effective strategy which enlists bystander cells or noninfected DCs to produce *endogenous* IFN to generate the antiviral state instead of through *exogenous* IFN. These noninfected cells have not had their innate immune response to virus blunted as in HCV infection or masked as in HBV infection. This opens up the possibility of therapeutics that can initiate the typical cascade of events that constitute an effective and physiologically optimal host antiviral response for that individual. This strategy could provide all the benefits observed with direct treatment with recombinant IFNα but could alleviate the problems that are associated with reduced effectiveness and increased side effects of IFN injections. For example many cells have type I IFN receptors and so systemic administration of recombinant IFNα can lead to unwanted side effects because of inappropriate IFN signaling. In contrast, endogenous IFN inducers primarily induce IFN production in immune cells such as DCs. This limits the amount of IFN produced and the localization of IFN signaling and its effects. A strategy to induce endogenous IFN may provide drug targets that could be developed as a new generation of antiviral therapeutics. A closer look at the endogenous IFN signaling pathway may reveal such potential drug targets.

3. Induction of Endogenous IFN Signaling

The overall scheme of innate IFN induction and subsequent IFN signaling is shown in Fig. 1. The process begins with the engagement of pathogen-sensing PRRs primarily on plasmacytoid DCs (pDCs) *(34)* or other cell types *(35)* by pathogen-associated molecular patterns (PAMPs) generated during virus infection. There are several virus-specific PRRs mediating the induction of IFNs including PKR and RIG-I responding to dsRNA *(36,37)*, and mannose receptors (MR) and related lectin-type receptors responding to viral glycoproteins (GP) *(38)*. However, by far the most studied PRRs (and the most likely drug targets) are the TLRs. TLRs are type 1 transmembrane proteins initially identified based on homology with the Toll receptor from *Drosophila (39)*. The broadest repertoire of TLRs has been detected on phagocytic cells, such as neutrophils, macrophages, and DCs *(40)*. The subset of TLRs involved in recognition of viruses and viral components include TLR3, TLR7, TLR8, and TLR9. The ectodomains of these virus-specific TLR directly interact with viral components whereas the cytoplasmic tail, termed TIR (toll and interleukin 1 receptor-like) domain is responsible for intracellular signal transduction *(4)*. TLR7, TLR8, and TLR9 are predominantly localized in endosomal compartments *(41)*. Cellular localization of TLR3 is cell type specific, for example in fibroblasts it is expressed on the cell surface, whereas in DCs it is mainly confined to intracellular compartments *(42)*.

After specific recognition of different viral components by different virus-specific TLRs, signaling is initiated by interaction of the cytosolic TIR domain with TLR-type-specific adaptor molecules. For TLR7, TLR8, and TLR9, the interaction with MyD88 (myeloid differentiation factor 88) transduces the signal to TRAF6 (TNF receptor-associated factor 6) via IRAK-1 (IL-1R-associated protein kinase 1). TLR3 uses a different primary adaptor molecule, TRIF (TIR domain-containing adaptor-inducing IFNβ) *(43)*. Like MyD88, TRIF is believed to interact with TRAF6. The interaction with adaptor molecules leads to phosphorylation (by IκB kinase kinases, IKKs) and subsequent activation of transcription factors including nuclear factor-κB (NFκB) and the interferon regulatory factors (IRFs). Specifically IKKβ phosphorylates IκB leading to release of transcriptionally active NFκB which translocates to the nucleus. Similarly, phosphorylation and subsequent activation of IRFs are mediated by IKKε and TBK1 (TRAF-associated NFκB activator-binding kinase 1) *(44)*. TLR3 signaling activates IRF3, IRF5, and IRF7 whereas signaling through TLR7, TLR8, and presumably TLR9 is restricted to activation of IRF5 and 7 but not IRF3 *(44,45)*.

On activation, these transcription factors bind to corresponding sites in the IFNα/β promoter regions *(35)*. NFκB binds to the κB site present in IFNβ promoter. The IRFs bind to IFN-stimulated response element (ISRE) sites,

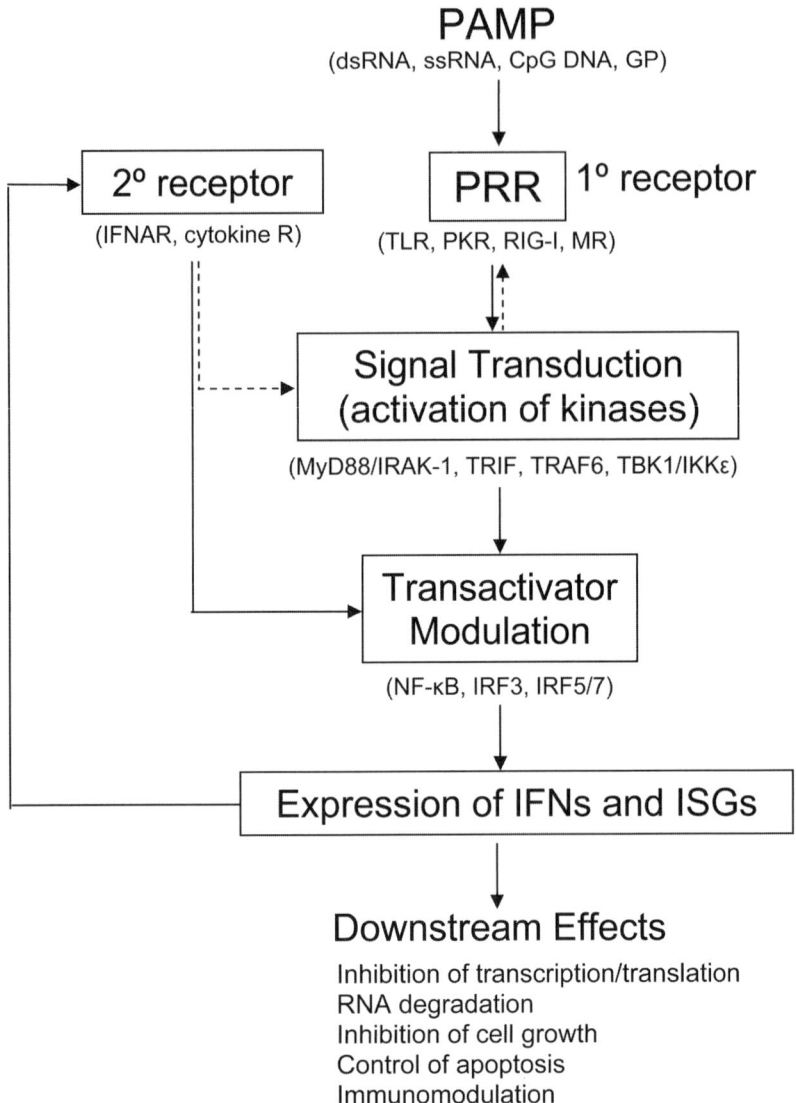

Fig. 1. Induction of innate IFN signaling. The process is initiated by PRR-mediated recognition of virus-presented PAMPs. PRR engagement stimulates molecular signal transduction, leading to activation of IFN- and ISG-specific transcription factors. Activated transcription factors translocate to the nucleus and bind to corresponding promoter regions, allowing the transcription and subsequent gene expression. Expressed IFNs and ISGs induce range of cellular effects, as well as provide a feedback signaling through a 2° (secondary) receptor. Dotted line indicates IFN/ISG-mediated regulation of PRR expression. PAMP, pathogen-associated molecular pattern. PRR, pattern recognition receptor. ISG, IFN-stimulated genes. Other abbreviations, *see* text.

present in promoters of all IFNβ and IFNα genes. Thus selective TLR-mediated signaling regulates endogenous expression of type I IFNs through transcriptional control by NFκB *(46)*, IRF3 *(47)*, IRF5 *(45)*, and IRF7 *(45,48)*.

Once expressed immediate-early IFNs (IFNβ and IFNα4 in mouse; IFNβ and IFNα1 in human) can provide autocrine and endocrine signals through type I IFN receptor (IFNAR; referred to as 2° receptor in Fig. 1) to further amplify the IFN response by inducing late phase IFNs, the IFNα subtypes. This second wave of IFN signaling via the IFN receptor stimulates JAK-STAT signaling pathways, leading to activation of ISGF3 (IFN-stimulated gene factor 3), a transcription factor that induces the expression of IRF7 *(12)*. Induction of IRF7 in cells that do not constitutively express this transcription factor leads to expression of IFNα *(49)*.

Type I IFNs, induced via TLR-mediated signaling, act in concert to promote expression of many ISGs that in turn stimulate a range of antiviral effector pathways that inhibit the replication of invading viruses *(12,50)*. ISGs trigger the inhibition of viral gene translation (by deactivation of the translation initiation factor eIF-2α and by activation of mRNA-hydrolyzing RNAse L) and viral transcription (by interfering with the trafficking and/or activity of viral polymerases through the action of Mx proteins *(51)*. In addition the action of IFNs and ISGs prevents further cellular infection which requires cell division by regulating cell growth and apoptosis (reviewed in *[50,51]*). Finally, amplification of the antiviral immune response is achieved through IFN-mediated activation of macrophages and other antigen-presenting cells, enhancement of antigen processing and presentation, regulation of induction of cellular and humoral responses *(50,52,53)*, as well as induction of proinflammatory cytokines (e.g., TNFα, IL-6, IL-12) *(4,41,54)*.

From our current understanding of the molecular mechanisms of IFN signaling (Fig. 1) it is clear that mimicking the action of viral PAMPs by engaging PRRs such as virus-specific TLRs can lead to endogenous induction of IFNs and the subsequent downstream ISG-mediated antiviral effects. Examination of the nature of recognition of viral PAMPs could lead to the generation of synthetic TLR ligands that can be exploited as antiviral therapeutics.

4. Virus-Specific PAMPs

Viruses require host cells to survive and propagate. During this process, they often generate unique forms of genetic materials which are distinguishable from host material. These viral products are viral PAMPs that can be recognized by host PRRs and subsequently initiate the host IFN response. Three virus-specific nucleic acid species (dsRNA, ssRNA, and unmethylated CpG-containing DNA) have been identified as natural virus PAMPs *(55)*.

4.1. dsRNA

The presence of dsRNAs is often associated with virus infection as they are produced during many RNA virus infections as a part of genomic RNA, a replicative intermediate or a unique stem and loop structure *(56)*. At least three distinct dsRNA-specific PRRs have been identified, PKR, RIG-I, and TLR3 (Table 3), but some data suggest the existence of others *(57)*. Unlike PKR or RIG-I, TLR3 is constitutively expressed in professional antigen-presenting cells (APCs) such as DCs and macrophages *(58)*. Thus synthetic dsRNA analogs that can engage TLR3 represent a viable strategy to induce type I IFNs. Synthetic analogs such as poly I:C have been generated and have been shown to mimic dsRNA-specific PRRs-mediated signaling and exhibit potent antiviral activity in vivo and in vitro *(59,60)*. Ampligen is another synthetic analogue of dsRNA. Structurally it is homologous to poly I:C with a single basepair mismatch (poly[I]:poly[C(12)U]). The biological action of ampligen closely resembles that of poly I:C. Thus, it has been reported to act as an immunomodulator inducing cytokine production (including IFNs) and as an antiviral, specifically activating PKR and $2'$-$5'$-OAS antiviral mechanisms *(61)*.

4.2. ssRNA

ssRNA derived from ssRNA viruses such as influenza virus, HIV, or vesicular stomatitis virus (VSV) is recognized by TLR7 in human and mice *(62)* or TLR8 in humans *(63)* (Table 3). TLR7 or TLR8 is preferentially expressed on professional APCs, in particular, pDCs and, on activation, transmit signals to produce type I IFN. Abbreviated RNA sequences derived from viral genome or synthetic ssRNA sequences can also trigger TLR7-dependent IFNα production from human or mouse DC thus providing opportunities for synthetic mimics of natural viral ssRNA *(64)*. Indeed TLR7 and TLR8 have been shown to respond to small synthetic molecules. This is a significant milestone in the validation of the pharmaceutical exploitation of TLR7 and TLR8 as therapeutic targets. At least two distinct classes of small synthetic molecules, guanosine analogs and imidazoquinoline-like molecules, have been identified that elicit signaling through TLR7 and/or TLR8. Guanosine analogs such as isatoribine (7-thia-8-oxoguanosine or TOG) or loxoribine (7-ally-7,8-dihydro-8-oxo-guanosine) have demonstrable antiviral activity and capacity to induce IFN *(65)*. Recent in vitro and in vivo studies demonstrated that these guanosine analogs trigger the innate immune response via TLR7 but not TLR8 in a MyD88-dependent manner *(66,67)*. Synthetic imidazoquinoline-like molecules can also act as TLR7 agonists (imiquimod) or both TLR7 and TLR8 agonists (resiquimod or R-848) *(66,68)* and exhibit potent antiviral and antitumor activity *(69)*. Compared to guanosine analogs, resiquimod exhibited superior potency in induction of IFN and other

Table 3
List of Viral PAMPs, their PRRs, Synthetic Analogs and Small Molecule Agonists

	Viral PAMPs		
	dsRNA	ssRNA	Unmethylated CpG DNA
PRRs **(Cellular localization)**	PKR, RIG-I (cytoplasmic) TLR3 (endosomal or extracellular)	TLR7 or 8 (endosomal)	TLR 9 (endosomal)
Source of **viral PAMPs**	Replicative intermediates of DNA and RNA viruses	ssRNA viruses (Influenza, VSV, HIV)	Viral DNA sequences (MCMV, HSV)
Synthetic PAMPs	Poly I:C	ssORN	CpG ODN
Synthetic small **molecule agonist**	None	Imiquimod (TLR7) Resiquimod (TLR7/8) Guanosine analogs Isatoribine and Loxoribine (TLR7)	None

PKR, dsRNA-activated protein kinase; ssORN, single stranded oligoribonucleotide, ODN, oligonucleotide.

inflammatory cytokines *(66)*. Whereas the nature of the promiscuity of resiquimod (being both TLR7 and 8 agonists) remains unclear, fundamental differences between TLR7 and TLR8 pathways have been revealed using exclusively TLR7-specific or TLR8-specific imidazoquinoline-like compounds *(70)*. TLR7 agonists were more proficient in induction of IFNα and IFN-regulated chemokines through selective activation of pDCs, whereas TLR8 agonists were more effective at producing proinflammatory cytokines such as TNFα, IL-12 and MIP-1α through myeloid DCs (mDCs) *(70)*. Based on these data, one may speculate that it is more desirable to develop TLR7 agonists rather than TLR8 agonists for more effective antiviral activity and less unwanted inflammatory responses.

4.3. Oligonucleotide DNA with Unmethylated CpG Motifs

Oligonucleotide DNA sequences (ODNs) containing unmethylated CpG motifs are frequently found in DNA sequences from DNA viruses and bacterias but not in mammalian DNA *(71)*. Such sequences have been shown to trigger the innate immune responses via TLR9 engagement (Table 3) *(72,73)*. On cellular uptake of CpG ODNs, TLR9 migrates from the endoplasmic reticulum to the endosomes, directly interacts with CpG ODN, and initiates signaling *(74)*. Expression of human TLR9 appears to be restricted to B cells and pDC *(75)* but TLR9 signaling in response to CpG ODNs induces type I IFN only in pDCs *(76)*.

Synthetic analogues of natural TLR9 agonists have been extensively investigated because of their immunomodulatory potential *(77)*. At least three distinct types of CpG sequences have been identified based on target effector cells and functional consequences. CpG-A (also known as D type) ODNs are composed of mixture of phosphorothioate (PS) and phosphodiester (PO) backbone and contain a single hexameric purine-pyrimidine-CG-purine-pyrimidine motif with flanking palindromic sequences that forms the stem–loop secondary structure. CpG-A appears to preferentially act on pDC, eliciting high amount of IFNα and subsequently activating NK cells to produce IFNγ. CpG-B (also known as K type) ODNs contain multiple CpG motifs exclusively on a PS backbone and preferentially but not exclusively induce proliferation and activation of B cells and subsequently IL-6 production. Owing to these characteristics, B-type CpG ODNs have been widely tested as vaccine adjuvants in animal models, particularly to enhance humoral immunity *(77)*. CpG-C (C type) ODNs consist entirely of PS backbone with multiple CpG motifs embedded in the palindromic sequence. The CpG-C sequences are capable of executing both CpG-A and CpG-B-specific functions; IFNα production as well as B-cell proliferation and IL-6 production.

In addition to these three types of CpG ODNs, unique ODN sequences with synthetic base replacements are also TLR9 agonists *(78,79)*. In these ODNs, the C was replaced with a bicyclic heterobase [1-(2'-deoxy-b-D-ribo-furasyl)-2-oxo-7-deaza-8-methyl purine; R], creating an ODN with the RpG motifs. Alternatively, G was replaced with 2'-deoxy-7-deazaguanosine (R), generating an ODN with the CpR dinucleoside motifs. These variations showed potent immunomodulatory activity via TLR9 in vitro and in vivo mouse models *(78,79)*.

It is currently unclear what determines the type and potency of a CpG ODN. However, it appears that at least three aspects of TLR9 binding to ODN sequences are involved: first, the efficiency of the uptake of the ODN by the target effector cells and the trafficking to the appropriate cellular compartment (delivery); second, the actual interaction with TLR9; and third, the capability of an ODN to multimerize and thus crosslink TLR9 receptors. Multimerization appears to be critical for TLR9 activity as the minimal ODN sequences are required to form a structural moiety that allows ODN multimerization *(80)*.

Moreover, the ODNs that do not aggregate were shown to act as TLR9 signaling antagonists *(80)*. In contrast to TLR7/8, no small synthetic molecules have been reported to have TLR9 agonist activity. However, a non-DNA-based TLR9 agonist, hemozoin, has been identified. This relatively large molecular weight detoxification product of heme, produced during malaria infection, may contain potential small molecule TLR9 recognition patterns *(81)*.

There appears to be certain degree of promiscuity in the recognition of viral PAMPs by the virus-specific PRRs so that agonists of different molecular compositions can trigger the pathways *(82)*. The necessity for therapeutic agents to act as selective or multiple TLR agonists is an important issue yet to be addressed. Acting to stimulate signaling through more than one TLR can be beneficial to the host as viruses can be detected through multiple PRRs. For example, ssRNA influenza virus can interact with TLR7 via ssRNA or TLR3 via the replicative intermediate dsRNA *(57)*. Similarly, TLR9 and TLR3 have been identified as essential components in defense against mouse cytomegalovirus (MCMV) *(83)*. In contrast, acting primarily on TLR8 signaling may lead to predominant proinflammatory cytokine effects and the potential for unwanted side effects *(70)*. In sum, the identification of synthetic analogs and small molecules that can mimic natural TLR ligands has provided the opportunity to explore their potential as therapeutic antiviral agents.

5. Therapeutic Strategies for Antiviral TLR Ligands

Because of their ability to impact the innate immune response, TLR agonists have been investigated as therapeutics against viral infections and as vaccine adjuvants for the treatment of cancer, allergies and for the prevention of viral diseases.

5.1. TLR Agonists as Antiviral Immunostimulators

Several preclinical proof-of-principle studies have demonstrated in vivo efficacy of TLR agonists in various viral infection models. The first generation of small molecule TLR agonists is selective for TLR7 or TLR8. The TLR7 agonist imiquimod is a synthetic imidazoquinoline-like molecule that has been approved to treat HPV infection-associated warts. There is ample evidence that the antiviral activity of imiquimod is dependent on TLR7 signaling (reviewed in *[84]*). Early studies showed that topical imiquimod administration induces IFNα and TNFα at the site of drug application *(85)*, and additional studies demonstrated a clinical response of genital warts to imiquimod *(86–88)*. In addition, imiquimod and a more potent analog, resiquimod, have shown promise for the treatment of herpesvirus infection. In guinea pigs, these agents can suppress genital HSV-2 recurrences and enhance long-lasting protective HSV-specific T-cell memory that lasts even after therapy is discontinued *(89,90)*. Inhibition of antiviral activity by imidazoquinolines in this animal model of HSV-2 correlated with the induction of 2′,5′-OAS activity, suggesting a role for IFN-regulated genes in viral clearance *(90)*. In addition to the imidazoquinolines, other TLR7 agonists, such as nucleoside analog isatoribine, induce cytokine production and NK cell activation and protect rodents against challenge with several different RNA viruses in an IFNα-dependent manner *(91,92)*.

The TLR3 agonists, poly I:C and ampligen, are dsRNA mimics that can protect mice against myocarditis induced by Coxsackie B3 virus *(93)*, mortality induced by West Nile virus *(94)*, and encephalitis caused by the Modoc flavivirus *(95)*. In animal models of infection, ampligen enhances the antiviral effects of exogenous IFNα *(93,95)*. Moreover, ampligen acts synergistically with anti-retroviral drugs to improve anti-HIV responses in vitro *(96)*. In mouse models of herpesvirus infection, locally delivered poly I:C is protective against a subsequent challenge with HSV-2 (prophylactic application) but was not efficacious to treat ongoing herpesviral infection (therapeutic application) *(97)*. The reason for this lack of efficacy may be related to the time of treatment, as has been previously shown for CpG ODNs *(98)*.

TLR9 agonists such as CpG ODNs administered intranasally can protect mice against a subsequent challenge with vaccinia virus, indicating efficacy when administered by a mucosal route *(99)*. In addition, CpG administered intra-vaginally not only protects mice against a subsequent challenge with HSV-2 but has therapeutic antiviral properties when administered a few hours after infection *(98)*.

A recent report demonstrates that several TLR agonists, CpG ODNs (TLR9 agonist), resiquimod (TLR7 agonist), and poly I:C (TLR3 agonist), potently inhibit HBV replication in HBV transgenic mice at doses that induce intra-hepatic IFNα/β (for all TLR agonists tested) and IFNγ (for CpG ODNs) *(23)*. These data are very encouraging because they provide evidence for TLR agonists as potential therapeutics for chronic HBV infection.

5.2. TLR Agonists as Antiviral Vaccine Adjuvants

One strategy to improve the efficacy of vaccines is to use TLR agonists as adjuvants that specifically target DCs and B cells. Recent preclinical data have shown that TLR7/8 agonists, imiquimod and resiquimod, and TLR9 agonists, CpG ODN, can markedly improve Ag-specific humoral and cellular antiviral immune responses.

In murine studies, resiquimod demonstrated moderate adjuvant activity for HIV-1 Gag DNA vaccine *(100)*. This activity was dramatically enhanced using a resiquimod–Gag protein conjugate which increased Th1 cytokines and Gag-specific CD8[+] T-cell responses in immunized mice *(101)*. Conjugation of an adjuvant to an Ag may enhance DC activation by virtue of simultaneous enhancement of antigen presentation and TLR stimulation *(101)*. Moreover resiquimod and imiquimod increased the HSV-2-specific CTL and antibody response and significantly reduced the occurrence of herpetic lesions of latently infected guinea pigs immunized with HSV-2 glycoprotein *(89,102)*.

There is evidence that two therapeutic CpG ODNs (1018 ISS and CpG 7909) administered in combination with HBsAg may lead to enhanced protective responses as exemplified by the induction of Th1 cytokines and Ag-specific T-cell responses in primates *(103,104)*. The adjuvanticity of CpG (1018 ISS) can be further enhanced by the formation of microparticles containing polyvalent CpG. An example of such a conjugate is CpG ODN complexed to polymyxin B. Such a conjugate can facilitate the clustering and more efficient triggering of TLR9-mediated signaling *(103)*. 1018 ISS has also been shown to enhance virus-specific antibody production, secretion of Th1 cytokines, and cell-mediated immunity in mice vaccinated with the HIV gp120 envelop protein *(105)*. Another study demonstrated that CpG ODN boosted the immunogenicity of HBV vaccines in healthy and SIV-infected primates *(106)*. These data suggest that there is potential for improvement of vaccine immunogenicity in poorly responding individuals such as those infected with HIV.

5.3. Clinical Experience of TLR Agonists

Based on the efficacy and reasonable toxicology profiles seen in preclinical studies, some TLR agonists have advanced into the clinic. Some have been evaluated as antiviral immunostimulators. Imiquimod is approved in the United States and in Europe as a 5% cream formulation for the treatment of external genital warts *(107)*. A more potent analog, resiquimod, has been evaluated in a randomized, placebo-controlled phase I study of healthy adults. This study demonstrated that resiquimod delivered topically is well tolerated and increases levels of mRNA for IFNα, IL-6, and IL-8 in dermal skin biopsies *(108)*. A subsequent phase II study for the treatment of HSV *(109)* showed a significant reduction in the frequency of HSV recurrences *(110)*. However, the following phase III trials were suspended because of lack of efficacy *(111)*.

Isatoribine is currently under investigation in clinical studies for the treatment of HCV. Data from a phase Ib study showed that intravenous administration of isatoribine is safe and well tolerated. Patients received seven daily injections of 200, 400, 600, or 800 mg isatoribine. All patients who received 800 mg exhibited a significant viral load reduction during treatment, with a median change of -0.94 $\log_{10}$ from baseline. This effect was associated with induction of the biomarker $2',5'$-OAS *(112)*. Recently, it has been reported that clinical trials for the potential treatment of HCV are underway for the orally bioavailable prodrugs of isatoribine, ANA-971 and ANA-975 *(113)*. According to a company press release in May 2005, interim data from phase I studies showed that conversion of ANA-975 to isatoribine in human plasma was effective, delivering levels of clinically relevant isatoribine.

Recent advances indicated that CpG oligonucleotides could be used to treat chronic viral infections. In a phase Ib clinical study in patients chronically infected with HCV, actilon (CpG 10101) was safe and efficacious, decreasing HCV viral load by at least 1 $\log_{10}$ (reported in Digestive Disease Week meeting in Chicago, IL, 2005). Of note, patients enrolled for this study were infected with genotype 1 HCV and previously failed to show a sustained response to standard therapy with IFNα plus ribavirin, suggesting that CpG may offer an improvement over the current anti-HCV therapy. Patients given a 20 mg subcutaneous injection of actilon twice weekly achieved a 96% decrease of viral load within 4 wk. Of the six patients receiving this dose, five achieved at least a 90% reduction of viral load. Mild-to-moderate injection site reactions and mild flu-like symptoms were observed in actilon-treated patients.

A phase II clinical study designed to test the effect of ampligen (400 mg intravenous injection twice weekly) in combination with HAART in multidrug-resistant HIV-infected patients is currently being evaluated. Interim data of this study showed that patients treated with ampligen plus HAART experienced a significant decrease of HIV-1 load (mean decrease of 0.5 $\log_{10}$) (reported in 14th International AIDS Conference, Barcelona, 2002). In another phase II trial, the effects of ampligen therapy in patients during HAART strategic treatment interruptions (STI) were evaluated. Interim clinical data from this study showed that ampligen significantly increases CD8$^+$ cell count during interruption of HAART treatment and significantly augments the mean time required to resume HAART treatment (reported in 16th International Conference on Antiviral Research, Savannah, GA, 2003).

Others CpG ODNs have been evaluated as vaccine adjuvants. In a phase I trial, the safety and adjuvanticity of CpG 7909 for the influenza vaccine fluarix was evaluated in a double-blinded randomized trial *(114)*. CpG 7909 proved to be safe as an adjuvant when one-tenth of the vaccine dose was administered *(114)*. In addition, in combination with HBV vaccine Engerix-B, CpG 7909 accelerated seroconversion and increased the percentage of subjects with protective levels of anti-HBsAg antibodies *(115)*. Similar findings were reported with the CpG ODN 1018 ISS. Administration of this adjuvant with the standard vaccine regimen generated anti-HBs antibodies in patients who had previously failed to respond to HBV vaccination *(116)*.

In summary, to date studies have shown that select TLR agonists are safe and well tolerated in humans. Moreover, interim results from clinical studies have demonstrated that TLR agonists are efficacious in a variety of viral infections and when administered by different routes, including intravenous, intranasal, and topical. The challenge remains to develop orally available TLR agonists and to further address the applicability of these novel immunomodulatory agents in the therapy of viral infections as well as fully evaluating the potential safety issues of their administration.

6. Potential Issues Related to Therapeutic Use of TLR Agonists

The early signs from the in vivo and initial clinical studies demonstrate that the use of TLR agonists as a strategy to enhance the antiviral innate immune response holds much promise. However, there may be potential drawbacks from the strategy of induction of innate antiviral response. These include the triggering of overt autoimmune disease, immunosuppression, increased susceptibility of the host to pathogenic agents that cause toxic shock and deleterious effects on cholesterol metabolism.

6.1. Autoimmunity

Every healthy person produces naturally occurring autoantibodies and T cells that react against self-antigens. In most cases contact between autoreactive T cells and their target antigen is not sufficient to induce and autoimmune disease. TLR stimulation has been shown in animal models of autoimmune disease to break this "ignorance" in autoreactive T cells, triggering overt organ-specific and systemic autoimmunity. This "innate autoimmunity" has been demonstrated with TLR3 (poly I:C) and TLR7/8 (resiquimod) agonists in models of autoimmune diabetes *(117)* and with TLR9 agonists (CpG) in models of multiple sclerosis and systemic lupus erythromatosus (SLE) *(118,119)*. Recent reports suggest involvement of TLR3 and TLR9 ligands in other autoimmune diseases such as glomerulonephritis and autoimmune hepatitis in animal models *(120,121)*. These observations support the notion that viral triggers could help set off autoimmune disease and that the TLR stimulation with agonists mimic this viral trigger at a molecular level. It has been demonstrated in some of these animal models that the inflammatory process associated with TLR stimulation is mediated by IFNα produced by mononuclear cells—again mimicking what viruses normally do *(117,122,123)*. IFNAR receptor-deficient mice did not develop autoimmune disease and IFNα substituted for the effect of TLR agonists *(117)*. Circulating IFNα levels are often increased in SLE patients and a strong IFNα signature is seen in gene array studies and the disease can be prevented in animal models by blockade of IFNα *(124,125)*. Data demonstrating the induction of autoimmune disease are compelling in the respective mouse models of disease, and the involvement of IFN makes for a rational mediator of these effects. However, the link in lupus is still speculative. Despite these data, autoimmunity is rare and even prevented by infection in other models *(126)*. Studies using an RIP-mOVA mouse model have demonstrated that DC activation by TLR ligands including CpG ODNs, poly I:C, and LPS was insufficient to break peripheral crosstolerance in the absence of specific CD4+ T-cell help *(127)*. These data support the view that nonspecific DC activation by TLR ligands generally does not compromise peripheral tolerance.

6.2. Immunosuppression

Daily injections of CpG ODNs cause mice to develop considerable pathology in their lymphoid organs with changes in both structure and function. Specifically destruction of the lymphoid-follicle architecture and suppression of follicular dendritic cells and germinal B lymphocytes was observed. Subsequently, multi-focal liver necrosis and hemorrhagic ascites developed after 3 wk. These effects were observed in mice treated with high doses of CpG ODNs and were dependent on TLR9 but not TLR3 or poly I:C *(128)*. The effect of CpG ODN-induced injury at lower dosages expected to have beneficial antiviral effects has not been evaluated. In addition, these adverse effects have not been reported with TLR7 ligands.

6.3. Toxic Shock

TLR9 ligands such as CpG ODNs when coadministered with sublethal doses of LPS or D-galactosamine have been shown to cause toxic shock by triggering the overproduction of TNFα leading to severe morbidity and mortality *(129)*. To examine whether such toxicities are likely to occur during treatment, CpG ODNs at doses equal to or exceeding those typically used in adjuvant experiments were injected weekly for 4 mo into normal Balb/c mice. All the animals remained physically fit and none showed macroscopic or microscopic evidence of tissue damage or inflammation *(130)*. Similar results have been reported in nonhuman primates and normal human volunteers administered with CpG-containing DNA on a weekly or monthly basis *(131)*.

6.4. Cholesterol Metabolism

The TLR3 ligand, poly I:C, has been demonstrated in C57Bl/6 mouse macrophages to block induction of liver X receptor (LXR) signaling and thus strongly inhibit cholesterol efflux from macrophages, an effect that is mediated by IRF3 *(132)*. This observation has pathophysiologic implications in athero-sclerosis as loss of macrophage LXR expression has been demonstrated to dramatically accelerate the disease *(132)*. However, further studies to demonstrate these effects in vivo have not been reported.

6.5. Clinical Safety

Many TLR agonists have advanced to the clinic providing information on the safety of these potential drugs.

For TLR9 agonists, CpG ODNs have been administered to more than 500 patients in more than a dozen clinical trials. Many of these are phase I studies that evaluate safety and immunomodulatory effects of CpG ODN delivered alone or in combination with vaccines, antibodies, or allergens. Phase II studies have been subsequently initiated to evaluate efficacy in the treatment of viral hepatitis as an antiviral immunomodulator or as a vaccine adjuvant. CpG ODNs being developed as vaccine adjuvants include CpG 7909 and ISS 1018. In phase I trials both CpG ODNs administered by coinjection with HBV vaccines demonstrated no clinically significant toxicities or serious adverse events for up to four doses separated by at least 1 wk and with a maximum dose of 1 mg/kg (CpG 7909) and 3 mg/kg (ISS 1018) *(115,116)*. As monotherapy for treating HCV infection, CpG 10101 injected subcutaneously at doses up to 20 mg twice weekly for 4 wk has been evaluated for safety in healthy volunteers and 18 patients with chronic HCV infection. In these studies, the drug was generally well tolerated and no serious side effects or dose-limiting toxicities were observed *(133)*.

For TLR7 agonists, resiquimod and isatoribine have been evaluated in phase I clinical trials. Resiquimod 0.01% as a topical application has been evaluated in healthy adults. A twice weekly dose regimen induced IFNα and was well tolerated *(108)* and was advanced to a phase II study that subsequently demonstrated efficacy in patients with genital herpesviral infection, increasing median time to recurrence from 57 d in vehicle control to 169 d in the resiquimod-treated patients *(121)*. No safety data for resiquimod delivered systemically have been reported. Imiquimod, an analog of resiquimod, previously approved for the topical treatment of HPV, has been evaluated for safety when administered orally in humans. To determine maximum tolerated dose, toxicity, and biological response in humans, a phase I clinical trial was conducted with 14 subjects who received 100–500 mg imiquimod p.o. either once or twice weekly. Imiquimod induced IFNα in serum in 10 of 19 doses of 200–300 mg. Dose-limiting side effects included fatigue, fever, headache, and lymphocytopenia; no hepatic or renal toxicity or other hematological changes exceeded the normal range. Twice-weekly doses up to 300 mg were well tolerated, with the longest twice-weekly treatments being 200 mg for 9 wk and 100 mg for 25 wk *(134)*. Isatoribine has been developed to treat patients chronically infected with HCV. In multiple dose studies in 32 HCV-infected patients, isatoribine was well tolerated up to 800 mg/kg injected intravenously once-a-day for 7 d. Adverse events were all mild to moderate and mostly related to typical flu-like injection site effects. Joint pain, insomnia, and headache were also observed in about 25% of patients receiving the 800 mg/kg dose *(135)*.

In summary, although in vivo studies in rodents, primarily transgenic mice, have shown that there may be potential safety issues using TLR agonists as therapeutics, none of the clinical studies in which safety of TLR agonists were evaluated has demonstrated any of the side effects observed in the rodent studies. This may be in part related to the dosing regimen in clinical studies and in most cases this is only once or twice a week. The most important observation was that in humans there was efficacy with these TLR agonists seen using these safe dose regimens. However, the lessons learned from in vivo studies in animals do provide awareness to monitor for potential drug liabilities in future clinical studies, particularly in patients that have ongoing autoimmune or cardiac disease.

7. Conclusion

Over the last 5 yr, there have been remarkable advances in our knowledge of the innate immune responses to viral infections. This has been fueled primarily by the identification and characterization of TLRs. Currently TLR3, TLR7, and TLR9 stand at the forefront of the host innate immune responses to viral PAMPs. The ability of these TLRs to recognize viral products, induce endogenous IFN signaling, and secrete type I IFNs makes TLR agonists an attractive new therapeutic approach against viral infection. This approach has the potential to emulate the success of recombinant IFN as a drug for treating chronic viral hepatitis but without the limitations of recombinant IFN. There are still many challenges ahead that current and future clinical studies should address: (a) Will the side effects seen in animal models manifest themselves in human trials to be insurmountable to clinical development (so far this has not been the case)? (b) What degree of selectivity is preferred for ligands targeting TLR3, TLR7, or TLR9-induced IFN signaling? (c) What is the most beneficial therapeutic approach of TLR agonists, as adjuvant to vaccines, as standalone immunomodulators or in combination with known antivirals? The battle between host and virus is constantly evolving but hopefully with the aid of new therapeutics targeting the innate immune response the balance can be effectively tipped to favor the host.

References

1. Janeway, C. A., Jr. (1989) Approaching the asymptote? Evolution and revolution in immunology. *Cold Spring Harb. Symp. Quant. Biol.* **54**(Pt 1), 1–13.
2. Medzhitov, R., Preston-Hurlburt, P., and Janeway, C. A., Jr. (1997) A human homologue of the Drosophila Toll protein signals activation of adaptive immunity. *Nature* **388**, 394–397.
3. Medzhitov, R. (2001) Toll-like receptors and innate immunity. *Nat. Rev. Immunol.* **1**, 135–145.
4. Takeda, K., Kaisho, T., and Akira, S. (2003) Toll-like receptors. *Annu. Rev. Immunol.* **21**, 335–376.
5. Le Bon, A. and Tough, D. F. (2002) Links between innate and adaptive immunity via type I interferon. *Curr. Opin. Immunol.* **14**, 432–436.

6. Weber, F., Kochs, G., and Haller, O. (2004) Inverse interference: how viruses fight the interferon system. *Viral. Immunol.* **17,** 498–515.

7. Isaacs, A. and Lindenmann, J. (1957) Virus interference. I. The interferon. *Proc. R. Soc. Lond. B Biol. Sci.* **147,** 258–267.

8. Samuel, C. E. (2001) Antiviral actions of interferons. *Clin. Microbiol. Rev.* **14,** 778–809, Table of contents.

9. Kotenko, S. V., Gallagher, G., Baurin, V. V., et al. (2003) IFN-lambdas mediate antiviral protection through a distinct class II cytokine receptor complex. *Nat. Immunol.* **4,** 69–77.

10. Dumoutier, L., Lejeune, D., Hor, S., Fickenscher, H., and Renauld, J. C. (2003) Cloning of a new type II cytokine receptor activating signal transducer and activator of transcription (STAT)1, STAT2 and STAT3. *Biochem. J.* **370,** 391–396.

11. de Veer, M. J., Holko, M., Frevel, M., et al. (2001) Functional classification of interferon-stimulated genes identified using microarrays. *J. Leukoc. Biol.* **69,** 912–920.

12. Sen, G. C. (2001) Viruses and interferons. *Annu. Rev. Microbiol.* **55,** 255–281.

13. Bonjardim, C. A. (2005) Interferons (IFNs) are key cytokines in both innate and adaptive antiviral immune responses—and viruses counteract IFN action. *Microbes. Infect.* **7,** 569–578.

14. Rehermann, B. and Nascimbeni, M. (2005) Immunology of hepatitis B virus and hepatitis C virus infection. *Nat. Rev. Immunol.* **5,** 215–229.

15. Su, A. I., Pezacki, J. P., Wodicka, L., et al. (2002) Genomic analysis of the host response to hepatitis C virus infection. *Proc. Natl Acad. Sci. U. S. A.* **99,** 15,669–15,674.

16. Katze, M. G., He, Y., and Gale, M., Jr. (2002) Viruses and interferon: a fight for supremacy. *Nat. Rev. Immunol.* **2,** 675–687.

17. Foy, E., Li, K., Wang, C., et al. (2003) Regulation of interferon regulatory factor-3 by the hepatitis C virus serine protease. *Science* **300,** 1145–1148.

18. Foy, E., Li, K., Sumpter, R., Jr., et al. (2005) Control of antiviral defenses through hepatitis C virus disruption of retinoic acid-inducible gene-I signaling. *Proc. Natl Acad. Sci. U. S. A.* **102,** 2986–2991.

19. He, Y., Nakao, H., Tan, S. L., et al. (2002) Subversion of cell signaling pathways by hepatitis C virus nonstructural 5A protein via interaction with Grb2 and P85 phosphatidylinositol 3-kinase. *J. Virol.* **76,** 9207–9217.

20. Jinushi, M., Takehara, T., Tatsumi, T., et al. (2004) Negative regulation of NK cell activities by inhibitory receptor CD94/NKG2A leads to altered NK cell-induced modulation of dendritic cell functions in chronic hepatitis C virus infection. *J. Immunol.* **173,** 6072–6081.

21. Wieland, S. F. and Chisari, F. V. (2005) Stealth and cunning: hepatitis B and hepatitis C viruses. *J. Virol.* **79,** 9369–9380.

22. Wieland, S., Thimme, R., Purcell, R. H., and Chisari, F. V. (2004) Genomic analysis of the host response to hepatitis B virus infection. *Proc. Natl Acad. Sci. U. S. A.* **101,** 6669–6674.

23. Isogawa, M., Robek, M. D., Furuichi, Y., and Chisari, F. V. (2005) Toll-like receptor signaling inhibits hepatitis B virus replication in vivo. *J. Virol.* **79,** 7269–7272.

24. Pestka, S., Krause, C. D., and Walter, M. R. (2004) Interferons, interferon-like cytokines, and their receptors. *Immunol. Rev.* **202,** 8–32.
25. Heathcote, J. and Main, J. (2005) Treatment of hepatitis C. *J. Viral. Hepat.* **12,** 223–235.
26. Arosemena, L. R., Cortes, R. A., Servin, L., and Schiff, E. R. (2005) Current and future treatment of chronic hepatitis B. *Minerva. Gastroenterol. Dietol.* **51,** 77–93.
27. Pawlotsky, J. M. (2005) Current and future concepts in hepatitis C therapy. *Semin. Liver Dis.* **25,** 72–83.
28. Park, W. and Keeffe, E. B. (2004) Diagnosis and treatment of chronic hepatitis B. *Minerva. Gastroenterol. Dietol.* **50,** 289–303.
29. Shepherd, J., Waugh, N., and Hewitson, P. (2000) Combination therapy (interferon alfa and ribavirin) in the treatment of chronic hepatitis C: a rapid and systematic review. *Health Technol. Assess.* **4,** 1–67.
30. Health, N. I. (2002). Management of hepatitis C. In *National Institutes of Health Consensus Development Conference Statement,* Vol. 36. Hepatology, pp. S3–S20.
31. Ong, J. P. and Younossi, Z. M. (2004) Managing the hematologic side effects of antiviral therapy for chronic hepatitis C: anemia, neutropenia, and thrombocytopenia. *Cleve. Clin. J. Med.* **71**(Suppl 3), S17–S21.
32. Crone, C. and Gabriel, G. M. (2003) Comprehensive review of hepatitis C for psychiatrists: risks, screening, diagnosis, treatment, and interferon-based therapy complications. *J. Psychiatr. Pract.* **9,** 93–110.
33. Kraus, M. R., Schafer, A., Bentink, T., et al. (2005) Sexual dysfunction in males with chronic hepatitis C and antiviral therapy: interferon-induced functional androgen deficiency or depression? *J. Endocrinol.* **185,** 345–352.
34. Siegal, F. P., Kadowaki, N., Shodell, M., et al. (1999) The nature of the principal type 1 interferon-producing cells in human blood. *Science* **284,** 1835–1837.
35. Malmgaard, L. (2004) Induction and regulation of IFNs during viral infections. *J. Interferon. Cytokine Res.* **24,** 439–454.
36. Der, S. D. and Lau, A. S. (1995) Involvement of the double-stranded-RNA-dependent kinase PKR in interferon expression and interferon-mediated antiviral activity. *Proc. Natl Acad. Sci. U. S. A.* **92,** 8841–8845.
37. Yoneyama, M., Kikuchi, M., Natsukawa, T., et al. (2004) The RNA helicase RIG-I has an essential function in double-stranded RNA-induced innate antiviral responses. *Nat. Immunol.* **5,** 730–737.
38. Milone, M. C. and Fitzgerald-Bocarsly, P. (1998) The mannose receptor mediates induction of IFN-alpha in peripheral blood dendritic cells by enveloped RNA and DNA viruses. *J. Immunol.* **161,** 2391–2399.
39. Rock, F. L., Hardiman, G., Timans, J. C., Kastelein, R. A., and Bazan, J. F. (1998) A family of human receptors structurally related to Drosophila Toll. *Proc. Natl Acad. Sci. U. S. A.* **95,** 588–593.
40. Zarember, K. A. and Godowski, P. J. (2002) Tissue expression of human Toll-like receptors and differential regulation of Toll-like receptor mRNAs in leukocytes in response to microbes, their products, and cytokines. *J. Immunol.* **168,** 554–561.

41. Takeda, K. and Akira, S. (2005) Toll-like receptors in innate immunity. *Int. Immunol.* **17**, 1–14.

42. Matsumoto, M., Funami, K., Tanabe, M., et al. (2003) Subcellular localization of Toll-like receptor 3 in human dendritic cells. *J. Immunol.* **171**, 3154–3162.

43. Yamamoto, M., Sato, S., Hemmi, H., et al. (2003) Role of adaptor TRIF in the MyD88-independent toll-like receptor signaling pathway. *Science* **301**, 640–643.

44. Pitha, P. M. (2004) Unexpected similarities in cellular responses to bacterial and viral invasion. *Proc. Natl Acad. Sci. U. S. A.* **101**, 695–696.

45. Schoenemeyer, A., Barnes, B. J., Mancl, M. E., et al. (2005) The interferon regulatory factor, IRF5, is a central mediator of toll-like receptor 7 signaling. *J. Biol. Chem.* **280**, 17,005–17,012.

46. Santoro, M. G., Rossi, A., and Amici, C. (2003) NF-kappaB and virus infection: who controls whom. *EMBO J.* **22**, 2552–2560.

47. Sato, M., Tanaka, N., Hata, N., Oda, E., and Taniguchi, T. (1998) Involvement of the IRF family transcription factor IRF-3 in virus-induced activation of the IFN-beta gene. *FEBS Lett.* **425**, 112–116.

48. Sato, M., Suemori, H., Hata, N., et al. (2000) Distinct and essential roles of transcription factors IRF-3 and IRF-7 in response to viruses for IFN-alpha/beta gene induction. *Immunity* **13**, 539–548.

49. Sato, M., Hata, N., Asagiri, M., Nakaya, T., Taniguchi, T., and Tanaka, N. (1998) Positive feedback regulation of type I IFN genes by the IFN-inducible transcription factor IRF-7. *FEBS Lett.* **441**, 106–110.

50. Levy, D. E., Marie, I., and Prakash, A. (2003) Ringing the interferon alarm: differential regulation of gene expression at the interface between innate and adaptive immunity. *Curr. Opin. Immunol.* **15**, 52–58.

51. Stark, G. R., Kerr, I. M., Williams, B. R., Silverman, R. H., and Schreiber, R. D. (1998) How cells respond to interferons. *Annu. Rev. Biochem.* **67**, 227–264.

52. Takaoka, A. and Taniguchi, T. (2003) New aspects of IFN-alpha/beta signalling in immunity, oncogenesis and bone metabolism. *Cancer Sci.* **94**, 405–411.

53. Uze, G., Lutfalla, G., and Mogensen, K. E. (1995) Alpha and beta interferons and their receptor and their friends and relations. *J. Interferon. Cytokine Res.* **15**, 3–26.

54. Akira, S. and Takeda, K. (2004) Toll-like receptor signalling. *Nat. Rev. Immunol.* **4**, 499–511.

55. Bowie, A. G. and Haga, I. R. (2005) The role of Toll-like receptors in the host response to viruses. *Mol. Immunol.* **42**, 859–867.

56. Balachandran, S., Roberts, P. C., Brown, L. E., et al. (2000) Essential role for the dsRNA-dependent protein kinase PKR in innate immunity to viral infection. *Immunity* **13**, 129–141.

57. Barchet, W., Krug, A., Cella, M., et al. (2005) Dendritic cells respond to influenza virus through TLR7- and PKR-independent pathways. *Eur. J. Immunol.* **35**, 236–242.

58. Muzio, M., Bosisio, D., Polentarutti, N., et al. (2000) Differential expression and regulation of toll-like receptors (TLR) in human leukocytes: selective expression of TLR3 in dendritic cells. *J. Immunol.* **164**, 5998–6004.

59. Ali, S. and Kukolj, G. (2005) Interferon regulatory factor 3-independent double-stranded RNA-induced inhibition of hepatitis C virus replicons in human embryonic kidney 293 cells. *J. Virol.* **79,** 3174–3178.

60. Ichinohe, T., Watanabe, I., Ito, S., et al. (2005) Synthetic double-stranded RNA poly(I:C) combined with mucosal vaccine protects against influenza virus infection. *J. Virol.* **79,** 2910–2919.

61. Adams, M., Navabi, H., Jasani, B., et al. (2003) Dendritic cell (DC) based therapy for cervical cancer: use of DC pulsed with tumour lysate and matured with a novel synthetic clinically non-toxic double stranded RNA analogue poly [I]:poly [C(12)U] (Ampligen R). *Vaccine* **21,** 787–790.

62. Lund, J. M., Alexopoulou, L., Sato, A., et al. (2004) Recognition of single-stranded RNA viruses by Toll-like receptor 7. *Proc. Natl Acad. Sci. U. S. A.* **101,** 5598–5603.

63. Heil, F., Hemmi, H., Hochrein, H., et al. (2004) Species-specific recognition of single-stranded RNA via toll-like receptor 7 and 8. *Science* **303,** 1526–1529.

64. Diebold, S. S., Kaisho, T., Hemmi, H., Akira, S., and Reis e Sousa, C. (2004) Innate antiviral responses by means of TLR7-mediated recognition of single-stranded RNA. *Science* **303,** 1529–1531.

65. Smee, D. F., Alaghamandan, H. A., Cottam, H. B., Jolley, W. B., and Robins, R. K. (1990) Antiviral activity of the novel immune modulator 7-thia-8-oxoguanosine. *J. Biol. Response Mod.* **9,** 24–32.

66. Lee, J., Chuang, T. H., Redecke, V., et al. (2003) Molecular basis for the immunostimulatory activity of guanine nucleoside analogs: activation of Toll-like receptor 7. *Proc. Natl Acad. Sci. U. S. A.* **100,** 6646–6651.

67. Heil, F., Ahmad-Nejad, P., Hemmi, H., et al. (2003) The Toll-like receptor 7 (TLR7)-specific stimulus loxoribine uncovers a strong relationship within the TLR7, 8 and 9 subfamily. *Eur. J. Immunol.* **33,** 2987–2997.

68. Hemmi, H., Kaisho, T., Takeuchi, O., et al. (2002) Small anti-viral compounds activate immune cells via the TLR7 MyD88-dependent signaling pathway. *Nat. Immunol.* **3,** 196–200.

69. Miller, R. L., Gerster, J. F., Owens, M. L., Slade, H. B., and Tomai, M. A. (1999) Imiquimod applied topically: a novel immune response modifier and new class of drug. *Int. J. Immunopharmacol.* **21,** 1–14.

70. Gorden, K. B., Gorski, K. S., Gibson, S. J., et al. (2005) Synthetic TLR agonists reveal functional differences between human TLR7 and TLR8. *J. Immunol.* **174,** 1259–1268.

71. Krieg, A. M. (2002) CpG motifs in bacterial DNA and their immune effects. *Annu. Rev. Immunol.* **20,** 709–760.

72. Krug, A., French, A. R., Barchet, W., et al. (2004) TLR9-dependent recognition of MCMV by IPC and DC generates coordinated cytokine responses that activate antiviral NK cell function. *Immunity* **21,** 107–119.

73. Hochrein, H., Schlatter, B., O'Keeffe, M., et al. (2004) Herpes simplex virus type-1 induces IFN-alpha production via Toll-like receptor 9-dependent and -independent pathways. *Proc. Natl Acad. Sci. U. S. A.* **101,** 11,416–11,421.

74. Latz, E., Schoenemeyer, A., Visintin, A., et al. (2004) TLR9 signals after translocating from the ER to CpG DNA in the lysosome. *Nat. Immunol.* **5,** 190–198.

75. Krieg, A. M. (2003) CpG motifs: the active ingredient in bacterial extracts? *Nat. Med.* **9,** 831–835.

76. Liu, Y. J. (2005) IPC: Professional Type 1 Interferon-Producing Cells and Plasmacytoid Dendritic Cell Precursors. *Annu. Rev. Immunol.* **23,** 275–306.

77. Klinman, D. M. (2004) Immunotherapeutic uses of CpG oligodeoxynucleotides. *Nat. Rev. Immunol.* **4,** 249–258.

78. Kandimalla, E. R., Bhagat, L., Zhu, F. G., et al. (2003) A dinucleotide motif in oligonucleotides shows potent immunomodulatory activity and overrides species-specific recognition observed with CpG motif. *Proc. Natl Acad. Sci. U. S. A.* **100,** 14,303–14,308.

79. Kandimalla, E. R., Bhagat, L., Li, Y., et al. (2005) Immunomodulatory oligonucleotides containing a cytosine-phosphate-2′-deoxy-7-deazaguanosine motif as potent Toll-like receptor 9 agonists. *Proc. Natl Acad. Sci. U. S. A.* **102,** 6925–6930.

80. Wu, C. C., Lee, J., Raz, E., Corr, M., and Carson, D. A. (2004) Necessity of oligonucleotide aggregation for toll-like receptor 9 activation. *J. Biol. Chem.* **279,** 33,071–33,078.

81. Coban, C., Ishii, K. J., Kawai, T., et al. (2005) Toll-like receptor 9 mediates innate immune activation by the malaria pigment hemozoin. *J. Exp. Med.* **201,** 19–25.

82. Ulevitch, R. J. (2004) Therapeutics targeting the innate immune system. *Nat. Rev. Immunol.* **4,** 512–520.

83. Zamanian-Daryoush, M., Mogensen, T. H., DiDonato, J. A., and Williams, B. R. (2000) NF-kappaB activation by double-stranded-RNA-activated protein kinase (PKR) is mediated through NF-kappaB-inducing kinase and IkappaB kinase. *Mol. Cell. Biol.* **20,** 1278–1290.

84. Stanley, M. A. (2002) Imiquimod and the imidazoquinolones: mechanism of action and therapeutic potential. *Clin. Exp. Dermatol.* **27,** 571–577.

85. Imbertson, L. M., Beaurline, J. M., Couture, A. M., et al. (1998) Cytokine induction in hairless mouse and rat skin after topical application of the immune response modifiers imiquimod and S-28463. *J. Invest. Dermatol.* **110,** 734–739.

86. Garland, S. M., Sellors, J. W., Wikstrom, A., et al. (2001) Imiquimod 5% cream is a safe and effective self-applied treatment for anogenital warts—results of an open-label, multicentre Phase IIIB trial. *Int. J. STD AIDS* **12,** 722–729.

87. Buck, H. W., Fortier, M., Knudsen, J., and Paavonen, J. (2002) Imiquimod 5% cream in the treatment of anogenital warts in female patients. *Int. J. Gynaecol. Obstet.* **77,** 231–238.

88. Beutner, K. R., Tyring, S. K., Trofatter, K. F., Jr., et al. (1998) Imiquimod, a patient-applied immune-response modifier for treatment of external genital warts. *Antimicrob. Agents Chemother.* **42,** 789–794.

89. Bernstein, D. I., Harrison, C. J., Tomai, M. A., and Miller, R. L. (2001) Daily or weekly therapy with resiquimod (R-848) reduces genital recurrences in herpes simplex virus-infected guinea pigs during and after treatment. *J. Infect. Dis.* **183,** 844–849.

90. Harrison, C. J., Miller, R. L., and Bernstein, D. I. (1994) Posttherapy suppression of genital herpes simplex virus (HSV) recurrences and enhancement of HSV-specific T-cell memory by imiquimod in guinea pigs. *Antimicrob. Agents Chemother.* **38,** 2059–2064.

91. Smee, D. F., Alaghamandan, H. A., Cottam, H. B., Sharma, B. S., Jolley, W. B., and Robins, R. K. (1989) Broad-spectrum in vivo antiviral activity of 7-thia-8-oxoguanosine, a novel immunopotentiating agent. *Antimicrob. Agents Chemother.* **33,** 1487–1492.

92. Smee, D. F., Alaghamandan, H. A., Jin, A., Sharma, B. S., and Jolley, W. B. (1990) Roles of interferon and natural killer cells in the antiviral activity of 7-thia-8-oxoguanosine against Semliki Forest virus infections in mice. *Antiviral Res.* **13,** 91–102.

93. Padalko, E., Nuyens, D., De Palma, A., et al. (2004) The interferon inducer ampligen [poly(I)-poly(C12U)] markedly protects mice against coxsackie B3 virus-induced myocarditis. *Antimicrob. Agents Chemother.* **48,** 267–274.

94. Morrey, J. D., Day, C. W., Julander, J. G., Blatt, L. M., Smee, D. F., and Sidwell, R. W. (2004) Effect of interferon-alpha and interferon-inducers on West Nile virus in mouse and hamster animal models. *Antivir. Chem. Chemother.* **15,** 101–109.

95. Leyssen, P., Drosten, C., Paning, M., et al. (2003) Interferons, interferon inducers, and interferon-ribavirin in treatment of flavivirus-induced encephalitis in mice. *Antimicrob. Agents Chemother.* **47,** 777–782.

96. Essey, R. J., McDougall, B. R., and Robinson, W. E., Jr. (2001) Mismatched double-stranded RNA (polyI-polyC(12)U) is synergistic with multiple anti-HIV drugs and is active against drug-sensitive and drug-resistant HIV-1 in vitro. *Antiviral Res.* **51,** 189–202.

97. Ashkar, A. A., Yao, X. D., Gill, N., Sajic, D., Patrick, A. J., and Rosenthal, K. L. (2004) Toll-like receptor (TLR)-3, but not TLR4, agonist protects against genital herpes infection in the absence of inflammation seen with CpG DNA. *J. Infect. Dis.* **190,** 1841–1849.

98. Sajic, D., Ashkar, A. A., Patrick, A. J., et al. (2003) Parameters of CpG oligo-deoxynucleotide-induced protection against intravaginal HSV-2 challenge. *J. Med. Virol.* **71,** 561–568.

99. Rees, D. G., Gates, A. J., Green, M., et al. (2005) CpG-DNA protects against a lethal orthopoxvirus infection in a murine model. *Antiviral Res.* **65,** 87–95.

100. Otero, M., Calarota, S. A., Felber, B., et al. (2004) Resiquimod is a modest adjuvant for HIV-1 gag-based genetic immunization in a mouse model. *Vaccine* **22,** 1782–1790.

101. Wille-Reece, U., Wu, C. Y., Flynn, B. J., Kedl, R. M., and Seder, R. A. (2005) Immunization with HIV-1 Gag Protein Conjugated to a TLR7/8 Agonist Results in the Generation of HIV-1 Gag-Specific Th1 and CD8+ T Cell Responses. *J. Immunol.* **174,** 7676–7683.

102. Bernstein, D. I., Harrison, C. J., Tepe, E. R., Shahwan, A., and Miller, R. L. (1995) Effect of imiquimod as an adjuvant for immunotherapy of genital HSV in guinea-pigs. *Vaccine* **13,** 72–76.

103. Marshall, J. D., Higgins, D., Abbate, C., et al. (2004) Polymyxin B enhances ISS-mediated immune responses across multiple species. *Cell. Immunol.* **229,** 93–105.

104. Davis, H. L., Suparto, II, Weeratna, R. R., et al. (2000) CpG DNA overcomes hyporesponsiveness to hepatitis B vaccine in orangutans. *Vaccine* **18,** 1920–1924.

105. Horner, A. A., Datta, S. K., Takabayashi, K., et al. (2001) Immunostimulatory DNA-based vaccines elicit multifaceted immune responses against HIV at systemic and mucosal sites. *J. Immunol.* **167,** 1584–1591.

106. Verthelyi, D., Wang, V. W., Lifson, J. D., and Klinman, D. M. (2004) CpG oligodeoxynucleotides improve the response to hepatitis B immunization in healthy and SIV-infected rhesus macaques. *Aids* **18,** 1003–1008.

107. Dockrell, D. H. and Kinghorn, G. R. (2001) Imiquimod and resiquimod as novel immunomodulators. *J. Antimicrob. Chemother.* **48,** 751–755.

108. Sauder, D. N., Smith, M. H., Senta-McMillian, T., Soria, I., and Meng, T. C. (2003) Randomized, single-blind, placebo-controlled study of topical application of the immune response modulator resiquimod in healthy adults. *Antimicrob. Agents Chemother.* **47,** 3846–3852.

109. Jones, T. (2003) Resiquimod 3M. *Curr. Opin. Investig. Drugs* **4,** 214–218.

110. Spruance, S. L., Tyring, S. K., Smith, M. H., and Meng, T. C. (2001) Application of a topical immune response modifier, resiquimod gel, to modify the recurrence rate of recurrent genital herpes: a pilot study. *J. Infect. Dis.* **184,** 196–200.

111. Wu, J. J., Huang, D. B., and Tyring, S. K. (2004) Resiquimod: a new immune response modifier with potential as a vaccine adjuvant for Th1 immune responses. *Antiviral Res.* **64,** 79–83.

112. Raney, A. K., Hamatake, R. K., and Hong, Z. (2004) HEP DART 2003: Frontiers in drug development for viral hepatitis. *Expert Opin. Investig. Drugs* **13,** 289–293.

113. Pawlotsky, J. M. and McHutchison, J. G. (2004) Hepatitis C. Development of new drugs and clinical trials: promises and pitfalls. Summary of an AASLD hepatitis single topic conference, Chicago, IL, February 27-March 1, 2003. *Hepatology* **39,** 554–567.

114. Cooper, C. L., Davis, H. L., Morris, M. L., et al. (2004) Safety and immunogenicity of CPG 7909 injection as an adjuvant to Fluarix influenza vaccine. *Vaccine* **22,** 3136–3143.

115. Cooper, C. L., Davis, H. L., Morris, M. L., et al. (2004) CPG 7909, an immunostimulatory TLR9 agonist oligodeoxynucleotide, as adjuvant to Engerix-B HBV vaccine in healthy adults: a double-blind phase I/II study. *J. Clin. Immunol.* **24,** 693–701.

116. Halperin, S. A., Van Nest, G., Smith, B., Abtahi, S., Whiley, H., and Eiden, J. J. (2003) A phase I study of the safety and immunogenicity of recombinant hepatitis B surface antigen co-administered with an immunostimulatory phosphorothioate oligonucleotide adjuvant. *Vaccine* **21,** 2461–2467.

117. Lang, K. S., Recher, M., Junt, T., et al. (2005) Toll-like receptor engagement converts T-cell autoreactivity into overt autoimmune disease. *Nat. Med.* **11,** 138–145.

118. Segal, B. M., Chang, J. T., and Shevach, E. M. (2000) CpG oligonucleotides are potent adjuvants for the activation of autoreactive encephalitogenic T cells in vivo. *J. Immunol.* **164,** 5683–5688.

119. Anders, H. J., Vielhauer, V., Eis, V., et al. (2004) Activation of toll-like receptor-9 induces progression of renal disease in MRL-Fas(lpr) mice. *FASEB J.* **18,** 534–536.

120. Anders, H. J., Banas, B., and Schlondorff, D. (2004) Signaling danger: toll-like receptors and their potential roles in kidney disease. *J. Am. Soc. Nephrol.* **15,** 854–867.

121. Sacher, T., Knolle, P., Nichterlein, T., Arnold, B., Hammerling, G. J., and Limmer, A. (2002) CpG-ODN-induced inflammation is sufficient to cause T-cell-mediated autoaggression against hepatocytes. *Eur. J. Immunol.* **32,** 3628–3637.

122. Theofilopoulos, A. N., Baccala, R., Beutler, B., and Kono, D. H. (2005) Type I interferons (alpha/beta) in immunity and autoimmunity. *Annu. Rev. Immunol.* **23,** 307–336.

123. Baechler, E. C., Gregersen, P. K., and Behrens, T. W. (2004) The emerging role of interferon in human systemic lupus erythematosus. *Curr. Opin. Immunol.* **16,** 801–807.

124. Kirou, K. A., Lee, C., George, S., Louca, K., Peterson, M. G., and Crow, M. K. (2005) Activation of the interferon-alpha pathway identifies a subgroup of systemic lupus erythematosus patients with distinct serologic features and active disease. *Arthritis Rheum.* **52,** 1491–1503.

125. Kono, D. H., Baccala, R., and Theofilopoulos, A. N. (2003) Inhibition of lupus by genetic alteration of the interferon-alpha/beta receptor. *Autoimmunity* **36,** 503–510.

126. Hron, J. D. and Peng, S. L. (2004) Type I IFN protects against murine lupus. *J. Immunol.* **173,** 2134–2142.

127. Hamilton-Williams, E. E., Lang, A., Benke, D., Davey, G. M., Wiesmuller, K. H., and Kurts, C. (2005) Cutting edge: TLR ligands are not sufficient to break cross-tolerance to self-antigens. *J. Immunol.* **174,** 1159–1163.

128. Heikenwalder, M., Polymenidou, M., Junt, T., et al. (2004) Lymphoid follicle destruction and immunosuppression after repeated CpG oligodeoxynucleotide administration. *Nat. Med.* **10,** 187–192.

129. Gao, J. J., Xue, Q., Papasian, C. J., and Morrison, D. C. (2001) Bacterial DNA and lipopolysaccharide induce synergistic production of TNF-alpha through a post-transcriptional mechanism. *J. Immunol.* **166,** 6855–6860.

130. Klinman, D. M., Takeno, M., Ichino, M., et al. (1997) DNA vaccines: safety and efficacy issues. *Springer Semin. Immunopathol.* **19,** 245–256.

131. Klinman, D. M., Currie, D., Gursel, I., and Verthelyi, D. (2004) Use of CpG oligodeoxynucleotides as immune adjuvants. *Immunol. Rev.* **199,** 201–216.

132. Castrillo, A., Joseph, S. B., Vaidya, S. A., et al. (2003) Crosstalk between LXR and toll-like receptor signaling mediates bacterial and viral antagonism of cholesterol metabolism. *Mol. Cell.* **12,** 805–816.

133. Schmalbach, E. S., Morris, M. L., Adhami, M. A., Laframboise, C., Davis, H. L., and Peese, P. (2004). CpG 10101 (Actilon) oligodeoxynucleotide TLR9 agonist: Pharmacokinetics and Pharmacodynamics in normal volunteers. In *Interscience Conference on Antimicrobial Agents and Chemotherapy.* (Abstract)

134. Witt, P. L., Ritch, P. S., Reding, D., et al. (1993) Phase I trial of an oral immunomodulator and interferon inducer in cancer patients. *Cancer Res.* **53,** 5176–5180.

135. Hormanns, Y. (2004). Isatoribine, a TLR7 agonist, significantly reduced plasma viral load in a clinical proof-of-concept study in patients with chronic HCV infection. In *AASLD meeting 2004.*

136. van Kooyk, Y. and Geijtenbeek, T. B. (2003) DC-SIGN: escape mechanism for pathogens. *Nat. Rev. Immunol.* **3,** 697–709.

137. Quesada, J. R., Hersh, E. M., Manning, J., et al. (1986) Treatment of hairy cell leukemia with recombinant alpha-interferon. *Blood* **68,** 493–497.

138. Eron, L. J., Judson, F., Tucker, S., et al. (1986) Interferon therapy for condylomata acuminata. *N. Engl. J. Med.* **315,** 1059–1064.

139. Davis, G. L., Balart, L. A., Schiff, E. R., et al. (1990) Treatment of chronic hepatitis C with recombinant alpha-interferon. A multicentre randomized, controlled trial. The Hepatitis Interventional Therapy Group. *J. Hepatol.* **11**(Suppl 1)**,** S31–S35.

140. Poynard, T., Leroy, V., Cohard, M., et al. (1996) Meta-analysis of interferon randomized trials in the treatment of viral hepatitis C: effects of dose and duration. *Hepatology* **24,** 778–789.

141. Poynard, T., Marcellin, P., Lee, S. S., et al. (1998) Randomised trial of interferon alpha2b plus ribavirin for 48 weeks or for 24 weeks versus interferon alpha2b plus placebo for 48 weeks for treatment of chronic infection with hepatitis C virus. International Hepatitis Interventional Therapy Group (IHIT). *Lancet* **352,** 1426–1432.

142. McHutchison, J. G., Gordon, S. C., Schiff, E. R., et al. (1998) Interferon alfa-2b alone or in combination with ribavirin as initial treatment for chronic hepatitis C. Hepatitis Interventional Therapy Group. *N. Engl. J. Med.* **339,** 1485–1492.

143. Manns, M. P., McHutchison, J. G., Gordon, S. C., et al. (2001) Peginterferon alfa-2b plus ribavirin compared with interferon alfa-2b plus ribavirin for initial treatment of chronic hepatitis C: a randomised trial. *Lancet* **358,** 958–965.

144. Fried, M. W., Shiffman, M. L., Reddy, K. R., et al. (2002) Peginterferon alfa-2a plus ribavirin for chronic hepatitis C virus infection. *N. Engl. J. Med.* **347,** 975–982.

145. Wong, D. K., Cheung, A. M., O'Rourke, K., Naylor, C. D., Detsky, A. S., and Heathcote, J. (1993) Effect of alpha-interferon treatment in patients with hepatitis B e antigen-positive chronic hepatitis B. A meta-analysis. *Ann. Intern. Med.* **119,** 312–323.

12

Pharmacogenetics in the Clinic

Kai I. Cheang

Abstract

Pharmacogenetics is the inherited basis of differences among individuals in their response to drugs. Genetic polymorphisms of drug-metabolizing enzymes may account for as much as 30% of interindividual differences in drug disposition and response. An increasing number of drug target polymorphisms have also been linked to differences in drug response. This chapter reviews some examples of the use of pharmacogenetics in clinical practice. Despite the increasing number of examples of genetic polymorphisms affecting drug response in the literature, pharmacogenetic data are rarely used in current clinical practice. The limitations that have prevented the use of pharmacogenetic testing in clinical practice are reviewed.

Key Words: Pharmacogenetics; pharmacogenomics; adverse drug events; efficacy; individualized therapy; drug–gene interactions; personalized medicine.

1. Introduction

Pharmacogenetics is the study of the genetic basis for interindividual differences in drug response. Genetic polymorphisms, defined as genetic variations occurring in at least 1% of the human population, form the genetic basis for variations in drug response. This field of investigation began half a century ago when hemolysis in some patients taking the antimalarial primaquine was shown to be caused by an inherited deficiency of glucose-6-phosphate dehydrogenase *(1)*. Since the sequencing of the human genome, there have been increasing examples in the literature documenting the association between genetic polymorphisms and drug response. These include reports on the genetic variability for drug-metabolizing enzymes, drug transporters, drug targets, and disease-modifying genes. However, translation of these pharmacogenetic observations into clinical practice has been limited by numerous hurdles. This chapter will first review some examples of how pharmacogenetic testing could aid in tailoring

From: *Biopharmaceutical Drug Design and Development*
Edited by: S. Wu-Pong and Y. Rojanasakul © Humana Press Inc., Totowa, NJ

pharmacotherapy to the individual patient. Limitations of using pharmacogenetic testing in actual clinical practice will also be reviewed.

2. Pharmacogenetics in Clinical Practice

2.1. Dose Adjustment

Genetic polymorphisms may partially account for variability in drug pharmaco-kinetics or disposition among individuals. In current medical practice, patients are usually treated initially with standard doses of a drug, and the clinician expects an "average" drug exposure and "average" drug response, similar to what would be reported in the clinical literature concerning the drug. Genetic differences in drug-metabolizing enzymes may lead to varying levels of plasma concentrations and drug exposure when the same dose of the drug is adminis-tered. As a result, clinical response to the drug may be altered, both in terms of drug efficacy and dose-dependent adverse events. This is especially important for drugs with narrow therapeutic index, such as mercaptopurine, warfarin, and others. More uniform and predictable drug exposure, which may be achieved by dosage adjustment according to the patient's genetic profile of their meta-bolizing enzymes, may aid in predicting drug response and reducing dose-related side effects.

2.1.1. Pharmacogenetics of Drug-Metabolizing Enzymes

Genetic polymorphisms have been reported for phase I, phase II, and nucleotide base metabolizing enzymes. An important clinical example of how pharmacogenetic differences in drug-metabolizing enzymes affect clinical out-comes is illustrated by the genetic polymorphism of thiopurine methyltrans-ferase (TPMT) *(2)*. TPMT catalyzes the S-methylation of azathioprine and 6-mercaptopurine. In patients treated for acute lymphoblastic leukemia (ALL), mercaptopurine is used as maintenance therapy. Mercaptopurine is metabolized to thioguanine, which is responsible for the bone marrow toxicity, as well as the therapeutic efficacy in ALL. TPMT catalyzes mercaptopurine's metabolism to an inactive metabolite. Patients with TPMT deficiency require a drastic reduction in mercaptopurine dose to prevent severe myelosuppression *(2)*. Certain TPMT genotypes are associated with reduced TPMT activity, and the TPMT genetic polymorphism is an important predictor of myelotoxicity associated with mer-captopurine and is clinically used to prevent myelotoxicity in ALL patients.

Most commercially available drugs are metabolized by cytochrome P450 enzymes. Polymorphisms of these P450 enzymes may also affect therapeutic responses to medications. For example, P450 2C9 is responsible for the meta-bolism of warfarin and phenytoin, both of which are drugs with narrow therapeutic indices. Clearances of these drugs are decreased in patients with 2C9 variant alleles. For example, certain 2C9 haplotypes are associated with reduced warfarin maintenance therapeutic doses *(3)*. Knowledge of 2C9 phenotypes, therefore, may aid in optimizing titration schedules to achieve therapeutic anticoagulant effect whereas minimizing the risk of bleeding when a patient is initiated on warfarin therapy. Establishment of stable maintenance warfarin doses currently necessitate frequent laboratory monitoring during the first month of therapy. It is estimated that at least 50% of the variability in maintenance warfarin doses is because of genetic polymorphisms of both the 2C9 metabolizing enzyme and the vitamin K epoxide reductase complex 1 (VKORC1), which affects clotting factor synthesis *(4–6)*. Although no genetic polymorphism-based dosing algorithm has been prospectively validated, genetic information on 2C9 and VKORC1 may lead to faster establishment of therapeutic and stable anticoagulation and a decrease in the risk of major bleeding events *(7)*. The change in warfarin's prescribing information to contain such pharmacogenetic information was the subject of discussion during a Food and Drug Administration (FDA) Clinical Pharmacology Subcommittee meeting in November 2005 *(4)*.

Another clinically relevant example of how polymorphisms in drug-metabolizing enzymes affect dosage selection can be illustrated by irinotecan in colorectal cancer. Irinotecan undergoes metabolism by carboxylesterases into an active metabolite, SN-38, which is then conjugated by the enzyme UDP-glucuronosyl transferase 1A1 (UGT1A1) to form a glucuronide metabolite. In individuals with the UGT1A1*28 genetic polymorphism, UGT1A1 enzyme activity is reduced. About 10% of the North American population is homozygous for the UGT1A1*28 polymorphism. Patients homozygous for UGT1A1*28 have a higher exposure to the SN-38 active metabolite than patients without the polymorphism when irinotecan was administered. This increased exposure may increase the risk of neutropenia and diarrhea. This increased drug exposure in patients with the polymorphism is recognized in the labeling of irinotecan, in which recommendations were given to reduce the starting dose by at least one level in patients known to be homozygous for the UGT1A1*28 allele *(8)*. Although a specific recommendation is given, the precise optimal dose modification is unknown and subsequent dose adjustments may be necessary depending on patients' tolerance.

2.1.2. Pharmacogenetics of Drug Transporters

In addition, polymorphisms in genes encoding drug transporters have also been identified. A prominent example of these drug transporters is the P-glycoprotein, a transmembrane efflux pump. P-glycoprotein is found in a wide variety of cells and tissues, including the intestinal enterocytes, renal proximal tubules, hepatocytes, and capillary endothelial cells of the blood–brain barrier. Therefore, P-glycoprotein plays an important role in drug disposition by reducing drug absorption, increasing elimination, and decreasing the ability of drugs to cross the blood–brain barrier. P-glycoproteins are also found in tumor cells. P-glycoproteins mediate an active efflux of chemotherapeutic agents from tumor cells, hence promoting multidrug resistance to anticancer agents. The dispositions of many drugs are affected by P-glycoproteins. These include cancer chemotherapeutic drugs, antiepileptic agents, protease inhibitors (for treating human immunodeficiency virus [HIV] infections), and cardiac drugs such as digoxin and diltiazem. The multidrug resistance 1 gene (MDR1) codes for P-glycoprotein. Numerous single-nucleotide polymorphisms (SNPs) have been identified in the MDR1 gene. These polymorphisms may influence drug pharmacokinetics, and hence drug response.

Numerous genetic polymorphisms exist for other metabolizing enzymes and drug transporters. These genetic polymorphisms may affect the drug's pharmacokinetic parameters in drug absorption, distribution, metabolism and excretion. Table 1 reviews some selected examples of the effect of genetic polymorphisms influencing pharmacokinetics. Interindividual differences in drug response because of variability in drug disposition resulting from genetic polymorphisms may be minimized by appropriate dose adjustment, if a dose-adjustment algorithm based on pharmacogenetic data is available. Importantly, there is now an FDA-approved pharmacogenomic microarray test designed for clinical applications (AmpliChip CYP450 Test®, Roche Diagnostics). The test provides information on 2D6 and 2C19 genes, which are involved in the metabolism of about 25% of prescription drugs. Although not widely used at the time of publication, the test will facilitate individualizing treatment doses for medications metabolized by 2D6 and 2C19.

2.2. Selection of Drug Therapy

Genetic differences may also affect drug selection. Polymorphisms in a gene coding for a drug receptor may render a drug acting via that receptor less (or more) effective. Although not usually considered as examples of pharmacogenetics because they do not involve genetic variations of the host, molecular diagnostics of tumor cells and HIV enable the selection of appropriate therapeutic agents and are important recent advances.

2.2.1. Drug Receptor Polymorphisms

Pharmacogenetic data may affect drug selection. For example, genetic polymorphisms in drug receptors may affect the degree of pharmacologic (either agonistic or antagonistic) actions via these receptors. The beta-1 receptor polymorphisms and antihypertensive response to beta-1 antagonists serve as an example. Beta-antagonists are commonly used for treating hypertension. The beta-1 receptor gene contains two common SNPs at codons 49 and 389. In a

Table 1
Selected Examples of Genetic Polymorphisms Affecting Pharmacokinetics

	Medications	Possible drug effects
P450 Drug-metabolizing enzymes		
CYP P450 2C9	Warfarin	Decreased metabolism may lead to supertherapeutic anticoagulant effect
	Phenytoin	Decreased metabolism may lead to toxicity
CYP P450 2D6	Selective serotonin reuptake inhibitors	Antidepressant toxicity because of supratherapeutic drug concentrations in poor metabolizer; poor treatment response in extensive metabolizers
	Tricyclic antidepressants	
	Codeine	Decreased analgesic effect of codeine in patients with reduced 2D6 metabolism of codeine to its active metabolite
	Metoprolol	Potential increased drug concentrations in poor metabolizers which may lead to increased side effects
	Thioridazine	Increased concentration in patients with reduced 2D6 metabolism, which may lead to life-threatening Torsade de pointes arrhythmia
Phase III metabolizing enzymes		
Uridine diphosphate glucuronyltransferase	Irinotecan	Decreased clearance leads to gastrointestinal toxicity
Nucleotide base metabolizing enzymes		
Thiopurine methyltransferase	Mercaptopurine	Low enzyme activity leads to severe myelosupression
Dihydropyrimidine dehydrogenase	Flurouracil	Deduced enzyme activity may lead to life threatening

(Continued)

Table 1 (*Continued*)

	Medications	Possible drug effects
		neurological, hematological, and gastrointestinal toxicity
Drug transporters P-glycoprotein (MDR-1)	Digoxin	Decreased digoxin bioavailability
	HIV protease inhibitors	Decreased response in CD4 count

clinical study with metoprolol, patients who are homozygous for Ser49 and Arg389 had the greatest blood pressure reduction, suggesting that the beta-1 receptor haplotype may be used as a predictor of response to beta-blockers *(9)*.

2.2.2. Molecular Diagnostics

The field of molecular diagnostics is rapidly advancing. Currently there are already specific clinical examples of how therapeutic agents are chosen based on information obtained with molecular diagnostics, regularly used in clinical practice. Examples of these include transtuzumab (Herceptin®) *(10)*, imatinib (Gleevec®) *(11)*, gefitinib (Iressa®) *(12)*, and HIV genetic testing.

2.2.2.1. MOLECULAR DIAGNOSTICS IN ONCOLOGY

HER2 (human epidermal growth factor receptor) is a protein that is overexpressed in up to 30% of invasive breast cancer cells *(10)*. A humanized monoclonal antibody against HER2, transtuzumab (Herceptin®), is the first of a class of chemotherapeutic agents whose design is directed to specific molecular targets, in this case the HER2 receptors. Transtuzumab has shown efficacy as monotherapy in patients with HER2 overexpression, as well as in combination with taxane-based chemotherapy in metastatic breast cancer overexpressing the HER2 receptor. The molecular diagnostic test for HER2 has received FDA approval.

Another drug that benefits from genetic testing is imatinib (Gleevec®) *(11)*. In chronic myelogenous leukemia, the definitive diagnosis is a cytogenetic analysis of bone marrow specimens for the Philadelphia chromosome, which is a translocation between chromosome 9 and chromosome 22, resulting in the oncogenic fusion protein BCR-ABL (BCR from chromosome 22 and ABL from chromosome 9). The molecular diagnostic test measures the presence of BCR-ABL and the level of overexpression. Imatinib is a tyrosine kinase inhibitor that inhibits BCR-ABL, an abnormal tyrosine kinase. Imatinib inhibits the proliferation and induces programmed cell death in BCR-ABL positive cells. The molecular diagnostic test for BCR-ABL has been approved by FDA.

Gefitinib (Iressa®) is another tyrosine kinase inhibitor used in the treatment of non-small-cell lung cancer (NSCLC), a leading cause of cancer death in the United States in both men and women. Gefitinib targets the epidermal growth factor receptor (EGFR) that is overexpressed in up to 80% of NSCLCs. Most patients with NSCLCs do not respond to gefitinib. However, the drug leads to rapid and dramatic clinical response in 10% of patients. Recently, specific mutations in the EGFR binding site for gefitinib were found to be more prevalent in patients responding to the drug than for nonresponders. It is thought that these mutations in the EGFR binding site mediate increased growth factor signaling which is susceptible to inhibition by gefitinib. Screening for EGFR mutations in NSCLCs may help identify patients who will have clinical response to gefitinib *(12)*. The screening test will also identify those who are not likely to benefit from gefitinib, and help avoid delaying other therapy and the expense of an ineffective drug. An FDA-approved test for EGFR mutations is not yet available at the time of writing, but some medical centers have begun offering molecular diagnostics developed in-house for EGFR mutations, and a clinically relevant test is likely to be developed in the future.

2.2.3. Drug Resistance Testing in HIV

Drug resistance testing for HIV antiretroviral therapy is now considered standard of care. Prospective and retrospective data suggest that the availability of antiretroviral genotypic resistance data is an important factor in achieving response to therapy *(13,14)*. These data have led to expert panels' recommendation of the use of resistance testing in HIV antiretroviral therapy *(15,16)*. The Panel on Clinical Practices of HIV Infection convened by the US Department of Health and Human Services recommends drug resistance testing in cases of virologic failure (to maximize the number of active drugs utilized in a new regimen after a patient has failed combination antiretroviral therapy) or when there is suboptimal suppression of viral load after initiation of antiretroviral therapy.

An FDA-approved test for HIV drug resistance is available. TruGene® (Bayer Diagnostics) involves sequencing of genes encoding the two main drug treatment targets, the protease and reverse transcriptase genes. The genetic information is accompanied by a constantly updated algorithm to guide treatment options. The report is clinician-friendly and points clinicians to resistance or sensitivity of the various antiretroviral therapies. A minimum HIV viral load of 1000 copies per mL is necessary for reliable amplification of the virus and the genotype to be detectable. The goal of HIV resistance testing is to maximize the number of active antiretroviral therapy used against the infection.

2.2.4. Drug-Metabolizing Enzyme Polymorphisms Affecting Choice of Therapy

Besides affecting the pharmacokinetic characteristics of drugs and therefore necessitating dosage adjustments, polymorphisms in drug-metabolizing enzymes may also affect the choice of therapeutic agents, when the algorithm for dosage adjustments may not be readily available, such as in the example of tamoxifen. This example also illustrates how genetic polymorphisms in P450 drug-metabolizing enzymes affect therapeutic efficacy by the important role they play in the bioformation of active metabolites, which may have important implications for certain pharmacologic agents. Tamoxifen, an antiestrogen agent used as an adjuvant in the treatment of estrogen-dependent breast cancer, is metabolized by CYP 2D6 to an active metabolite with about 100 times the antiestrogen potency of the parent compound. Breast cancer patients who were homozygous for CYP 2D6 *4 (poor metabolizers) had the highest risk of relapse and worst disease-free survival, even when nodal involvement and tumor size were accounted for, when compared to patients who were heterozygous (*4/wt) or noncarriers of the allele (wt/wt). The poor prognosis of patients who were poor metabolizers was presumably because of a low concentration of the active metabolite with potent antiestrogen effects *(17)*. Hence, differences in response to tamoxifen may be explained by CYP 2D6 genotype. Genotypic information on CYP 2D6 may help clinicians in selecting optimal therapy for estrogen-dependent breast cancer.

Numerous other examples of genetic polymorphisms affecting drug selection exist. Table 2 reviews some selected examples of how genotypic information may aid the clinician in choosing the optimal therapy for different disease states.

3. Hurdles in Translating Pharmacogenetics to the Clinic

Despite great advances in the discovery of how genotypic information could affect drug dosage modifications and initial drug selections, pharmacogenetic information is rarely used in clinical practice. A recent survey suggested that even though knowledge of drug-metabolizing polymorphisms have existed since a half-century ago, pharmacogenetic tests for drug-metabolizing enzymes are rarely used by clinicians *(18)*. This is despite continual claims in the literature that pharmacogenetic testing for drug-metabolizing enzymes may prevent adverse clinical outcomes. Because evidence for some of the drug-metabolizing enzymes have been existent for almost five decades, the reasons for the lack of utilization of pharmacogenetics in the clinic seem to be beyond the natural "lag time" for translation of scientific findings to clinical practice. What are the possible reasons for the low clinical utilization for pharmacogenetic tests?

Table 2
Selected Examples of Genetic Polymorphisms Affecting Selection of Drug Therapy

	Medications	Possible drug effects
Beta-1 adrenergic receptor	Beta-1 antagonist (e.g., metoprolol)	Decreased blood pressure and cardiovascular response to beta-1 blockers
Beta-2 adrenergic receptor	Beta-2 agonists (e.g., albuterol)	Decreased bronchodilation effect in asthma
Serotonin transporter	Serotonin reuptake inhibitors (antidepressants)	Decreased antidepressant response
Human major histocompatibility complex (HLA)	Antiretroviral agent, abacavir	Increased risk of hypersensitivity reaction
Prothrombin and factor V	Oral contraceptives	Increased risk of venous thromboembolism (stroke or deep vein thrombosis)

3.1. Inadequacy of Evidence Relevant to the Clinician

3.1.1. Lack of Studies with Clinical Outcomes as Endpoints

With few exceptions, there is a lack of prospective hypothesis-driven clinical studies that demonstrate pharmacogenetic-based individualizations of drug therapy leading to improved clinical outcomes. For example, most pharmacogenetic studies with drug-metabolizing enzymes link genotypic findings to drug concentrations or exposure. Studies with clinical outcomes as endpoints are much more difficult to conduct because nongenetic confounders, such as compliance, health habits (smoking, diet), concomitant drugs, and disease states, are difficult to control. In addition, responses to most drugs are likely to be multigenic in nature. A well-controlled trial with definitive clinical endpoints will require a large number of study subjects and will be very expensive.

In the case of drug-metabolizing enzymes, differences in drug exposure because of metabolizing enzyme genetic polymorphisms do not necessarily translate into differences in therapeutic efficacy or adverse effects. For example, metoprolol, a beta-1 antagonist, is metabolized by CYP 2D6. Individuals who are poor metabolizers of CYP 2D6 have increased exposure to the drug. However, even when CYP 2D6 genotype correlates with variations in pharmacokinetics, these genetic and pharmacokinetic variations are not associated with efficacy or adverse effects because of adrenergic blockade *(19)*. Hence, not all genetic variations associated with a surrogate phenotype will be associated with differences in clinically meaningful outcomes.

In addition to the lack of studies with clinically important outcomes, cost–benefit analyses of pharmacogenetic-based individualizations of pharmacotherapy will also add to the evidence supporting their use.

3.1.2. Lack of Information Regarding Pharmacogenetic-Based Dosing Algorithms

Dosage modifications of 6-mercaptopurin based on TPMT genotypic information serves as an important example of how genotypic information can facilitate optimal dosing of therapeutic agents *(2)*. In this example, patients who are homozygous for the allele resulting in reduced TPMT activity will receive 5–10% of the standard dose of 6-mercaptopurin for ALL maintenance therapy *(2,20)*. However, for many other drugs, information on dosing algorithms that facilitate appropriate and precise pharmacogenetic-based dosage adjustments recommendations does not exist. Most of the studies lack specific explanation on how to translate study findings for use in clinical situations. Specific and straightforward recommendations on how to individualize therapy based on the results of pharmacogenetic tests are critical if current pharmacogenetic knowledge is to be applied in the clinic to assist in rational drug choices and improve the benefit–risk ratio of drug therapy.

3.1.3. Inadequate Pharmacogenetic-Based Prescribing Information

If pharmacogenetics were to be commonly used in clinical therapeutic decision-making, they should be available in prescribing information widely accessible to health-care providers. A recent analysis of drug package inserts (prescribing information) revealed that only 35% of package inserts contained pharmacogenetic data *(21)*. Of these, only about 10% of those contain adequate information to guide therapeutic decisions. Importantly, pharmacogenetic-based dosing recommendations were only available for two drugs. This deficiency in useful prescribing information may reflect the relatively recent widespread interest in the field of pharmacogenetics in the effort to capitalize the human genome. In addition, only recently have pharmacogenetic assays with clinically relevant turnaround time become available. Nonetheless, the lack of clinician-friendly pharmacogenetic information in the labeling of medications will hamper the clinical applicability of pharmacogenetic information.

Clinician-friendly, straightforward, and actionable recommendations can also prevent the misinterpretation of pharmacogenetic information. For example, consider the antidepressant Venlafaxine. Venlafaxine is metabolized by CYP

2D6 to equipotent active metabolite. Given the equipotent nature of the active metabolite, CYP 2D6 genotypic differences do not change the exposure to the active drug moiety because both the parent drug and the metabolite are equally potent. Regular practicing clinicians are likely not aware of a drug's specific pharmacokinetic and metabolite characteristics and will only be able to properly adjust drug dosages if a straightforward recommendation is given for a specific drug.

In addition, because of the lack of molecular genetic training in health-care providers, even when pharmacogenetic information is available, genetic information may be misunderstood even by health-care practitioners. In one study, more than 31% of physicians were reported to misinterpret genetic results *(22)*. This could be because of the lack of systemic education on pharmacogenetics of health-care professionals currently in practice. The re-education of the medical community on these pharmacogenetic relationships to drug response may prove to be challenging. As the complexity of genetic information grew because of a greater understanding of the nature of polygenic drug response, matching a patient's genotype to a specific therapy or dose may be beyond the capability of individual clinicians. Advancement of existing information technology in health-care delivery will be essential for pharmacogenetics to be utilized in clinical practice *(23)*.

3.2. Genotypic Vs Phenotypic Identification

Sometimes, knowledge of a patient's phenotype may render it unnecessary to identify the exact genotype, at least in the clinic. For example, in the above TPMT scenario, one could test for TPMT activity instead of TPMT genotype. The availability of other validated biomarker assays, combined with the lack of third-party reimbursement of genetic tests *(24)*, and the relatively high current cost of these genetic tests may hinder the translation of genetically guided information into clinical practice.

3.3. Polygenic Traits

For the most part, the pharmacogenetic information currently available (such as those in Tables 1 and 2) represents examples of monogenetic determinant of drug response. Many of these polymorphisms are also highly penetrant, with clearly identifiable phenotypes, such as drug concentrations. However, drug treatment outcomes represent a complex phenotype, possibly encoded by dozens to hundreds of genes, influenced by numerous environmental factors, and the interaction between these various gene and environmental variables *(25)*. The term "pharmacogenomics" illustrates these polygenic determinants of drug response. Detailed knowledge to associate specific genetic sequence variants to drug outcome is often not available. Therefore, it will be crucial to screen genotypic markers across the whole genome systematically and combine them with extensive phenotypic characterization of therapy response. Such studies have yet to be performed.

It has only very recently become technologically possible to perform such whole genome association studies. Although genome-wide association scans offer the potential to discover genetic variants governing drug response, several technological and methodological problems exist. First, in these studies, large samples are necessary to find markers with definite effects on drug response while controlling for false positives. Secondly, currently even with the least expensive genotyping technology and strategy, genome-wide studies are still extremely expensive. The Nation Institutes of Health has issued a request for proposal to encourage the development of appropriate methodologies.

3.4. Standardization and Availability of Genetic Tests with Clinically Relevant Turnaround Time

For pharmacogenetics to be successfully used in the clinic, genetic tests need to be available, with clinically relevant turnaround time. Currently, only few FDA-approved paramagnetic tests are available. Examples of these include BCR-ABL genetic test for imatinib, AmpliChip CYP450® for 2D6 and 2C19 metabolizing enzyme, and TruGene® for testing HIV drug resistance.

Most genetic tests bypass the FDA because they are categorized as services, which are not regulated by the FDA. Laboratories at major academic centers often offer genetic tests that have been developed "in-house." Many of these laboratory-based techniques may be too time intensive for routine clinical use. In addition, although these academic clinical laboratories are Clinical Laboratory Improvement Amendments (CLIA) certified and their "analyte-specific reagents" used in these assays are regulated by the FDA, different laboratories may utilize different test systems such that results may not be generalizable across various platforms *(26)*. In addition, laboratory reporting may also need to be standardized. Andersson et al. *(27)* reported a study evaluating factor V Leiden genetic testing and the clinical utility of different report formats to physicians when interpreting results of the genetic test. It was found that there was a considerable degree of variability in the contents of these reports, with some lacking information deemed critical by the standard of professional guidelines and recommendations.

4. Conclusions

Pharmacogenetics may facilitate rational therapeutics both in dosage adjustments and in initial drug selection for treatment. However, pharmacogenetic-based individualized therapy is yet to be widely translated into clinical practice. Hurdles for translating pharmacogenetics from the laboratory to the clinic include: lack of clinical studies with clinically relevant outcomes, few widely available standardized genetic assays, lack of information on polygenic drug response, and finally, the lack of straightforward, clinician-friendly pharmaco-genetic information at the point of prescribing. The above problems will serve as future investigative directions of translating genetic information into clinical practice. Pharmacogenetics and pharmacogenomics have great potential to facilitate improved therapeutics.

References

1. Beutler, E. (1969) Moderne Probleme der Humangenetik. *Ergebn. Inn. Med. Kinderheilk.* **12,** 52–125.
2. Relling, M. V., Hancock, M. L., Rivera, G. K., et al. (1999) Mercaptopurine therapy intolerance and heterozygosity at the thiopurine S-methyltransferase gene locus. *J. Natl Cancer Inst.* **91**(23)**,** 2001–2008.
3. Veenstra, D. L., Blough, D. K., Higashi, M. K., et al. (2005) CYP2C9 haplotype structure in European American warfarin patients and association with clinical outcomes. *Clin. Pharmacol. Ther.* **77**(5)**,** 353–364.
4. Clinical Pharmacology Subcommittee of the Advisory Committee for Pharmaceutical Science, Food and Drug Administration. November 14, 2005 (http://www.fda.gov/ohrms/dockets/ac/05/slides/2005-4194S1_Slide-Index.htm).
5. Rieder, M. J., Reiner, A. P., Gage, B. F., et al. (2005) Effect of VKORC1 haplotypes on transcriptional regulation and warfarin dose. *N. Engl. J. Med.* **352**(22)**,** 2285–2293.
6. Sconce, E. A., Khan, T. I., Wynne, H. A., et al. (2005) The impact of CYP2C9 and VKORC1 genetic polymorphism and patient characteristics upon warfarin dose requirements: proposal for a new dosing regimen. *Blood* **106**(7)**,** 2329–2333.
7. Higashi, M. K., Veenstra, D. L., Kondo, L. M., et al. (2002) Association between CYP2C9 genetic variants and anticoagulation-related outcomes during warfarin therapy. *JAMA* **287**(13)**,** 1690–1698.
8. Camptosar Prescribing Information, Pfizer.
9. Johnson, J. A., Zineh, I., Puckett, B. J., McGorray, S. P., Yarandi, H. N., and Pauly, D. F. (2003) Beta 1-adrenergic receptor polymorphisms and antihypertensive response to metoprolol. *Clin. Pharmacol. Ther.* **74**(1)**,** 44–52.
10. De Laurentiis, M., Cancello, G., Zinno, L., et al. (2005) Targeting HER2 as a therapeutic strategy for breast cancer: a paradigmatic shift of drug development in oncology. *Ann. Oncol.* **16**(Suppl 4)**,** iv7–iv13.
11. Gorre, M. E., Mohammed, M., Ellwood, K., et al. (2001) Clinical resistance to STI-571 cancer therapy caused by BCR-ABL gene mutation or amplification. *Science* **293**(5531)**,** 876–880.
12. Lynch, T. J., Bell, D. W., Sordella, R., et al. (2004) Activating mutations in the epidermal growth factor receptor underlying responsiveness of non-small-cell lung cancer to gefitinib. *N. Engl. J. Med.* **350**(21)**,** 2129–2139.
13. Bossi, P., Peytavin, G., Ait-Mohand, H., et al. (2004) GENOPHAR: a randomized study of plasma drug measurements in association with genotypic resistance testing and expert advice to optimize therapy in patients failing antiretroviral therapy. *HIV Med.* **5**(5)**,** 352–359.

14. Quirk, E., McLeod, H., and Powderly, W. (2004) The pharmacogenetics of anti-retroviral therapy: a review of studies to date. *Clin. Infect. Dis.* **39**(1), 98–106.

15. US Department of Health and Human Services Panel on Clinical Practices for Treatment of HIV Infection A. (2005) Guidelines for the use of antiretroviral agents in HIV-1-infected adults and adolescents. October 6, 2005 (http://aidsinfo.nih.gov).

16. Hirsch, M. S., Brun-Vezinet, F., D'Aquila, R. T., et al. (2000) Antiretroviral drug resistance testing in adult HIV-1 infection: recommendations of an International AIDS Society-USA Panel. *JAMA* **283**(18), 2417–2426.

17. Goetz, M. P., Rae, J. M., Suman, V. J., et al. (2005) Pharmacogenetics of tamoxifen biotransformation is associated with clinical outcomes of efficacy and hot flashes. *J. Clin. Oncol.* **23**(36), 9312–9318.

18. Gardiner, S. J. and Begg, E. J. (2005) Pharmacogenetic testing for drug metabolizing enzymes: is it happening in practice? *Pharmacogenet. Genomics* **15**(5), 365–369.

19. Zineh, I., Beitelshees, A. L., Gaedigk, A., et al. (2004) Pharmacokinetics and CYP2D6 genotypes do not predict metoprolol adverse events or efficacy in hypertension. *Clin. Pharmacol. Ther.* **76**(6), 536–544.

20. Evans, W. E., Hon, Y. Y., Bomgaars, L., et al. (2001) Preponderance of thiopurine S-methyltransferase deficiency and heterozygosity among patients intolerant to mercaptopurine or azathioprine. *J. Clin. Oncol.* **19**(8), 2293–2301.

21. Zineh, I., Gerhard, T., Aquilante, C. L., Beitelshees, A. L., Beasley, B. N., and Hartzema, A. G. (2004) Availability of pharmacogenomics-based prescribing information in drug package inserts for currently approved drugs. *Pharmacogenomics J.* **4**(6), 354–358.

22. Giardiello, F. M., Brensinger, J. D., Petersen, G. M., et al. (1997) The Use and Interpretation of Commercial APC Gene Testing for Familial Adenomatous Polyposis. *N. Engl. J. Med.* **336**(12), 823–827.

23. Ginsburg, G. S., Konstance, R. P., Allsbrook, J. S., Schulman, K. A. (2005) Implications of pharmacogenomics for drug development and clinical practice. *Arch. Intern. Med.* **165**(20), 2331–2336.

24. Logue, L. J. (2003) Genetic testing coverage and reimbursement: a provider's dilemma. *Clin. Leadersh. Manag. Rev.* **17**(6), 346–350.

25. Evans, W. E. and McLeod, H. L. (2003) Pharmacogenomics—drug disposition, drug targets, and side effects. *N. Engl. J. Med.* **348**(6), 538–549.

26. US Food and Drug Administration. (2003) Guidance for Industry: Pharmacogenomic Data Submissions 2003 (http://www.fda.gov/cber/gdlns/pharmdtasub.pdf).

27. Andersson, H. C., Krousel-Wood, M. A., Jackson, K. E., Rice, J., and Lubin, I. M. (2002) Medical genetic test reporting for cystic fibrosis (deltaF508) and factor V Leiden in North American laboratories. *Genet. Med.* **4**(5), 324–327.

13

Molecular Modeling: Considerations for the Design of Pharmaceuticals and Biopharmaceuticals

Philip D. Mosier and Glen E. Kellogg

Abstract

In this chapter, we provide the reader with a broad overview of the tools and techniques commonly used in the fields of molecular modeling and computer-assisted drug design (CADD) and at the same time present context with respect to the discovery, design, and development of therapeutic agents. Note is also made to some of the challenges that lie at the cutting edge of these disciplines. We highlight a number of techniques and related software programs as they apply to the development of therapeutic pharmaceutical and biopharmaceutical agents.

Key Words: Molecular modeling; computer-aided drug design; ligand-based drug design; structure-based drug design; QSAR; virtual screening; *de novo* design; docking; scoring; homology modeling.

1. Introduction

Molecular modeling and computer-aided drug design (CADD) have had a significant impact on the understanding of biological structure and function over the last two decades. The ability to "visualize" a model has been a tremendous boon to medicinal chemistry and in many other areas of chemistry, particularly towards the development of new therapeutic agents. Companies and academic research groups involved in the design and development of pharmaceuticals now routinely employ computational methods including molecular modeling. Whereas *bio*pharmaceutical design and development have perhaps not benefited from computational techniques as much as the more traditional, chemically derived therapeutic moieties, the potential for this technology to have a significant impact is still quite large.

From: *Biopharmaceutical Drug Design and Development*
Edited by: S. Wu-Pong and Y. Rojanasakul © Humana Press Inc., Totowa, NJ

It is our intent to provide the reader of this chapter with a broad overview of the tools and techniques commonly used in the field of molecular modeling and CADD and at the same time present some important historical background and also some of the challenges that lie at the cutting edge of these disciplines as they apply to the development of pharmaceutical and biopharmaceutical agents. We will highlight a number of techniques and related software programs, but certainly cannot be exhaustive in this brief chapter. It is hoped that the interested reader will follow up by perusing some of the references we provide.

Let us first define some terms that are commonly used to describe the methods outlined in this chapter. Molecular modeling is the process by which representations of chemical entities are constructed (usually in a computer) in order to examine their structure, to observe their behavior, and/or to determine their properties. Three other terms, CADD, cheminformatics, and bioinformatics, are also frequently used to collectively describe the various tasks computational chemists employ during the course of the drug design and development process. Despite the bevy of names used, there is much overlap in the methods these terms describe. CADD is a more general term that refers collectively to the computational processes and methods used to generate potential pharmaceutically relevant molecules (lead generation) and to optimize the properties of known pharmaceutical agents (lead optimization). Cheminformatics encompasses CADD and is generally used to refer to the storage, retrieval, and processing of chemical information and is not restricted to the realm of biopharmaceuticals and other drug compounds. In contrast, bioinformatics refers to the storage, retrieval, and processing of biological information like the kind that would be found, for example, in genomic databases. In this brief chapter, we will focus primarily on molecular modeling and CADD.

Two parallel tracks have traditionally been taken regarding the development of computational methods related to the design and development of biopharmaceuticals; these have recently converged into the discipline known as CADD. On one of these tracks, medicinal chemists and pharmacologists have developed computational tools to facilitate the design and development of small molecule pharmaceutical agents based on traditional medicinal chemistry structure–activity principles. These are the methods that comprise *ligand-based design*. In a similar fashion, biochemists and biophysicists have developed tools to aid in the elucidation or characterization of the biological macromolecules that serve as targets for small molecule and biopharmaceutical agents. These are the methods that comprise *structure-based design*. Each of these two paradigms has advantages and disadvantages. The advantage of ligand-based design is that meaningful results may be derived in the absence of structural or other information regarding the drug target. The principal disadvantage of ligand-based design is that the drug target is not known. In the absence of this information, guesses must be made about the nature of the drug binding site based on the similarity of the ligands known to bind to a particular target. The advantage of structure-based design is that this structural information is known and can be used to design high-affinity biopharmaceutical agents that have optimal complementarity to the binding site. The disadvantage of structure-based design is that experimentally derived structural information regarding the target is difficult and time-consuming to obtain: the target (typically a protein) must be cloned, expressed, and purified in sufficient quantity to allow its determination by X-ray diffraction crystallography, nuclear magnetic resonance (NMR) spectroscopy, or electron diffraction methods. Methods have been developed (*vide infra*) to model a particular macromolecular drug target based on its similarity to other related targets (i.e., homology modeling), but these are not as reliable as direct structure determination because assumptions regarding the conformation of the desired target must be made. In the following sections, we will describe some important molecular visualization methods, followed by discussions of the major methods and tools used in both ligand- and structure-based design.

2. Molecular Visualization and Representation

Although it seems obvious, it is worth reiterating that almost certainly the most valuable result of molecular modeling is the ability to actually see and explore, in three dimensions, the molecule's size, shape, and structure. In addition, a number of fundamental tools that facilitate the accurate representation of molecules in a computer will be described.

2.1. Visualization

The ability to visualize and interact with molecular models is one of the most useful and fundamental features of CADD applications. Computers have been used to display molecular graphics even in the earliest days of computational chemistry. As early as 1966, Cyrus Levinthal at MIT had been using a primitive visualization system (nicknamed "Kluge") comprised of an oscilloscope fitted to an early time-sharing computer to interactively display and model the folding mechanisms of proteins *(1)*. This initial work inspired many others to develop new software to take advantage of the advances being made in the field of computer science, and the fledgling discipline known as interactive molecular graphics was off to a running start. Early adopters of this new technology included Garland Marshall *(2)*, who wrote molecular modeling software designed to explore conformational analysis that would eventually become the heart of SYBYL, the flagship application of Tripos, Inc., a company which he founded in 1979. At the same time, Todd Wipke recognized the need for an efficient chemical storage and retrieval system and developed the MACCS (Molecular

ACCess System) database software and who cofounded the software company Molecular Design Limited (MDL), now owned by Elsevier.

As computer graphics hardware became more capable (and at the same time less costly) the ways in which molecules could be displayed also became more sophisticated. An important step in the representation of higher-order protein structure was the ribbon plots that were hand-drawn by Jane Richardson *(3)*. These were later adapted to computer display systems by Carson and Bugg *(4)*. A particularly useful way to render a molecule in general is to display its solvent-accessible surface. This is because it not only depicts the molecule's bulk size and shape as "seen" by neighboring molecules, but also allows properties such as electrostatic potential or hydrophobicity to be mapped onto the rendered surface, allowing the user to, for example, visualize and identify regions of the molecule well suited to certain types of inter- and intramolecular interactions or to assess the chemical reactivity of a molecule. The methods developed by Michael Connolly *(5)* are now widely used for this purpose. Figure 1 shows six different common rendering schemes.

New and useful ways of displaying molecules and their associated properties are routinely being developed by research groups such as those of Brickmann *(6)* at Darmstadt Technical University who developed the MOLCAD software package. The release of the application programming interface (API) known as OpenGL *(7)* by Silicon Graphics, Inc. (SGI), and its adoption as a graphical display standard by companies producing powerful and low-cost consumer graphics cards (fueled primarily by the video game industry) have made it possible for virtually every personal computer made today to display sophisticated graphics in an interactive way. Most of the CADD software packages in use today are designed to run on UNIX machines (and SGI in particular). However, the computer hardware and software industries are very volatile and the recent trend, which can be expected to become dominant in the next few years, is for migration of cutting-edge software to more commodity platforms and operating systems such as Intel/Linux and Macintosh/OS X. Four of the more commonly utilized commercially available general molecular modeling software packages which boast advanced graphics capabilities are SYBYL (Tripos, Inc., St Louis, MO), Cerius2 and InsightII (Accelrys, Inc., San Diego, CA), and MOE (Chemical Computing Group, Inc., Montreal, Quebec, Canada). Some, if not all, of these programs are available or are in the process of being ported to PCs and other kinds of computers running the Linux operating system. Others, too numerous to review here, but which have somewhat diminished feature sets, are designed to run under Microsoft Windows-based personal computers.

2.2. Representation of Molecular Structure

Before any meaningful molecular modeling may be accomplished, the representation of the molecules in the computer must themselves be valid.

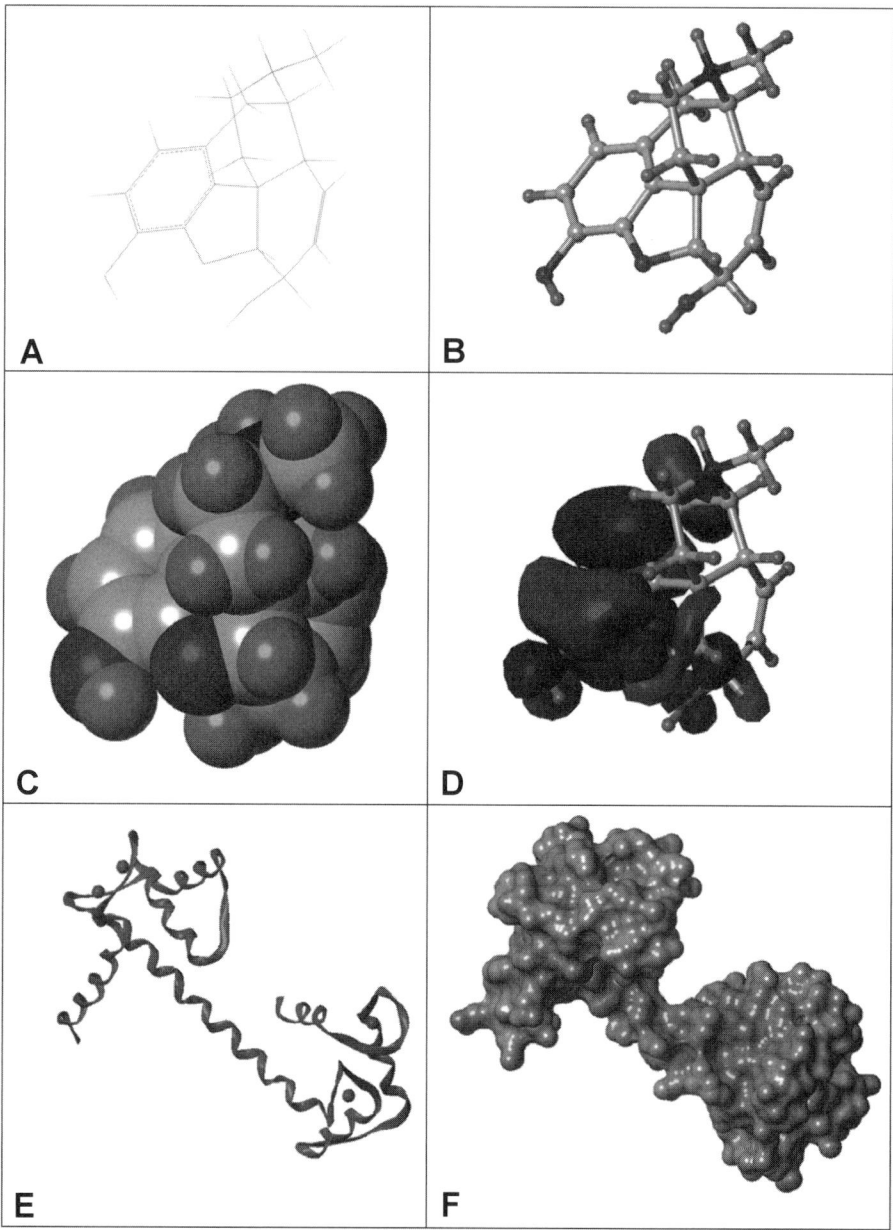

Fig. 1. Six different ways to display (render) a molecule: (**A**) morphine: wire frame; (**B**) morphine: ball-and-stick; (**C**) morphine: Corey-Pauling-Koltun (CPK) space-filling; (**D**) morphine: highest occupied molecular orbital (HOMO) determined using a semi-empirical QM routine; (**E**) calmodulin: Richardson-style ribbon (calcium ions rendered as CPK); (**F**) calmodulin: Connolly solvent-accessible surface.

The degree of accuracy to which structural and chemical characteristics can be reliably represented depends on the way in which the molecules are processed. Typically, molecules are stored in electronic form in structure files. Common file formats that are highly portable, i.e., transferable from one modeling program or one computer to another, are MACCS (.sdf), SYBYL (.mol2), and Brookhaven (.pdb). The pdb format is most useful for biomacromolecules and the sdf format is most useful for small molecules, whereas the mol2 format can be effectively used with either. The information contained in the structure files includes the positions of the atomic nuclei (given as X, Y, and Z coordinates) as well as a connection table specifying the atoms that are connected and the bond types connecting them. The differences between the file formats have to do with the information they are capable of storing; for example, the sdf format cannot code substructure (residue level) information and is thus not useful to represent proteins, whereas the pdb format in its native form cannot represent bond order and/or atom hybridization.

As the 3D coordinates may in general take on any value assigned to them, they must be assigned meaningful values. Depending on the level of accuracy desired and the amount of available time and computational power, varying degrees of accuracy may be achieved using different computational methods. Rule-based "expert" systems are the least computationally demanding and are used to assign plausible 3D conformations to molecules based on their 2D structure (i.e., atom types and connection table only). Such methods include CONCORD *(8)* and CORINA *(9)*. The advantage of using these methods is speed: thousands of compounds may be processed per minute with current technology, making these methods ideally suited to the processing of large databases prior to virtual screening. Alternatively, molecular mechanics (MM) routines to calculate 3D structure from atom coordinates and connection tables are more computationally expensive (though quite facile with the currently available computer hardware).

MM methods are based on Newtonian physics ("balls attached to springs") and employ force fields coupled with multivariate minimization algorithms to calculate an enthalpy-related energy for a particular molecular conformation. Equation 1 shows the form of a typical MM force field. Individual terms in the equation account for the contribution to the total energy arising from the non-ideality of bond lengths and bond angles, as well as from van der Waals and electrostatic contributions.

$$E = \frac{1}{2} \sum_{bonds} k_b (b - b_0)^2 + \frac{1}{2} \sum_{angles} k_\theta (\theta - \theta_0)^2 +$$

$$\frac{1}{2} \sum_{\substack{dihedral \\ angles}} k_\phi [1 + \cos(n\phi - \delta)] + \sum_{\substack{nonbonded \\ pairs}} \left[\frac{A}{r^{12}} - \frac{C}{r^6} + \frac{q_1 q_2}{Dr} \right] \tag{1}$$

Popular MM programs include Allinger's MM series *(10–12)*, the Merck Molecular Force Field *(13)*, and the Tripos Force Field *(14)*, which are well suited to the conformational optimization of small molecules. Force fields optimized for macromolecules have also been developed and include the programs AMBER *(15)* and CHARMM *(16)*.

MM routines can be very effectively used when a fast but accurate assignment of molecular conformation is necessary. If other molecular properties such as atomic charges and polarizabilities are needed, then more sophisticated (and computationally more expensive) quantum mechanical (QM) methods are required. The most rigorous of the QM methods are known as *ab initio* methods, because the calculated molecular properties are based on "first principles" and the solution of the Schrödinger equation in particular. These methods determine the electronic occupancy of molecular orbitals. *Ab initio* software packages in common use include Gaussian (Gaussian, Inc., Wallingford, CT), Spartan (Wavefunction, Inc., Irvine, CA), and GAMESS (Gordon Research Group, Ames Laboratory/Iowa State University, Ames, IA). In order to improve the performance of these methods, some approximations regarding the way atomic orbitals are represented have been developed which significantly decrease the CPU time needed, resulting in what are known as *semiempirical* molecular orbital methods. Commonly employed software packages include MOPAC (Quantum Chemistry Program Exchange (QCPE), Indiana University, Bloomington, IN), AMPAC (SemiChem, Inc., Kansas City, MO), and Spartan (Wavefunction, Inc., Irvine, CA), which provide interfaces to QM routines for small molecules. QM methods have traditionally been limited to the optimization of small molecules, but have also more recently been applied (through the use of extremely large and highly parallelized computer systems and highly optimized computer code) to macromolecular systems *(17)*.

2.3. Conformational Analysis

The MM and QM energy minimization methods described in the last section will likely find a relatively low-energy (and thus probable) conformation for a molecule from an arbitrary starting set of coordinates. This relatively low-energy value is known as a *local minimum*. However, if the initial coordinates correspond to a high-energy state, then the resulting local minimum sometimes corresponds to a conformation that is also relatively high in energy. One way to find the *global minimum*, or absolute lowest-energy conformation, is to perform a *systematic search*. In the systematic search, rotatable bonds are systematically adjusted in an incremental fashion so that the *conformational space* defined by the rotatable bonds is completely sampled. An energy value is calculated for each conformation and the conformation that results in the lowest energy is selected as the global minimum. Alternatively, a *grid search* may be performed, in which a geometry optimization is run at each incremental conformation change. Systematic and grid searches are generally too computationally demanding to be used on larger molecules, and for these molecules, a *random search* may be used in which the rotatable bonds are randomly adjusted so that the molecule's conformational space is randomly sampled.

Another technique used to avoid local minima and to guide the molecule into a lower-energy state is molecular dynamics (MD). MD simulations combine a MM force field with the classical equations of motion to simulate the actual movement of the atoms in the molecule. The amount of atomic motion experienced by the atoms is usually specified by the user in terms of an absolute temperature. MD is useful, for example, when one wishes to observe the behavior of a molecule over an extended period of time to find new local minima or when vibrational spectra are desired, as this information may also be derived from an MD "run". MD simulation is used quite extensively to model the behavior of proteins. However, at the present time, these simulations rarely exceed what represent a few actual nanoseconds of molecular movement. The *Lecture Notes in Computational Science and Engineering* series *(18)* includes reviews of novel MD methodologies used to address the simulation of macromolecules.

3. Ligand-Based Drug Design

Ligand-based design comprises those methods used to find and/or optimize lead drug compound in the absence of knowledge about the drug target. As such, ligand-based design methods must *infer* the structure of the target from information gleaned from the ligands that are known to have affinity for a given target.

3.1. Quantitative Structure–Activity Relationships

One of the earliest and most widely used modeling techniques is the Quantitative Structure–Activity Relationship or QSAR. The fundamental axiom in the development of QSAR models is that the observed biological activities of molecules can be described as a function of the molecules' structure. This paradigm is not limited to the prediction of biological activities. Other variants of the QSxR paradigm include Quantitative Structure–Property Relationships (QSPRs) to predict physical properties and Quantitative Structure–Toxicity Relationships (QSTRs) to predict chemical toxicity.

One of the first successful attempts to correlate molecular structure with a chemical property was achieved by Hammett *(19)*, who linked the electron withdrawing or donating character of a set of benzoate substituents with both the acid dissociation constant of substituted benzoic acids and the rate of hydrolysis for a similarly substituted set of ethyl benzoates. This type of QSPR is known as a *linear free energy relationship* (LFER) because a linear relationship exists between the Hammett sigma values (in this case) and properties related to Gibb's free energy. The well-known Hammett sigma parameters were derived from that QSPR model and are fundamental in understanding the relative chemical reactivity of similar compounds. In the mid-1950s, Taft *(20)* introduced the E_s parameter to take into account the effect that steric bulk had on a set of α-substituted acetates. Later, in what developed into the Pomona Medicinal Chemistry Project, Hansch and Fujita *(21,22)* built on Hammett's idea by developing a QSPR model that resulted in the assignment of a substituent-based lipophilicity parameter π to molecular fragments that differentially modified the octanol–water partition coefficient.

The measurement of lipophilicity is essential in understanding the pharmacokinetic and pharmacodynamic properties of drug molecules. The pioneering work of Hansch, Fujita, and many others has been incorporated into the popular CLOGP/BioLoom programs (Biobyte, Inc., Claremont, CA) used to accurately calculate log P values. Along the way, many other similar parameters have been developed for which we shall use the general term "S". Today, *Hansch analysis* or the *extrathermodynamic method* refers to any QSxR that conforms to equation 2, in which multiple linear regression (MLR) is used to correlate the parameters discussed above with a biological activity by determining the adjustable parameters *a* through *f.*

$$\log(1/C) = -a\pi^2 + b\pi + c\sigma + dE_s + eS + f \tag{2}$$

The Hansch analysis has been shown to work well for small sets of homologous molecules. If the data set consists of a more diverse or noncongeneric group of molecules, then a more generalized version of the extrathermodynamic Hansch approach may be used to correlate the desired property or biological activity with the topological, geometric, and electronic properties of the data set. This is the approach taken by QSxR modeling packages such as ADAPT *(23,24)* and MDL QSAR (Elsevier MDL, Inc.).

The general methodology used to develop the "2D" QSxR models described above consists of three distinct steps. First, a set of molecules for which the desired biological activity or physical property is known is selected and assigned appropriate 3D conformations. This set of compounds is subdivided into a *training set* that is used to generate the model and a *prediction set* that is used to validate the model and ensure that the model is able to generalize or accurately predict the properties of molecules not used in the model-building process. Next, a set of molecular *descriptors* is calculated for each of the selected compounds. Many hundreds of descriptors have been devised *(25,26)* to encode various aspects of molecular structure.

In general, any descriptor may be classified as belonging to one of four major classes: topological, geometric, electronic, or hybrid. Topological descriptors are derived from the 2D connection table of a molecule and are graph-theoretical in nature. Examples of topological descriptors are the Randic connectivity indices *(27)*, the Kier-Hall-Murray χ branching indices *(28–34)*, and the electrotopological state indices *(35)*. Geometric descriptors encode the overall bulk size and shape of a molecule. Geometric descriptors include moments of inertia *(36)* and solvent-accessible surface areas *(37)*. Electronic descriptors convey information about the distribution of electrons in the molecule. Examples of electronic descriptors include polarizability *(38)*, dipole moment, the highest occupied and lowest unoccupied molecular orbital energies derived from QM methods. Hybrid descriptors combine elements of one or more of the three basic descriptor types. Examples of hybrid descriptors include charged partial surface areas *(39)* and BCUT descriptors *(40)*.

Typically, many more descriptors will be generated than are relevant to the property being studied. Thus, the third step in developing a QSxR model is to select the most pertinent descriptors from the large pool of those calculated, a process sometimes referred to as *feature selection*. Feature selection is most commonly performed using MLR or neural networks combined with a stochastic optimization method like a genetic algorithm or simulated annealing. MLR has the advantage of being statistically interpretable. The effect of each descriptor in the model can be easily determined and statistical significance of each one calculated individually. However, MLR is restricted in that it does not allow for nonlinear contributions from the independent variables. Neural networks (*see* Fig. 2), on the other hand, are quite adept at handling nonlinear phenomena. The disadvantage of neural networks is that they are "black boxes" in that the contribution and statistical significance of the individual descriptors are not always evident.

Finally, after a QSxR model has been built, it must be validated. That is, the model must be shown to accurately predict the properties of molecules that were not used in the development of the model. The predictions for the compounds that were left out of the model-building process (i.e., the prediction set) are computed at this point and the errors of prediction for both the training and prediction sets give an overall measure of the quality of the model. Scatter plots of the calculated vs observed values for the dependent variable are examined for outliers, and residual plots are examined to ensure that the error of the model is distributed uniformly. Other validation techniques may also be used, such as the leave-*n*-out cross-validation procedure.

The methodology used to develop "3D" QSxR models is much the same as for 2D models, with some important differences. The theory behind 3D QSxR is that an "impression" of the receptor site may be inferred by aligning common features of the molecules in the data set. The most common 3D QSxR methods are the grid-based CoMFA *(41)* and CoMSIA *(42,43)*; other methods include the distance geometry approach of Crippen and coworkers *(44–46)* and Hopfinger's molecular shape analysis *(47–49)*. The first step in a 3D QSxR study is to align

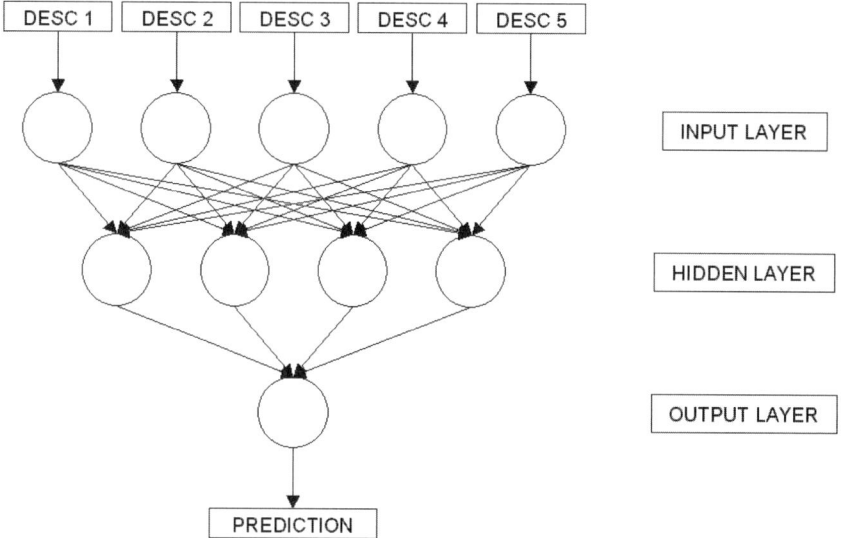

Fig. 2. A generalized schematic of a neural network. Each circle represents an individual neuron in the net. Each neuron accepts data from one or more inputs and outputs a value based on a nonlinear transformation of the input values. In the most common type of neural net, the information is processed in three steps. Neurons in the input layer accept descriptor values and send their transformed output to the hidden layer, which subsequently provides data for the output layer. The node(s) of the output layer then produce predicted values for the property of interest.

the molecules based on shared features such as a common core that serves as a scaffold for substituent groups and other pharmacophoric elements, e.g., hydrogen bond donating and accepting sites and hydrophobic regions. If an actual 3D pharmacophore is known, then this may be used to guide the alignment of the molecules. Tools such as GALAHAD (Tripos, Inc., St Louis, MO), Catalyst (Accelrys, Inc., San Diego, CA), FlexX (BioSolveIT, Inc., Sankt Augustin, Germany) and field-fit methods can simplify the alignment process. The set of aligned molecules is then suspended in and is surrounded by a regularly spaced grid in 3D space (X, Y, and Z directions). Electrostatic, steric, and (optionally) hydrophobic fields are then calculated for each molecule via a "probe" atom located at the intersections of the grid.

The electrostatic, steric, and hydrophobic values at each grid point are the independent variables in a 3D QSxR study. Because there are usually thousands of these "descriptors", partial least squares methods are used to extract the relevant information from the aligned fields. Validation in 3D QSxR methods is usually achieved using a leave-*n*-out method, in which the data set is divided into *m* parts, each part consisting of about *n* molecules. Each of the *m* groups in turn is left out of the model-building process and is assigned a prediction based on the remaining *m*−1 groups of the data set. Commonly, the leave-one-out method, in which each group consists of only one molecule, is used. 3D QSxR can be a powerful tool, particularly when the molecules of the data set are fairly similar and/or fairly rigid and an unambiguous alignment can be realized. If the molecules are flexible and/or too dissimilar, then this approach becomes much more speculative. Doweyko *(50)* has outlined and given examples of cases in which 3D QSAR results can be misleading. An advantage of 3D QSxR is that it lends itself well to interpretation by the lay chemist. Steric, electrostatic, and hydrophobic maps are generated that locate regions of space representing various regions of the molecules where more or less steric bulk, more positive or more negative charge, or more hydrophobic or more hydrophilic character is predicted to result in greater activity values.

QSxR methods have traditionally been shown to be very useful, particularly for the prediction of physical properties such as log *P* (2D QSPR) and in some cases biological activities (3D QSAR). 2D QSAR methods in particular have recently found new applicability in the modeling of ADME/Tox properties such as intestinal permeability, aqueous solubility, oral bioavailability, active transport, P-glycoprotein efflux, blood–brain barrier permeation, plasma protein binding, metabolic stability, interaction with cytochrome P-450 enzymes, and toxicity *(51)*.

3.2. Database Searching

A discussion of ligand-based drug design would not be complete without briefly mentioning the important role that chemical databases play. Modern chemical database systems like UNITY (Tripos, Inc., St Louis, MO) allow the user to search for molecules that have similar physiochemical properties to a *query* molecule, which is very often a promising lead compound or one of its substructures. The query often consists of a set of pertinent structural fragments or groups of fragments arranged in a specific 3D pharmacophoric pattern. A commonly used free and large database of commercially available compounds is ZINC, maintained at the University of California, San Francisco. Databases play a prominent role in structure-based drug design also and will be discussed further below.

3.3. Challenges and Opportunities

Some of the most exciting aspects of QSxR research involve the development of new descriptors and the application of new statistical methods as feature selection routines to chemical problems. Two of the newer feature selection methods that have been studied in this respect are support vector machines *(52)* and probabilistic/general regression neural networks *(53)*. Nevertheless, two issues continue to present challenges in the world of ligand-based drug design and QSxR. First, the interpretability of some (particularly topological) descriptors remains a significant challenge. This becomes an issue in the so-called inverse-QSxR problem, in which values for descriptors are sought that correspond to a desired activity. It might not be evident how a given (sometimes artificially defined) descriptor relates to more intuitive physiochemical properties. However, Kier and Hall *(54)* have provided rationale for understanding the connectivity indices. Second, there are opportunities for further understanding the issue of model applicability. Eriksson et al. *(55)* and Guha and Jurs *(56)* have addressed the question of how much confidence can be placed in a QSxR prediction. It is generally thought that by providing molecules to the QSxR model that are sufficiently similar to those of the training set, the prediction errors may be minimized. The question then becomes, "how similar is similar?" Frederique and Dragos *(57)* and Lipinski *(58)* have reviewed the efforts to understand and apply the similarity principle; Ghose and Viswanadhan *(59)* have reviewed its use in designing combinatorial libraries.

4. Structure-Based Drug Design

The advent of what is termed "structure-based drug design" was probably nearly coincident with the discovery that the 3D structure of biomacromolecules could be determined experimentally. The knowledge that there was a *molecular* basis for many diseases, and that this could potentially be gleaned from the structure of the normal or abnormal biomacromolecule, set into motion an entire scientific discipline (structural biology), numerous specialty pharmaceutical companies, and probably countless research projects in and out of academia. More recently, the US National Institutes of Health (NIH) has identified structural biology and protein structure determination as a key goal of the "NIH Roadmap". In fact, there are now over 44,000 unique entries of biomacromolecular structures in the "Protein Data Bank" (PDB) *(60)*, a publicly available database maintained by the Research Collaboratory for Structural Biology and hosted by Rutgers University.

Many of these entries include a bound ligand or inhibitor at the protein active site. This turns out to be data of enormous value for structure-based drug design, as it indicates first where the active site is, and also that this ligand is a more than plausible starting place for molecular design. One of the strategies of recent discovery programs based on structure-based design is to periodically crystallize and analyze structures of the protein complexed with the current "best" ligand to aid in fine-tuning the ligand's properties with respect to binding. A few of the structures in the PDB are the final products of this research effort—the commercially available drugs bound in their target protein.

In this section we intend to briefly review the process by which a highly active ligand can be designed from the known structure of a target protein or other biomacromolecule, e.g., DNA or RNA.

4.1. Modifying an Existing Substrate in Context

The most basic structure-based design technique is to use traditional medicinal chemistry design strategy linked with the 3D structure of a protein–ligand complex to optimize the binding of the ligand within its site. The key here is that the placement and interactions of the ligand can be easily visualized in three dimensions with molecular modeling software. One can identify the protein–ligand interactions that are energetically favorable for the binding and those that are energetically unfavorable. These latter cases immediately suggest chemical modifications that can be exploited, e.g., replace an amine with a carboxylate or a methyl with an amine. Also important are void spaces where there are no interactions—these represent additional opportunities for improving the ligand's binding efficacy. Figure 3 schematically illustrates this process.

By necessity, designing/optimizing drugs by this process is an iterative affair involving a close collaboration between the computational chemist and the synthetic chemist. Imagining compounds that cannot be made is of no value. Nevertheless, a number of computer programs have been created to totally automate this approach, i.e., the program suggests molecules that ideally fill the binding cavity with optimized properties for binding complementarity. Applying this approach with these programs is often termed "*de novo*" design. Several programs designed for this purpose have been described, e.g., LUDI *(61)* (Accelrys, Inc., San Diego, CA), LEAPFROG *(62,63)* (Tripos, Inc., St. Louis, MO), and AlleGrow (Bohacek, unpublished), but to date none have been more than modestly successful in designing new, useful, and accessible drug leads. The core technologies of these programs are generally based on a somewhat random process of iteratively building and scoring putative ligands; however, molecules thus created may end up being neither thermodynamically reasonable nor synthetically accessible. In an attempt to circumvent this problem, Paul Bartlett and coworkers *(64,65)* have designed 3D libraries of tricyclic hydrocarbons (TRIAD) and acyclic molecules (ILIAD) that are chemically reasonable starting points along with a complementary software package called CAVEAT. To date many success factors have remained unresolved for fully exploiting *de novo* ligand design techniques, but this remains an active area of research.

4.2. Docking and Scoring a New Ligand

The next most difficult computational modeling task is to "dock", i.e., place in the protein model, a ligand that is known to bind, but whose position and/or orientation in the site is unknown. As part of this exercise some metric of goodness of binding—the score—is calculated. The score is used in two ways: first to provide feedback as the ligand docking is optimized and secondly, at

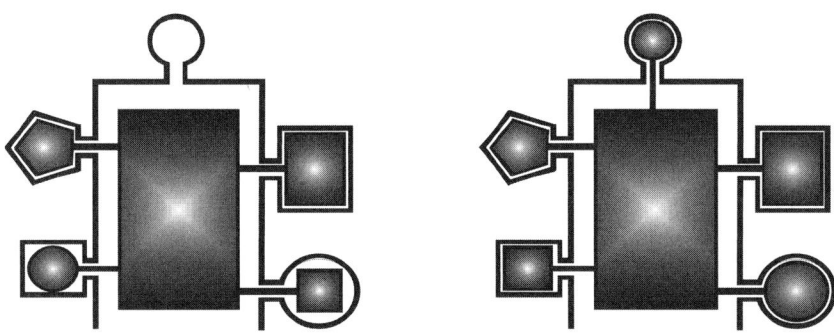

Fig. 3. Schematic representation of the process of optimizing a known ligand in the receptor site. The figure on the left depicts a ligand in the receptor site with optimal ligand–receptor interactions at some (but not all) possible interaction sites. The figure on the right shows the modified ligand with optimized ligand–receptor interactions at all possible interaction sites in the receptor.

completion, as a tool to compare the docking (poses) of multiple instances of the same ligand or similar ligands.

DOCK, the original docking program, was developed by Kuntz and associates at the University of California—San Francisco during the past two decades *(66–68)*. It uses a cavity search algorithm based on the Connolly surface code *(5)* to locate and characterize active sites that are then, in turn, filled with spheres with a range of radii. Atoms of the putative ligands are then fit to the centroids of the spheres to generate poses. Other algorithms and programs have gained popularity in recent years, e.g., the FlexX algorithm of Rarey, Kramer, and Lengauer *(69–71)* that rebuilds the ligand in situ; AutoDock from the laboratory of Olson at The Scripps Research Institute that uses atomic affinity grids and genetic algorithms for docking *(72,73)*; the GOLD (Genetic Optimisation for Ligand Docking) code of Jones et al. *(74,75)* that uses a genetic algorithm to generate flexible poses; and SLIDE (Screening for Ligands by Induced-fit Docking) by Schnecke and Kuhn *(76)*. FRED (Fast Rigid Exhaustive Docking) of McGann et al. *(77)* is a new program from OpenEye Scientific Software.

The sheer number of in-use scoring methods precludes even a brief synopsis of all of them here. The wide array of scoring algorithms ranges from simple contact scores based on (only) the number of atom–atom contacts, to surface area scores, to electrostatic-based scores, to grid-based scores, to MM energy scores, to knowledge-based scoring functions such as Potentials of Mean Force (PMF) *(78)*, and to Piecewise Linear Potential scoring functions *(79)*.

Interestingly, the current, still evolving, trend is for "consensus" scoring, i.e., using a combination of available scoring methods *(80)*. Böhm and Stahl have nicely reviewed the empirical scoring functions *(81)*. These methods use simple sums of easily calculated interaction features, e.g., Coulombic energies, van der Waals energies, number of rotatable bonds, contact surface areas between species, to calculate a score for the ligand–macromolecule interaction. Considerable effort is expended in tuning these functions and the overwhelming benefit is speed. Of the knowledge-based methods, the PMF score approach of Muegge and Martin *(82)* has received the most attention and has been the most successful. It converts extracts of experimental protein–ligand interaction distances and the like from the PDB *(60)* into Helmholtz free energies for ligand–protein atom pairs and is similar to the PMF approach used by Sippl *(83)* in protein folding studies. The OWFEG method *(84)* of Pearlman and Charifson was shown to be superior to a handful of other methods in predicting the binding efficacy for a set of 16 ligands bound to the p38 MAP kinase protein *(85)*, but still inferior to a thermodynamic integration method in predicting free energies of binding for members of the ligand series.

The scoring part of docking and scoring turns out to be the more difficult issue. Much current research is being devoted to being able to predict (quickly) the free energy of binding for a bound ligand. Most of the computational tools available are better suited to calculating the enthalpic contribution of free energy as opposed to the entropic contribution. Thermodynamic integration-type methods do yield free energy predictions of relatively high quality, but are costly. More empirical approaches, such as the HINT model *(86)* developed in our laboratory, have been shown to yield accuracy approaching the level of the costlier methods *(87,88)*. The bases of the HINT empirical model are experimental measurements of Log $P_{o/w}$ (partition coefficient for solute–solvent transfer between 1-octanol and water) and the well-established computational tools to predict this parameter. Because Log $P_{o/w}$ is a thermodynamic quantity that is related to free energy, the HINT model yields binding "scores" that correlate well with measured free energies of binding.

Molecular models derived from docking experiments can be extremely good, i.e., very similar to crystallographic results on the same complexes when applied to retrospective analyses of known experimental structures. However, there are often multiple models that are "scored" very similarly, and the actual model may be difficult to identify without ambiguity if the actual structure is not known. In fact, on some occasions, ligands may appear to bind in two or more very distinct modes with little score difference to differentiate the modes. It must also be noted that the crystallographic result is, itself, a model that has its own set of errors and uncertainties, especially if the crystallographic resolution is poor.

4.3. Database Searches (a.k.a. Virtual Screening)

Perhaps one can say that the next extension of structure-based drug design is to attempt identification of new ligands by "screening" them through rapid docking and scoring experiments. Yvonne Martin first described database searching in drug design well over a decade ago *(89)*. Her landmark program, ALADDIN, was the basis for the wide range of software packages available today for database searching or "mining". New ligands are housed in a database of chemical structures that most importantly includes their 3D coordinates. Each molecule is first rapidly examined to see if it has the basic pharmacophoric features matching the active site requirements (search query) and whether it is of obviously inappropriate size. Then, the surviving molecules are placed in the site either with docking tools as described above or by simply matching the pharmacophore features to the search query. These models can then be scored to highlight lead compounds for further computational or experimental study. This combined screening approach has been successfully used to identify a neurokinin-1 antagonist with submicromolar affinity (New Reference #1). We have also found *(90)* in several independent studies that active compounds can be identified in a similar way by utilizing the UNITY program of Tripos and searching through the National Cancer Institute (NCI) database and compound repository.

An alternative designation for this computational procedure is "virtual screening", i.e., it is the virtual analog of binding assays, particularly of the high-throughput (HTS) type. The analogy goes further in that, when a large database or library is examined, the expectation value of finding a high "hit" rate with either virtual screening or HTS is rather low.

One obvious and serious limitation of the database search is that most chemical entities have several to hundreds of energetically accessible conformations and the search methodology with its simplest implementation will only examine the conformation(s) stored in the database. Neither of the complete solutions to this problem, including all of the conformations in the database (too much storage space) or performing in situ conformational searches (too computationally expensive), are practical if one wishes to examine thousands to millions of possible ligands. One solution, flexible searching with directed tweak *(91)* optimization, has been implemented in UNITY and yields much higher hit rates as the conformational space of the putative ligands is explored during the search process. However, conformational flexibility of the active site residues, which may be just as important, has not been successfully and efficiently implemented to date.

4.4. Target Structure Prediction

A key limitation of the tools discussed above is that they require a structurally well-characterized target biomacromolecule. Whereas the number of structures available in the PDB is ever expanding, there are still a large number of proteins for which no experimental structural data is available. The technique known as *homology modeling* (or *comparative modeling*) is used whenever the 3D coordinates of the desired target macromolecule are not known, and the model is instead generated from a model of a related, homologous protein target. A homology modeling study typically consists of several steps. First, amino acid sequences of homologous proteins are sought. Several important computational tools have made it possible to navigate through the sea of data generated by genomics and proteomics projects. The Needleman–Wunsch algorithm *(92)* made it possible to compare two amino acid sequences and assign a measure of similarity between the sequences based on the similarity of the individual amino acids contained in each sequence. Subsequent improvements to the Needleman–Wunsch algorithm resulted in the FASTA *(93)* and the Basic Local Alignment Search Tool (BLAST) *(94)*, two algorithms that provide the basis for tools that are routinely used to search sequence databases like those at the National Center for Biotechnology Information (NCBI) (http://www.ncbi.nlm.nih.gov/) and at the Swiss-Prot database (http://us.expasy.org/sprot/) maintained by the Swiss Institute of Bioinformatics.

Once one or more sequences related to the protein to be modeled (the reference structures) are obtained, alignment tools such as ClustalW *(95)* are used to optimally align the amino acid sequences of the reference and target proteins based on the physiochemical properties of the individual residues. The gaps that arise in regions where amino acid insertions (where the target protein has more residues than the reference protein(s)) or deletions (where the target protein has fewer residues than the reference protein(s)) occur are modeled with a *loop searches*. The loop searches find suitable replacements for the nonhomologous regions of the protein typically by searching a library of reference protein structures taken from the PDB. The conformations of the side chains are often subsequently optimized (based on libraries of preferred conformations that are derived from protein crystal structures in the PDB) using a rotamer database-enabled program like SCWRL *(96–98)*. The target protein is further refined by energy-minimizing the structure using a force field to remove excess strain. Finally, the stereochemical integrity of a modeled target protein may be assessed using programs such as PROCHECK *(99)*, WHATIF *(100)*, or the ProTable facility within SYBYL. Examples of computer programs which can be used to automate the steps involved in homology modeling include the MODELLER program of Šali and Blundell *(101)*, the Advanced Protein Modeling capabilities within SYBYL, and the MODELER module of InsightII. Homology modeling has proven to be extremely useful, but it is not necessarily a perfect solution; it is problematic in that structural features found in the reference protein structures do not always translate in a meaningful way to the modeled target protein structure.

4.5. Challenges, Uncertainties, and Future Developments

The main challenges for structure-based drug discovery will continue to center on two issues: better scoring of binding to yield better predictions of free energy and exploiting and accurately translating available structural information to new proteins that are difficult to work with experimentally. One factor that always should be considered is that computer technology both in terms of capabilities and price continues to favor, with time, the increasing use of more exhaustive and expensive computational tools. Thus, calculations that are inaccessible today will likely be possible in the future, and calculations that are too slow for real-time analyses today may be quite reasonable in the future.

However, we would like to point out a few commonly used shortcuts and their consequences in the current state of the art in structure-based analyses. First, the effects of water, both displaced by the incoming ligand or co-occupying the active site with the ligand, are very often ignored. Water can play a number of roles, even in a static structure, that are difficult to precisely discern *(87)*. In the milieu surrounding the actual binding process the solvent is ubiquitous in virtually every stage of the event. In analyzing a bound ligand crystal structure, we are left with trying to unravel the entire process, sort of like detectives examining clues at a crime scene. This much is clear: some water molecules are absolutely energetically involved in binding the ligand in place; other water molecules act more like part of the protein, serving to "shape" the active site; still other water molecules, while observed in the crystal structures, appear to be "casual" waters of hydration; and, lastly, some water molecules were displaced by the binding and contributed to entropy—unfortunately, there is no way to identify these important water molecules! A second factor often ignored in virtual screening and structure-based drug design is consideration of the ionization states for the ionizable residues of the protein and functional groups of the ligand. The local "pH" may differ considerably from the global (solution) pH within active sites or at structurally important interfaces of the protein itself. If all ionizable groups are treated as if they were at, e.g., pH 7, obviously spurious molecular models are constructed. Consider that a single protonation of an acid group may reform an unfavorable interaction between two acids into an overall favorable interaction. Unfortunately, only *very* high resolution crystallography is able to locate the electron density for hydrogens and, even then, protonation assignments are not unambiguous.

5. Conclusions

It is not difficult to envision that structure-based drug discovery will continue to play a large part in future drug discovery. As the number of experimentally known protein structures increases, the tools and basis data to predict similar and homologous structures will accordingly improve in accuracy. We are also poised to develop "designer" drugs based on single point mutations in target proteins, as an end result of pharmacogenomics, or as a result of drug resistance in targets. The integration of ligand-based and structure-based drug design will further enhance our ability to develop clinically useful pharmaceutical agents with favorable ADME/Tox properties. We hope that we have piqued your interest in molecular modeling and CADD, and that you may even be interested in applying some of the methods presented here to your own research. All it takes to get started is a trip to your friendly neighborhood computational chemist!

References

1. Levinthal, C. (1966) Molecular model-building by computer. *Sci. Am.* **214,** 42–52.
2. Barry, C. D., Ellis, R. A., Graesser, S., and Marshall, G. R. (1969) Display and manipulation in three dimensions, In: *Pertinent Concepts in Computer Graphics* (Faiman, M. and Nievergelt, J., eds.), University of Illinois Press, Chicago, pp. 104–153.
3. Richardson, J. R. (1981) The anatomy and taxonomy of protein structure. *Adv. Protein Chem.* **34,** 167–339.
4. Carson, M. and Bugg, C. E. (1986) Algorithm for ribbon models of proteins. *J. Mol. Graph.* **4,** 121–122.
5. Connolly, M. L. (1985) Molecular surface triangulation. *J. Appl. Cryst.* **18,** 499–505.
6. Brickmann, J., Exner, T. E., Keil, M., and Marhöfer, R. J. (2000) Molecular graphics-trends and perspectives. *J. Mol. Model.* **6,** 328–340.
7. Shreiner, D. (2004) *OpenGL Programming Guide: The Official Reference Document to OpenGL*, version 1.4. 4th ed., Addison-Wesley, Boston.
8. Pearlman, R. S. (1987) Rapid Generation of High Quality Approximate 3D Molecular Structures. Chem. Des. Automat. News 2, 5–7.
9. Klebe, G. and Meitzner, T. (1994) A fast and efficient method to generate biologically relevant compounds. *J. Comput. Aided Mol. Des.* **8,** 583–606.
10. Allinger, N. L. (1977) Conformational Analysis. 130. MM2. A hydrocarbon force field utilizing V_1 and V_2 torsional terms. *J. Am. Chem. Soc.* **99,** 8127–8134.
11. Allinger, N. L., Kuohsiang, C., and Lii, J.-H. (1996) An improved force field (MM4) for saturated hydrocarbons. *J. Comput. Chem.* **17,** 642–668.
12. Allinger, N. L., Yuh, Y. H., and Lii, J. -H. (1989) Molecular mechanics. The MM3 force field for hydrocarbons. *J. Am. Chem. Soc.* **111,** 8551–8566.
13. Halgren, T. A. (1996) Merck molecular force filed. I. Basis, form, scope, parameterization and performance of MMFF94. *J. Comput. Chem.* **17,** 490–519.
14. Clark, M., Cramer, III R. D., and van Opdenhosch, N. (1989) Validation of the general purpose tripos 5.2 force field. *J. Comput. Chem.* **10,** 982–1012.
15. Pearlman, D. A., Case, D. A., Caldwell, J. W., et al. (1995) AMBER, a package of computer programs for applying molecular mechanics, normal mode analysis, molecular dynamics and free energy calculations to simulate the structural and energetic properties of molecules. *Comput. Phys. Commun.* **91,** 1–41.

16. MacKerell, A. D., Jr., Bashford, D., Bellott, M., et al. (1998) All-atom empirical potential for molecular modeling and dynamics studies of proteins. *J. Phys. Chem. B* **102,** 3586–3617.
17. Gogonea, V. and Merz, K. M. Jr. (1999) Fully quantum mechanical description of proteins in solution. Combining linear scaling quantum mechanical methodologies with the Poisson-Boltzman equation. *J. Phys. Chem. A* **103,** 5171–5188.
18. Schlick, T. and Gan, H. H., eds. (2002) Computational methods for macromolecules: challenges and applications: proceedings of the 3rd international workshop on algorithms for molecular modeling, New York, October 12–14, 2000, Springer, New York.
19. Hammett, L. P. (1940) *Physical Organic Chemistry*, McGraw-Hill, New York.
20. Taft, R. W. (1956) Chapter 13. In: *Steric Effects in Organic Chemistry* (Neuman, M. S., ed.),Wiley, New York, p. 556.
21. Hansch, C. and Fujita, T. (1964) Rho-sigma-pi Analysis. A method for the correlation of biological activity and chemical structure. *J. Am. Chem. Soc.* **86,** 1616–1626.
22. Hansch, C., Maloney, P. P., Fujita, T., and Muir, R. M. (1962) Correlation of biological activity of phenoxyacetic acids with Hammett substituent constants and partition coefficients. *Nature (London)* **194,** 178–180.
23. Jurs, P. C., Chow, J. T., and Yuan, M. (1979) Studies of chemical structure-biological activity relations using pattern recognition. In: *Computer-Assisted Drug Design* (Olson, E. C. and Christoffersen, R. E., eds.), The American Chemical Society, Washington, DC, pp. 103–129.
24. Stuper, A. J., Brugger, W. E., and Jurs, P. C. (1979) *Computer-Assisted Studies of Chemical Structure and Biological Function.* John Wiley & Sons, New York.
25. Karelson, M. (2000) *Molecular Descriptors in QSAR/QSPR*, John Wiley and Sons, Inc., New York.
26. Todeschini, R. and Consonni, V. (2000) *Handbook of Molecular Descriptors.* Wiley-VCH Verlag GmbH, Weinheim.
27. Randic, M. (1975) On characterization of molecular branching. *J. Am. Chem. Soc.* **97**(23)**,** 6609–6615.
28. Hall, L. H., Kier, L. B., and Murray, W. J. (1975) Molecular connectivity 2: relationship to water solubility and boiling point. *J. Pharm. Sci.* **64**(12)**,** 1974–1977.
29. Kier, L. B. and Hall, L. H. (1976) Molecular Connectivity 7: Specific Treatment of Heteroatoms. *J. Pharm. Sci.* **65**(12)**,** 1806–1809.
30. Kier, L. B., Hall, L. H., Murray, W. J., and Randic, M. (1975) Molecular connectivity 1: relationship to nonspecific local anesthesia. *J. Pharm. Sci.* **64**(12)**,** 1971–1974.
31. Kier, L. B., Murray, W. J., and Hall, L. H. (1975) Molecular connectivity. 4. relationships to biological activities. *J. Med. Chem.* **18**(12)**,** 1272–1274.
32. Murray, W. J., Hall, L. H., and Kier, L. B. (1975) Molecular connectivity 3: relationship to partition coefficients. *J. Pharm. Sci.* **64**(12)**,** 1978–1981.
33. Murray, W. J., Hall, L. H., and Kier, L. B. (1976) Molecular connectivity 5: connectivity series applied to density. *J. Pharm. Sci.* **65**(8)**,** 1226–1230.
34. Murray, W. J., Kier, L. B., and Hall, L. H. (1976) Molecular connectivity. 6. examination of the parabolic relationship between molecular connectivity and biological activity. *J. Med. Chem.* **19**(5)**,** 573–578.

35. Kier, L. B. and Hall, L. H. (1990) An electrotopological state index for atoms in molecules. *Pharm. Res.* **7**(8), 801–807.

36. Goldstein, H., Poole, C. P., and Safko, J. L. (2002) *Classical Mechanics*, 3rd ed., Addison Wesley, San Francisco.

37. Pearlman, R. S. (1980) Molecular surface areas and volumes and their use in structure/activity relationships. In: *Physical Chemical Properties of Drugs* (Yalkowsky, S. H., Sinkula, A. A., and Valvani, S. C., eds.), Marcel Dekker, New York.

38. Miller, K. J. and Savchik, J. A. (1979) A new empirical method to calculate average molecular polarizibilities. *J. Am. Chem. Soc.* **101**(24), 7206–7213.

39. Stanton, D. T. and Jurs, P. C. (1990) Development and use of charged partial surface area structural descriptors in computer-assisted quantitative structure-property relationship studies. *Anal. Chem.* **62**, 2323–2329.

40. Pearlman, R. S. and Smith, K. M. (1998) Novel software tools for chemical diversity. *Perspect. Drug Discov. Des.* **9**, 339–353.

41. Cramer, III R. D., Patterson, D. E., and Bunce, J. D. (1988) Comparative molecular field analysis (CoMFA). 1. Effect of shape on binding of steroids to carrier proteins. *J. Am. Chem. Soc.* **110**, 5959–5967.

42. Kearsley, S. K. and Smith, G. M. (1990) An alternative method for the alignment of molecular structures: maximizing electrostatic and steric overlap. *Tetrahedron Comput. Methodol.* **3**(6), 615–633.

43. Klebe, G., Abraham, U., and Mietzner, T. (1994) Molecular similarity indices in a comparative analysis (CoMSIA) of drug molecules to correlate and predict their biological activity. *J. Med. Chem.* **37**(24), 4130–4146.

44. Crippen, G. M. (1981) *Distance Geometry and Conformational Calculations*. Research Studies Press, New York.

45. Crippen, G. M. (1979) Distance geometry approach to rationalizing binding data. *J. Med. Chem.* **22**(8), 988–997.

46. Ghose, A. K. and Crippen, G. M. (1982) Quantitative structure-activity relationship by distance geometry: quinazolines as dihydrofolate reductase inhibitors. *J. Med. Chem.* **25**(8), 892–899.

47. Hopfinger, A. J. (1981) Inhibition of dihydrofolate reductase: structure-activity correlations of 2,4-diamino-5-benzylpyrimidines based upon molecular shape analysis. *J. Med. Chem.* **24**(7), 818–822.

48. Hopfinger, A. J. (1980) A QSAR investigation of dihydrofolate reductase inhibition by Baker Triazines based upon molecular shape analysis. *J. Am. Chem. Soc.* **102**(24), 7196–7206.

49. Hopfinger, A. J. (1983) Theory and application of molecular potential energy fields in molecular shape analysis: a quantitative structure-activity relationship study of 2,4-diamino-5-benzylpyrimidines as dihydrofolate reductase inhibitors. *J. Med. Chem.* **26**(7), 990–996.

50. Doweyko, A. (2004) 3D-QSAR Illusions. *J. Comput. Aided Mol. Des.* **18**, 587–596.

51. Clark, D. E. (2005) Computational prediction of ADMET properties: recent developments and future challenges. In: *Annual Reports in Computational Chemistry*

(Spellmeyer, D. C., Carlson, H., Crawford, T. D., et al., eds.), Elsevier, Amsterdam, pp. 133–151.

52. Burbidge, R., Trotter, M., Buxton, B., and Holden, S. (2001) Drug design by machine learning: support vector machines for pharmaceutical data analysis. *Comput.Chem.* **26,** 5–14.

53. Mosier, P. D. and Jurs, P. C. (2002) QSAR/QSPR Studies using probabilistic neural networks and generalized regression neural networks. *J. Chem. Inf. Comput. Sci.* **42,** 1460–1470.

54. Kier, L. B. and Hall, L. H. (2002) The meaning of molecular connectivity. *Croat. Chem. Acta* **75,** 371–382.

55. Eriksson, L., Jaworska, J., Worth, A. P., Cronin, M. T. D., McDowell, R. M., and Gramatica, P. (2003) Methods for reliability and uncertainty assessment and for applicability evaluations of classification- and regression-based QSARs. *Env. Health Perspect.* **111,** 1361–1375.

56. Guha, R. and Jurs, P. C. (2005) Determining the validity of a QSAR Model - a classification approach. *J. Chem. Inf. Model* **45,** 65–73.

57. Frederique, B. and Dragos, H. (2004) Molecular similarity and property similarity. *Curr. Topics Med. Chem.* **4,** 589–600.

58. Lipinski, C. A. (2005) Filtering in drug discovery. In: *Annual Reports in Computational Chemistry* (Spellmeyer, D. C., Carlson, H., Crawford, T. D., et al., eds.), Elsevier, Amsterdam, pp. 155–168.

59. Ghose, A. K. and Viswanadhan, V. N., eds. (2001) *Combinatorial Library Design and Evaluation.* Marcel Dekker, New York/Basel.

60. Berman, H. M., Battistuz, T., Bhat, T. N., et al. (2002) The protein data bank. *Acta Cryst. D* **58,** 899–907.

61. Böhm, H. -J. (1995) Site-directed structure generation by fragment-joining. *Perspect. Drug Discov. Des.* **3,** 21–33.

62. Dixon, J. S., Blaney, J. M., and Weininger, D. (1993) Characterizing and satisfying the steric and chemical restraints of binding sites. In: *The Third York Meeting,* pp. 29–30.

63. Payne, A. W. R. and Glen, R. C. (1993) Molecular recognition using a binary genetic search algorithm. *J. Mol. Graph.* **11,** 74–91.

64. Bartlett, P. A. (1986) Design of enzyme inhibitors: answering biological questions through organic synthesis. In: *Organic Synthesis, From Gnosis to Prognosis (NATO Advanced Study Institute)* (Chatgilialoglu, C. and Snieckus, V., eds.), Kluwer Academic Publishers, Dordrecht, pp. 137–173.

65. Lauri, G. and Bartlett, P. A. (1994) CAVEAT - a program to facilitate the design of organic molecules. *J. Comput. Aided Mol. Des.* **8,** 51–66.

66. DesJarlais, R. L., Sheridan, R. P., Seibel, G. L., Dixon, J. S., Kuntz, I. D., and Venkataraghavan, R. (1988) Using shape complementarity as an initial screen in designing ligands for a receptor binding site of known three-dimensional structure. *J. Med. Chem.* **31,** 722–729.

67. Kuntz, I. D., Blaney, J. M., Oatley, S. J., Langridge, R., and Ferrin, T. E. (1982) A geometric approach to macromolecule-ligand interactions. *J. Mol. Biol.* **161,** 269–288.

68. Meng, E. C., Schoichet, B. K., and Kuntz, I. D. (1993) Automated docking with grid-based energy evaluation. *J. Comput. Chem.* **13,** 505–524.
69. Kramer, B., Metz, G., Rarey, M., and Lengauer, T. (1999) Ligand docking and screening with FlexX. *Med. Chem. Res.* **9,** 463–478.
70. Rarey, M., Kramer, B., and Lengauer, T. (1997) Multiple automatic base selection: protein-ligand docking based on incremental construction without manual intervention. *J Comput. Aided Mol. Des.* **11,** 369–384.
71. Rarey, M., Kramer, B., Lengauer, T., and Klebe, G. (1996) A fast flexible docking method using an incremental construction algorithm. *J. Mol. Biol.* **261**(3), 470–489.
72. Morris, G. M., Goodsell, D. S., Halliday, R. S., et al. (1998) Automated docking using a Lamarckian genetic algorithm and empirical binding free energy function. *J. Comput. Chem.* **19,** 1639–1662.
73. Morris, G. M., Goodsell, D. S., Huey, R., and Olson, A. J. (1996) Distributed automated docking of flexible ligands to proteins: parallel applications of AutoDock 2.4. *J. Comput. Aided Mol. Des.* **10,** 293–304.
74. Jones, G., Willett, P., and Glen, R. C. (1995) Molecular recognition of receptor sites using a genetic algorithm with a description of solvation. *J. Mol. Biol.* **245,** 43–53.
75. Jones, G., Willett, P., Glen, R. C., Leach, A. R., and Taylor, R. (1997) Development and validation of a genetic algorithm for flexible docking. *J. Mol. Biol.* **267,** 727–748.
76. Schnecke, V. and Kuhn, L. A. (2000) Virtual screening with solvation and ligand-induced complimentarity. *Perspect. Drug Discov. Des.* **20,** 171–190.
77. McGann, M. R., Almond, H. R., Nicholls, A., and Brown, F. K. (2003) Gaussian docking functions. *Biopolymers* **68,** 76–90.
78. Muegge, I. and Martin, Y. C. (1999) A general and fast scoring function for protein-ligand interactions: a simplified potential approach. *J. Med. Chem.* **42,** 791–804.
79. Gelhaar, D. K., Verkhivker, G. M., Rejto, P. A., Sherman, C. J., Fogel, L. J., and Freer, S. T. (1995) Molecular recognition of the inhibitor AG-1343 by HIV-1 protease: conformationally flexible docking by evolutionary programming. *Chem. Biol.* **2,** 317–324.
80. Charifson, P. S., Corkery, J. J., Murcko, M. A., and Walters, W. P. (1999) Consensus scoring: a method for obtaining improved hit rates from docking databases of three-dimensional structures into proteins. *J. Med. Chem.* **42,** 5100–5109.
81. Böhm, H. -J. and Stahl, M. (1999) Rapid empirical scoring functions in virtual screening applications. *Med. Chem. Res.* **9,** 445–462.
82. Muegge, I., Martin, Y. C., Hajduk, P. J., and Fesik, S. W. (1999) Evaluation of PMF scoring in docking weak ligands into the FK506 binding protein. *J. Med. Chem.* **42,** 2498–2503.
83. Sippl, M. J., Ortner, M., Jaritz, M., Lackner, P., and Flöckner, H. (1996) Helmholtz free energies of atom pair interactions in proteins. *Folding Des.* **1,** 289–298.
84. Pearlman, D. A. and Charifson, P. S. (2001) Improved scoring of ligand-protein interactions using OWFEG free energy grids. *J. Med. Chem.* **44,** 502–511.

85. Pearlman, D. A. and Charifson, P. S. (2001) Are free energy calculations useful in practice? A comparison with rapid scoring functions for the p38 MAP kinase protein system. *J. Med. Chem.* **44,** 3417–3423.

86. Kellogg, G. E. and Abraham, D. J. (2000) Hydrophobicity: Is $LogP_{o/w}$ more than the sum of its parts? *Eur. J. Med. Chem.* **35,** 651–661.

87. Fornabaio, M., Spyrakis, F., Mozzarelli, A., Cozzini, P., Abraham, D. J., and Kellogg, G. E. (2004) Simple, intuitive, calculations of free energy of binding for protein-ligand complexes. 3. Including the free energy contribution of structural water molecules in HIV-1 protease-ligand complexes. *J. Med. Chem.* **47,** 4507–4516.

88. Gussio, R., Zaharevitz, D. W., McGrath, C. F., et al. (2000) Structure-based design modifications of the Paullone molecular scaffold for cyclin-dependent kinase inhibition. *Anti-Cancer Drug Des.* **15,** 53–66.

89. Evers, A. and Klebe, G. (2004) Successful virtual screening for a submicromolar antagonist of the neurokinin-1 receptor based on a ligand-supported homology model. *J. Med. Chem.* **47,** 5381–5392.

90. Fornabaio, M., Rastinejad, F., Kharalkar, S., Safo, M., Abraham, D. J., and Kellogg, G. E. (2005) Virtual screening approach: application of a hydropathic forcefield to 3D database searches. In: *229th ACS National Meeting March 13–17,* San Diego, CA.

91. Hurst, T. (1994) Flexible 3D searching: The directed tweak technique. *J. Chem. Inf. Comput. Sci.* **34,** 190–196.

92. Needleman, S. B. and Wunsch, C. D. (1970) A general method applicacable to the search for similarities in the amino acid sequence of two proteins. *J. Mol. Biol.* **48,** 443–453.

93. Lipman, D. J. and Pearson, W. R. (1985) Rapid and sensitive protein similarity searches. *Science* **227,** 1435–1441.

94. Altschul, S. F., Gish, W., Miller, W., Myers, E. W., and Lipman, D. J. (1990) Basic local alignment search tool. *J. Mol. Biol.* **215,** 403–410.

95. Thompson, J. D., Higgins, D. G., and Gibson, T. J. (1994) CLUSTAL W: improving the sensitivity of progressive multiple sequence alignment through sequence weightning, position-specific gap penalties and weight matrix choice. *Nucleic Acids Res.* **22**(22)**,** 4673–4680.

96. Canutescu, A. A., Shelenkov, A. A., and Dunbrack, R. L. Jr. (2003) A graph-theory algorithm for rapid protein side-chain prediction. *Protein Sci.* **12,** 2001–2014.

97. Dunbrack, R. L. Jr. and Cohen, F. E. (1997) Bayesian statistical analysis of protein side-chain Rotamer preferences. *Protein Sci.* **6,** 1661–1681.

98. Dunbrack, R. L. Jr. and Karplus, M. (1993) Backbone-dependent Rotamer library for proteins. *J. Mol. Biol.* **230,** 543–574.

99. Morris, A. L., MacArthur, M. W., Hutchinson, E. G., and Thornton, J. M. (1992) Stereochemical quality of protein structure coordinates. *Proteins* **12,** 345–364.

100. Vriend, G. (1990) WHAT IF: a molecular modeling and drug design program. *J. Mol. Graph.* **8,** 52–56.

101. Šali, A. and Blundell, T. L. (1993) Comparative protein modelling by satisfaction of spatial restraints. *J. Mol. Biol.* **234,** 779–815.

14

Macromolecular Drug Delivery

Neelam Azad and Yon Rojanasakul

Abstract

Advances in molecular biotechnology coupled with novel technologies such as combinatorial chemistry and high-throughput screening have led to the discovery of a large number of drugs with macromolecular properties. Macromolecular therapeutics encompasses a variety of approaches including recombinant proteins, genes, antisense, and small interfering RNAs, all of which have larger molecular dimensions than conventional drugs. These macromolecules have emerged as a powerful class of drugs for a wide range of therapeutic indications mainly on the basis of their site-specific activity and reduced side effects. However, these drugs present an enormous challenge for non-invasive delivery as they are poorly absorbed and rapidly metabolized in the body. To surmount these obstacles, either the other aspects of the drug may be exploited or novel delivery systems may be developed. Cost-effectiveness suggests that developing an improved delivery system for an existing drug is a better alternative than modifying the chemical structure of the drug or discovering new drug entities. This chapter focuses on macromolecular drugs and their delivery systems including liposomes, polymers, peptides, and nanoparticles.

Key Words: Macromolecules; drug delivery; proteins; DNA; RNA; oligonucleotides; liposomes; polymers; nanodelivery.

1. Introduction

Targeted delivery of therapeutic molecules is the most desirable feature of an effective drug therapy. The plethora of knowledge amassed in the field of biomedical sciences, particularly in molecular biology, has provided a deeper understanding of pathogenesis at the cellular as well as subcellular level. Low molecular weight conventional chemotherapeutic agents easily traverse different organs and enter various cells and subcellular organelles. This property makes them a very good candidate for disease treatment; however, it also imparts nonspecificity

From: *Biopharmaceutical Drug Design and Development*
Edited by: S. Wu-Pong and Y. Rojanasakul © Humana Press Inc., Totowa, NJ

and increases the incidence of side effects *(1)*. Additionally, low molecular weight drugs are rapidly eliminated from the body requiring frequent dosing to attain therapeutic levels *(2)*. This adversely affects patient compliance and cost-effectiveness of the drug.

Several approaches have been developed to circumvent the problems associated with nonspecific drug delivery. Recent advances in biotechnology coupled with novel technologies such as combinatorial chemistry and high-throughput screening have led to the discovery of a large number of drugs with macromolecular properties. "Macromolecule" is basically a biological term for molecules of high molecular mass, comprising many small organic molecules that are often referred to as monomers, e.g., carbohydrates, lipids, proteins, and nucleic acids. Synthetic macromolecules include polymers, liposomes, peptides, and other molecules that are composed of several smaller molecules. In contrast to the conventional drug therapy that utilizes only xenobiotics as medicines; endogenous macromolecules are now being commonly used as therapeutic modalities *(3)*. Macromolecular therapeutics encompasses a variety of approaches including recombinant proteins, peptides, antisense oligonucleotides (ONs), genes, and small interfering RNA (siRNA). Akin to most of the therapeutic systems, macromolecular therapeutics has its own advantages as well as drawbacks. It has emerged as a powerful class of drugs for a wide range of therapeutic indications solely on the basis of site-specific activity and reduced side effects. However, these drugs present an enormous challenge for noninvasive delivery as they are poorly absorbed in the gastrointestinal tract because of their hydrophilic nature, charged structure, and large molecular mass *(4)*. Moreover, these drugs are rapidly metabolized and are biologically as well as chemically unstable. It is well established that proteins and ONs are unstable and are rapidly cleared from the bloodstream *(4,5)*. To surmount these obstacles, either the other aspects of the drug may be exploited or novel delivery systems may be developed. Cost-effectiveness suggests that developing an improved delivery system for an existing drug is a better alternative than modifying the chemical structure of the drug or discovering a new drug. The increasingly appealing option is an appropriate delivery system tailored to enhance the efficacy of the drug, thus, emphasizing more on targeting the drug efficiently to the proper cellular and/or subcellular target.

In vivo delivery of macromolecules involves transport of the drug from the site of administration through various physiological barriers to the target site. First, drugs have to survive through the hostile extracellular conditions such as extreme pH, immune defense, enzymatic degradation, and scavenger systems *(6)*. All living cells are protected by a double-layered plasma membrane that selectively allows the uptake of smaller molecules. Similarly, the nucleus is guarded by a double-layered nuclear membrane as it contains important genetic material. Molecules with a size of 40–50 kDa or less can freely diffuse through plasma membrane and nuclear pores. However, large and charged molecules can enter the cell or nucleus only by active transport posing a problem for the entry of macromolecules *(7)*. In order to deliver macromolecules to the target site, therapeutic strategies directed at intracellular targets have to be developed.

Several factors have to be considered before selecting an agent for macromolecular drug delivery such as its chemical composition, biodegradability, hydrophilicity, and biocompatibility *(8)*. The delivery system should have a suitable functional group that allows it to bind to the macromolecular drug or targeting moiety. It should be inert, nonimmunogenic, water soluble, and biodegradable so that it is easily excreted from the body. It should also be reproducibly manufactured and conveniently administered to patients. Macromolecules are also used as drug carriers because of the diversity in their physico-chemical properties *(3)*. The discovery that macromolecules can localize into subcellular compartments known as lysosomes portended their development as drug carriers *(9)*. Macromolecular delivery systems generally enter cells by endocytosis and can be designed in such a way that the drug is released in response to the physiological or diseased conditions such as high temperature and acidic environment *(1)*. Primarily, in endocytosis the drug molecules are internalized into the cell in the form of membrane vesicles. In receptor-mediated endocytosis, the efficiency of internalization depends on the concentration and type of the ligand on the membrane. Even freely soluble macromolecules with no membrane affinity can enter the cell as a part of the fluid in endocytic vesicles by absorptive endocytosis, although this process is inefficient *(10)*. Inside the cell, the delivery systems are transported from the early and late endosomes, endolysosomes, and eventually to lysosomes where they are degraded and the drug molecules are released in the cytoplasm or the nucleus (Fig. 1). In order to achieve successful delivery of the drug into the cytosol it is necessary to devise delivery systems that will allow easy escape of the macromolecule from the endosome. Certain viruses, plant and bacterial toxins gain entry into the cytosol by the acidification related to endosomal maturation *(10,11)*. In a fusion process the toxins and viral proteins are transported across the endosomal membrane because of acid-induced conformational changes. Various strategies such as acid-sensitive liposomes and peptides have been used to mimic this natural process for the delivery of macromolecular drugs to their cytoplasmic targets.

The rationale for encapsulating or associating a drug with a macromolecular carrier is that it provides the following key advantages. (1) There is a resultant increase in the molecular weight of the macromolecular drug complex that significantly alters its biodistribution and even prevents its uptake into the nonspecific cells by endocytosis, thus, targeting the drug at the site of action. This improves the therapeutic index of an otherwise toxic drug by reducing the incidence of systemic side effects. (2) The encapsulated drug can be released over

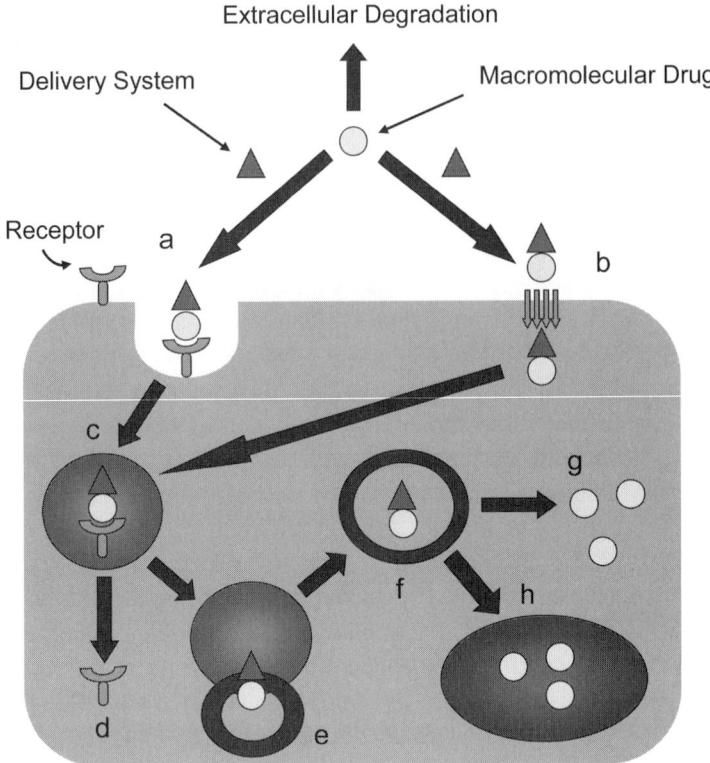

Fig. 1. Endocytosis-mediated uptake of macromolecular drugs. (a) Receptor-mediated endocytosis, (b) adsorptive endocytosis, (c) endosome, (d) receptor recycling, (e) endolysosome, (f) lysosome, (g) and (h) drug release into the cytosol and nucleus, respectively.

an extended period of time. This not only obviates the need for repeated high doses but also improves patient compliance *(1)*. (3) Macromolecular delivery systems can be used for the delivery of labile macromolecules such as proteins and ONs as these systems can protect the drug from the harmful physiological milieu, thus improving its stability in vivo *(1)*.

Several delivery strategies for effective macromolecular drug delivery have been developed such as high molecular weight carriers encompassing synthetic macromolecules such as liposomes, polymers, nanoparticles, peptides, and others. Lack of suitable drug delivery systems is not only affecting the conventional drugs but also hampering the progress of novel therapeutic strategies like siRNA and gene therapy. Delivery systems that are target specific and safe can improve the performance of some conventional medicines as well as have implications

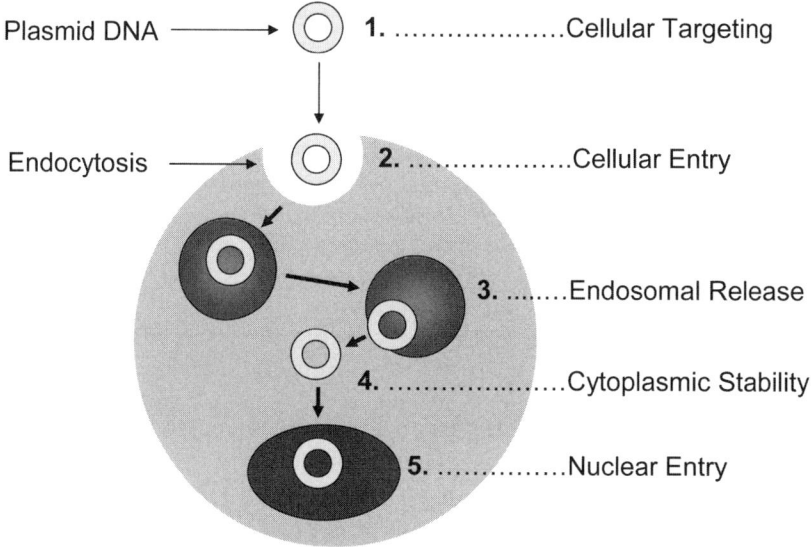

Fig. 2. Barriers to plasmid DNA delivery.

in the development and success of macromolecular therapeutic strategies. Because successful therapeutic delivery can be achieved by subcellular targeting of macromolecular drugs, this chapter focuses on the various macromolecular drugs and their delivery systems that can be modified and manipulated to achieve site-specific drug delivery for effective therapeutic applications.

2. Macromolecular Drugs

2.1. Deoxyribonucleic Acid

With the human genome sequenced, gene or DNA delivery is becoming the main focus of gene therapy and other DNA-based therapy. In gene therapy, genetic material is introduced into the cells either to block the expression of harmful proteins or to produce therapeutic proteins. To access the nuclear transcription machinery, therapeutic DNAs must cross the extracellular and intracellular barriers to reach their target site in the nucleus (Fig. 2). DNA is easily degraded by nucleases extracellularly and, hence, to begin with, DNA must be condensed with vectors that protect it from nuclease degradation *(6)*. This is followed by endocytosis which is a multistep process involving binding, internalization, formation of endosomes, fusion with lysosomes, and release of the DNA molecule *(12)*. DNA is generally degraded in this process because of low pH and enzymatic action. Nonetheless, the DNA molecules that survive finally enter the nucleus through nuclear pores or during cell division when the nuclear envelope ruptures *(13)*.

Table 1
Characteristics of an Ideal DNA Delivery System

It should protect the DNA from enzymatic degradation
It should be nontoxic, nonimmunogenic, and biocompatible
It should contain DNA of diameter less than 500 nm to facilitate endocytosis
It should easily penetrate the cell and target the site of action
It should protect the DNA from phagocytic system that removes it from the blood
circulation and thereby increases its in vivo bioavailability
It should allow easy escape of the encapsulated DNA into the nucleus for appropriate
transgene expression

As a result of all these barriers the number of DNA molecules decreases at each step of their passage from the cytoplasm to the nucleus.

Appropriate delivery systems are required to protect DNA from nuclease degradation and enhance its therapeutic efficacy. Amongst the DNA delivery systems developed so far, viral vectors are one of the most widely used strategies for clinical applications owing to their inherent ability to transport genetic material into the cell and nucleus, leading to enhanced gene expression. A desirable feature of long-term therapeutic effects is the integration of extraneous DNA into host cell chromatin structure *(6)*. This can be achieved by using retroviral vectors that can efficiently integrate into the host genome; however, there is risk of malignant transformation of the cell with these vectors. Furthermore, large-scale production of viral vectors is difficult and some viruses, e.g., adenoviruses, can also cause immunogenic and inflammatory reaction *(14)*. Owing to these safety issues, the search for alternative nonviral gene delivery systems has been growing intensely. Since 1950s, nonviral delivery systems such as cationic lipids and polymers have been studied for the delivery of nucleic acids as their structure can be widely manipulated enabling the investigation of structure–function relationships *(15,16)*. Some of the advantages of nonviral vectors over the more popular viral vectors include the possibility of transfecting cells with large DNA molecules, low cytotoxicity, less immunogenic reactions, reduced cost, and reproducibility *(17,18)*. Many of these delivery systems are discussed in greater details in subsequent sections and in a separate chapter. The ideal characteristics of DNA delivery systems are listed in Table 1.

2.2. Ribonucleic Acid

RNA helps in the transfer of genetic information from DNA to the protein-producing system of the cell. Currently, RNA interference (RNAi) is considered one of the most effective therapies for various genetic and acquired diseases. RNAi is a normal cellular process that causes posttranscriptional gene silencing in eukaryotic organisms *(19)*. Intracellular RNAi machinery can be easily activated against a target gene by utilizing siRNAs. When double-stranded RNA is injected into the cell or produced intracellularly, it is spliced by Dicer (RNase III nuclease) enzyme into smaller duplex pieces of 20–25 nucleotide length. Then these siRNA duplexes are guided by the RNA-induced silencing complex (RISC) present in the cytosol to the complementary sequences in the target mRNA *(20)* (Fig. 3). Essentially, siRNA duplexes are used as templates to cleave and downregulate complimentary mRNAs catalyzed by enzymes *(21)*. siRNA molecules have great potential as therapeutic agents for selective inhibition of genes causing various diseases. RNAi involves nucleic acids and hence is considered as the best option to treat diseases as these are less likely to be rejected by the body or cause toxic side effects *(22)*.

Although delivering siRNAs into living organisms is very intricate, two basic methods have been engineered. The first method of exogenous delivery involves direct insertion of synthetic siRNAs into the cell, whereas the second method of endogenous expression utilizes a plasmid or virus encoding a gene sequence that produces the desired siRNA *(19)* (Fig. 3). Exogenous delivery involves only the cytoplasm whereas endogenous method requires nuclear delivery for the transcription of the encoded siRNA. The second method may involve viral or nonviral vectors. Unlike gene therapy, RNAi may not include viral vectors that induce toxic responses because they can be introduced directly into the tissues *(19)*. However, this has its own implications such as the degradation of the siRNAs. Even though RNA duplexes are more stable than its single-stranded counterparts, siRNAs are modified chemically to enhance their biological stability *(23–27)*. The mechanism of siRNA uptake is not clear; however, it is observed that naked siRNAs are poorly transported into the cell and the negligible amount of nucleic acid that enters the cells remains sequestered within the endosomal vehicle where it is degraded by nucleases *(19,28)*. Moreover, pharmacokinetic study of siRNAs suggests that they accumulate in the organs of reticuloendothelial system (RES). This is good for the siRNAs targeted to such organs; however, targeting non-RES organs becomes challenging *(19)*. In addition, siRNAs are rapidly degraded and eliminated in vivo requiring repeated dosage. An appropriate delivery system that would improve the cellular uptake and intracellular trafficking of siRNAs is required.

Initially, two reports provided proof of the application of siRNAs as potential therapeutic agents. The first report demonstrated that high-pressure tail-vein injection of siRNA to mice silenced the vector expressing human placental alkaline phosphatase as well as the cotransfected reporter gene *(29)*. The other group directly injected siRNA into the liver of mice leading to the silencing of hepatitis C NS5B/luciferase fusion mRNA along with the

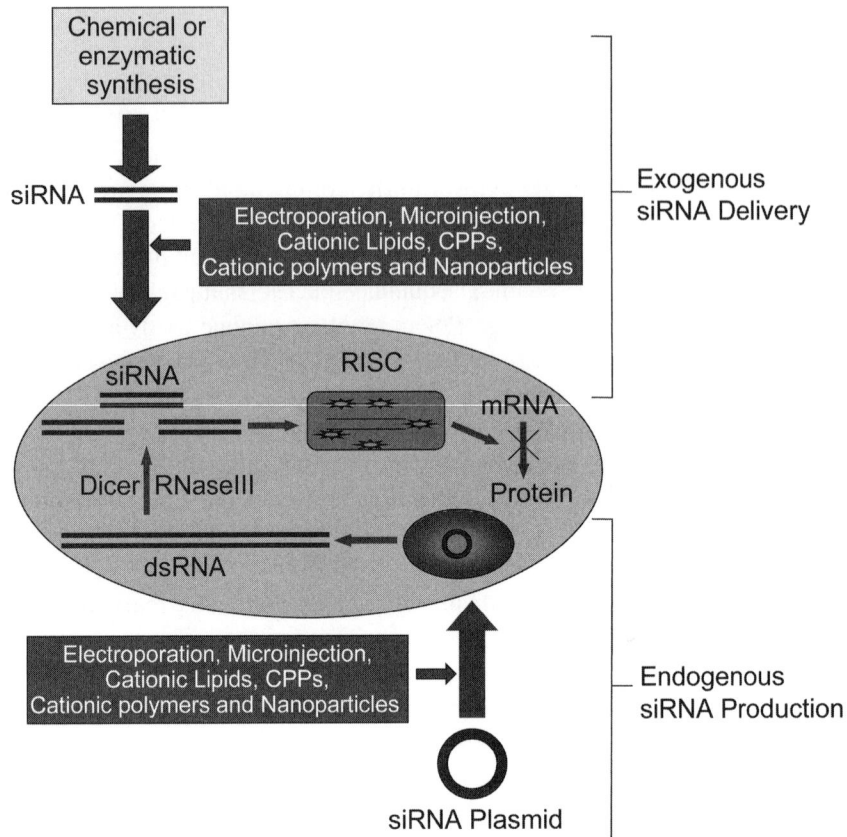

Fig. 3. Strategies for siRNA delivery. siRNA can be delivered exogenously into the cytoplasm or produced endogenously by delivering siRNA plasmid to the nucleus where it is transcribed and exported to the cytoplasm. RISC complexes with siRNA in the cytoplasm and guides it to target-specific mRNA degradation which results in inhibition of protein synthesis. Similar delivery strategies but with different optimal requirements can be used for both strategies.

luciferase gene *(30)*. Local delivery of siRNA to specific organs including brain *(31)*, eye *(32)*, and the lung *(33)* has also been tried successfully in animal models. One of the reports also claims that nasal delivery of siRNA can be used for systemic delivery including heart and kidney, in addition to the lung, as it causes protein knockout in all these organs *(19)*. Systemic administration using intravenous injection of naked siRNA to the liver and kidneys has also been reported. Although the delivery of siRNA is challenging, successful results with animal models have prompted phase I clinical trials with siRNAs *(19)*.

2.3. Oligonucleotides

ONs are small, single-stranded segments of nucleic acids ranging from 20 to 30 nucleotide bases in size. ONs are used as therapeutic agents in antisense therapies based on the ability of DNA or RNA ONs to downregulate the expression of disease-causing genes in a sequence-specific manner *(34)*. Antisense ONs have several properties that make them attractive therapeutic agents including their unique design that allows them to selectively bind by Watson–Crick hydrogen bonding to the complementary sequences in target mRNA. This prevents formation of the gene product by several presumed mechanisms including mRNA cleavage by the activation of RNase-H enzyme or translation arrest caused by blocking ribosomal reading *(35)*. The cytoplasm and nucleus are the main targets of antisense ONs and thus the cellular uptake of these molecules is very important. Plasma membrane restricts the transport of large molecules such as ONs that range from 3000 to 7500 Da in size *(36)*. Therefore, the cellular uptake of naked ONs is generally nonproductive with very few ONs entering the cells. In addition to poor cellular uptake, ONs also get trapped in endosomal compartments thereby affecting their intracellular distribution and availability at the target site. A negligible amount of ONs are also compartmentalized in other cellular organelles such as golgi complex and endoplasmic reticulum *(35)*. Generally DNA and RNA ONs are biologically unstable. ONs with an unmodified phosphodiester backbone are rapidly degraded in biological fluids by both endonucleases and exonucleases *(37–39)*. For effective therapy, ONs should be biologically stable and should possess appropriate chemistry for effective hybridization with the target mRNA. Copious chemically modified ONs have been developed that have improved stability and target binding affinity.

It is generally accepted that ONs are taken up by cells via endocytosis and finally enter the nucleus. This is based on the report where ONs injected into the cell directly accumulated in the nucleus, bypassing the endocytic pathway *(40)*. However, entry of ONs into the nucleus is cell-type specific with keratinocytes being the major type that can internalize ONs *(13)*. This is advantageous over DNA therapy where DNA of all sizes are immobilized in the nucleus and the cytosol-to-nuclear migration of the larger DNA constructs may be a rate-limiting step. In vitro and in vivo analyses along with pharmacokinetic distribution studies suggest that ON administered by almost any route appears to end up in the organs of the RES system such as liver, spleen, kidney, and lungs *(34)*. ONs within RES organs may induce adverse effects as these are known to exhibit some nonsequence-specific effects. Local delivery of antisense ONs has been reported to act on the brain of rats. Such studies indicate that antisense ONs may be more effective in some tissues than that predicted from cell culture studies *(34)*. However, in general, pharmacokinetic studies of ONs suggest that multiple parenteral dosing is required to obtain biological efficacy because of rapid pharmacokinetic distribution and elimination of ONs. Even orally delivered ONs are topic of debate owing to limited studies on the mechanism and effectiveness of these ONs *(35)*. In order to improve the cellular uptake and therapeutic activity of ONs, optimum delivery systems need to be developed. Several delivery systems for ONs are currently available including cationic liposomes and polymers, branched dendrimers, mini-osmotic pumps, and nanoparticles. These systems are discussed in subsequent sections.

2.4. Proteins and Peptides

Proteins are a class of organic macromolecules that perform a wide range of cellular functions. Proteins are polymers of amino acids, and peptides are short proteins. Proteins and peptides possess and control biological activities that mark them as therapeutic agents. Some of the examples include antimicrobial peptides, anti-inflammatory proteins, and antioxidant enzymes such as catalase and superoxide dismutase *(41)*. However, most proteins and peptides are rapidly eliminated from the circulation by renal filtration, proteolytic degradation, receptor-mediated clearance, and accumulation in unspecific organs and tissues *(42)*. High molecular weight proteins may generate neutralizing antibodies leading to an immune response. Additionally, most of them have limited biodistribution in vivo as they are unable to cross biological membranes in the absence of specific transport systems. Clinical application of proteins and peptides is further limited by their short biological half lives owing to low solubility or poor stability *(3)*. Short peptides have lower molecular weights and are thus degraded rapidly requiring frequent dosing. For instance, a lot of commercial peptides including insulin for diabetes, Fuzeon® for HIV, and Forteo® for osteoporosis require regular injections. Peptide and protein drugs normally exert their action either extracellularly by interacting with receptors on the cell surface or they may have targets inside the cell. In the case of intracellularly acting peptides and proteins, low permeability of cell membrane to these macromolecules presents an additional obstacle for the development of proteins and peptides as therapeutics. Additionally, the large size, hydrophilicity, physical and chemical lability of these macromolecules should be taken into account while preparing proteins/peptides for therapeutic purposes *(41)*. Endogenous proteins that are about 40 kDa in size cannot easily diffuse into the nucleus via the nuclear pore complex (NPC). Oral delivery of drugs is the most desirable route of administration; however, proteins and peptides administered by the oral

Table 2
Advantages of Encapsulating or Conjugating Protein and Peptide Drugs with Delivery Systems

1. Protects the drugs against enzymatic degradation and reduces renal filtration of the drug.
2. Improves the in vivo stability of these otherwise labile drugs.
3. Decreases nonspecific drug delivery.
4. Improves the drug transport and increases the amount of protein or peptide drug reaching its intracellular or extracellular site of action.
5. Improves the transport of the drug to its site of action.
6. Reduces drug clearance and thus increases the residence time of the drug at its site of action leading to a better therapeutic response.
7. Reduces the immunogenicity of proteins.
8. Reduces toxicity because of the high initial doses of the drug alone.

route show poor bioavailability because of their rapid degradation by proteolytic enzymes in the gastrointestinal tract *(43)*. Parenteral delivery of proteins and peptides is the most popular route as it bypasses the biological barriers that deter the passage of proteins and also leads to pharmacologic levels of proteins in a relatively short time *(44)*. Local delivery of proteins to the gut, sinus, and lungs has been attempted. However, these formulations require a delivery system that would protect protein and peptide drugs from enzymatic degradation and deliver them to the target site.

To address the problems associated with peptide and protein delivery, several approaches have been explored. Initially, the chemistry and amino acid sequence of these drugs were altered to decrease enzymatic degradation and antigenic side effects. Then methods were developed to fuse the peptide and protein drugs to immunoglobulins or albumin so as to improve their biological half-life *(45)*. Conjugating proteins and peptides with polymers or liposomes is another simple and effective way of prolonging their blood circulation by reducing glomerular filtration. In these systems the protein drug is encapsulated in or attached to the polymeric or liposomal matrix that releases the protein in a controlled manner by undergoing enzymatic digestion or hydrolysis. The advantages of encapsulating or associating proteins and peptides in delivery systems are listed in Table 2. With the exceptions of BioPORTER® which is a cationic lipid-based carrier *(46)* and TransIT® which is a histone-based polyamine *(47)*, most of these systems do not efficiently deliver proteins to their intracellular targets. However, liposomal systems have their own limitations. For example, liposomes lack site-specificity and are errantly compartmentalized in the liver, spleen, kidneys, and other RES systems causing unwanted side effects. Therefore, advanced drug delivery technologies such as PEGylation that involves linking specific poly-ethylene glycol (PEG) polymers to protein and peptide drugs are developed. Currently, PEG is the most widely used polymer for modifying therapeutically active proteins because it is less toxic, economical, and many of its molecular weight variants are commercially available. Some of the advantages of the PEGylation technique include optimized pharmacokinetics, decreased dosing frequency, enhanced efficacy, bioavailability, solubility, and stability. Two approved PEGylated interferon-α products are now available for hepatitis C treatment, viz., PEGinterferon α-2a (PEGASYS®—Roche) and PEGinterferon α-2b (PEG-INTRON®—Schering-Plough). Some other protein delivery strategies include linking of the protein to transduction domains of bacterial or plant toxins as well as to hydrophobic cell-penetrating peptides (CPPs) *(48,49)*. Endogenous proteins are generally imported into the nucleus via the NPC in a nuclear locali-zation signal (NLS)-dependent and ATP-dependent manner *(10)*. Attaching a ligand to the protein or its carrier facilitates its cellular uptake via several different pathways. It has been demonstrated that attaching NLS peptides to certain proteins increases their nuclear uptake via a receptor-mediated process. For example, coupling multiple copies of the NLS SV40 motif to proteins was demonstrated to enhance their nuclear uptake *(50,51)*. Some proteins and peptides such as protein transduction domains (PTDs) are naturally internalized by cells. Few PTDs which appear to translocate lipid membranes by a passive process have been identified; the most widely reported being HIV TAT protein, herpes simplex virus protein VP22, and the *Drosophila* homeoprotein antennapedia (Antp) *(52–54)*. It has been shown that these peptides can be used as vectors for the delivery of other macromolecules, including proteins.

Numerous proteins and peptides including growth factors, hormones, mono-clonal antibodies, and cytokines are undergoing clinical investigation for a range of clinical conditions. A major hindrance in the development of protein-based drugs is the cost as large quantities of proteins are required for thorough formulation and bioavailability studies *(41)*. In the future, there will be more protein and peptide drugs that will have similar problems leading to an increased demand for site-specific delivery systems.

3. Delivery Systems

Targeted delivery of high molecular weight macromolecules is a major hindrance in their successful clinical application. Delivery systems that target these macromolecular drugs to the specific site of action are the major deter-minants of the drug efficacy. Employing physical or chemical techniques such as electroporation, microinjection, and the use of detergents for the delivery of macromolecular drugs has its own advantages. However, the advantages are overshadowed by the shortcomings that include invasive delivery, effect on cell

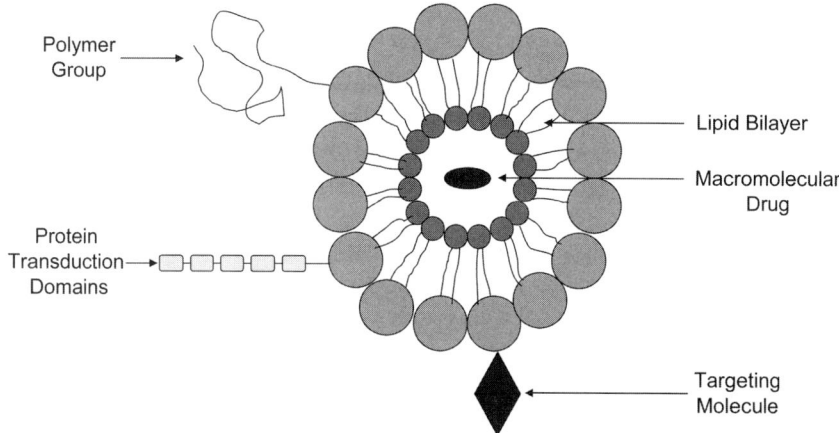

Fig. 4. Liposomal delivery for macromolecular drugs. Hydrophilic polymers such as PEG increase the circulation time of the drug by modifying the liposomal surface. PTDs facilitate intracellular delivery whereas targeting molecules such as antibodies assist in site-specific delivery.

viability, and limited clinical applications *(49)*. An ideal pharmacokinetic profile has the drug concentration in the therapeutic range, lower than the maximum tolerable dose, and it remains at that level for extended periods of time till the desired therapeutic effect is reached. Encapsulating the drug in delivery systems such as polymers, liposomes, and others is one such way of achieving an ideal pharmacokinetic profile. This not only protects the drug but also transports it to the target site. Delivery systems such as liposomes, polymers, peptides, and nanoparticles are gaining increasing popularity because of their higher specificity and applicability.

3.1. Liposomal Delivery

Liposomes are spherical phospholipid vesicles typically ranging from 100 to 200 nm in size with an inner compartment that can be used for the encapsulation of drugs *(55)* (Fig. 4). As mentioned earlier, because of the disadvantages associated with viral vectors, nonviral delivery systems such as liposomes are receiving increasing attention as these systems can protect the macromolecular drug from the harsh biological milieu and enhance its cellular delivery. Liposomes are inert, biocompatible with little toxicity and antigenic reactions. Although the encapsulation of the macromolecular drug in liposomes is ineffi-cient, liposomal design can be easily manipulated to provide protection to the enclosed macromolecular drug from enzymatic degradation and to increase its cellular uptake *(36)*. Anionic and neutral liposomes have received limited attention because of their poor encapsulation or association efficiency of macromolecules *(35)*. Cationic liposomes are the most widely used system owing to the electro-static interaction between the negatively charged macromolecules and the positively charged lipid carrier that results in a more neutral and stable complex. Typically these complexes have a net positive charge that interacts with the negative charge of the plasma membrane facilitating internalization of the complex into the cell by endocytosis *(56)*. The mechanism of action of these liposomes is incompletely understood and significant optimization is required as mentioned earlier. However, it is suggested that on membrane destabilization by the cationic liposome, anionic lipids from the cytoplasm-facing lipid layer tumble to the lumenal layer diffusing into the liposome. These anionic lipids then replace cationic lipids releasing the drug moiety into the perinuclear area and finally into the nucleus *(57)*. In order to further facilitate the release of nucleic acids from the endosomal or lysosomal compartment, helper lipids such as dioelyl-phosphatidylethanolamine (DOPE) are used. These helper molecules undergo a conformational change that destabilizes the vesicle membrane releasing the nucleic acid *(58)*.

Cationic liposomes are good when used in in vitro models, yet in vivo models may show nonspecific binding to the cellular components, blood plasma components, or the endothelial lining of the vessel, resulting in a short biological half-life *(56)*. Successful clinical application of these cationic reagents depends on a number of factors such as its chemical structure, target cell type, length, and the method of complex formation and the charge ratio *(59)*. The major problem with the use of liposomes is the toxicity associated with high charge ratio of cationic lipid species and the drug. Therefore, liposomes should be delivered to the target site so as to minimize side effects. Adding a targeting ligand to liposomes such as an antibody facilitates specific cell targeting (Fig. 4). Several studies have demonstrated that antibody-associated liposomes can augment cell-specific delivery and therapeutic activity of nucleic acids *(60,61)*. Another concern is the size of the liposomes. Small liposomes have longer circulation half-life than large liposomes. It has been reported that liposomes get trapped in the lung microvasculature of mice after tail-vein injections *(13)*. Liposomes administered systemically are rapidly cleared in vivo from the normal blood circulation by the RES system *(62–64)*. However, modifying liposomes with polymers such as PEG prevents the uptake by the RES resulting in a long-circulating drug–liposome complex *(65)* (Fig. 4). In cationic lipid–drug complex the linker group is the most important parameter that determines the chemical stability and biodegradability of a cationic lipid. Incorporating amide and carbamoyl bonds provides the most chemically stable but easily biodegradable cationic lipids *(66,67)*. However, taking into consideration the facts that liposomes can be administered in vivo by vascular system and can be easily stored in water for an extended period of time makes them interesting candidates for macromolecular drug delivery *(57,68)*.

The application of cationic liposomes has been found in several studies where it has been reported to deliver the drug selectively to the diseased tissue. Liposomes possess suitable characteristics for peptide and protein encapsulation. Long-circulating liposomes can be easily manipulated for the delivery of peptide and protein drugs. Encapsulating doxorubicin in PEG-coated liposomal systems demonstrated excellent electron paramagnetic resonance-based tumor therapy results and reduced toxicity of the original drug *(69)*. Coupling DNA with liposomes increases its circulation time and nuclear targeting. It has been observed that following microinjection DNA is rapidly dissociated from the cationic lipid in both cytoplasm and nucleus *(70)*. However, another group reported that the injection of naked DNA alone into the nucleus induces much higher and efficient gene expression as compared to the direct microinjection of DNA–cationic lipid complexes suggesting that DNA plasmid is released inefficiently from its carrier in the nucleus *(71)*. As cationic lipids are designed mainly on trial-and-error basis using the transgene expression as the end point, there is an inadequate understanding of the factors leading to transfection efficiency *(18)*.

Positively charged cationic lipids are the delivery systems of choice for siRNAs because of their high delivery efficiency *(25,72)*. Some of the commercial liposome formulations used for siRNA transfection are Lipofectin *(73)*, DMRIE-C *(74)*, and GeneTrans II *(75)*. Cationic lipid complexes have also been shown to successfully deliver siRNA after intraperitoneal injection *(76)*. Cationic lipids, lipid derivatives of polyamines, and cationic derivatives of cholesterol-diacyl glycerol have been shown to improve the cellular uptake of ONs *(77,78)*. However, the toxicity and serum sensitivity of the ON–liposome complexes is a major hindrance in their widespread application in vivo *(79)*. A number of techniques have been developed to circumvent these problems. The use of pH-sensitive fusogenic liposomes as well as antibody-associated liposomes have demonstrated cell-specific delivery and enhanced efficacy of antisense ONs *(35,36)*. Recently, cationic amphipiles have been demonstrated to increase the cellular uptake of ONs by 250-fold as compared to naked ON. These molecules consist of a hydrophobic cholic acid group covalently linked to spermine and/or spermidine groups which allows for association with nucleic acids. Unlike other cationic lipids, this compound is also active in the presence of high concentrations of serum and does not require a neutral lipid to facilitate ON delivery to cells *(80)*.

Liposomes are good in vitro delivery systems; however, there is still a limited understanding of their behavior in vivo and particularly inside cells. For successful clinical application, a more lucid correlation between the drug–liposome complex and all the steps that lead to efficient transfection needs to be established.

3.2. Polymer Delivery

In 1975, Ringsdorf formulated a polymer-based targeted drug delivery system *(1)*. Since then natural and synthetic polymers have been widely used for the delivery of low as well as high molecular weight drugs *(1,81)*. To develop a successful drug delivery system an appropriate polymer matrix should be selected. Polymers used for macromolecular drug delivery must be biocompatible lacking antigenicity, immunogenicity, and cytotoxic side effects. Polymers can be degradable or nondegradable. Developing biodegradable polymer systems is more advantageous than nondegradable systems as it eliminates the need of postdelivery removal of the polymer as the polymer is naturally resorbed by the body. Because biodegradable polymers degrade to smaller absorbable molecules, it is important to use monomers that are nontoxic in nature. Polylactide and poly (lactide-co-glycolide) (PLGA) are currently the most commonly used polymers as they are known to be biodegradable, biocompatible, and nontoxic. Naturally occurring polymers such as starch and gelatin are not very popular owing to their instability, rapid biodegradation, immunogenicity, and insufficient functional groups available for drug modification *(82)*. Successful therapeutic application of synthetic polymers depends primarily on the ability of the polymer to enhance transfection efficiency, reduce toxicity, and/or overcome biological barriers achieving targeted drug delivery.

Similar to cationic lipids, cationic polymers have gained increased popularity because of their electrostatic interaction with the negatively charged macromolecules. Macromolecular drugs are generally exposed to neutral or negatively charged environment because the luminal surfaces of blood vessels is covered with negatively charged moieties such as sialyl residues, chondroitin sulfate, and heparan sulfate *(56)*. The interaction between cationic charge of the polymer and the negative charge of the plasma membrane facilitates internalization of the macromolecular drug into the cytoplasm by pinocytosis or endocytosis or a combination of both. This is followed by the transport of the macromolecular drug through a sequence of vesicle fusion events into lysosomes. The active compound then diffuses into the cytoplasm or the nucleus, where it exerts its effect (Fig. 5). In this delivery system, the macromolecular drug is covalently linked as a unit to the polymeric backbone or to the polymeric carrier side chain via a spacer molecule as a pendant group *(1)*. The characteristics of the chemical bond used to link the macromolecular drug to the polymer including ester, imine, amide, hydrazide, azide, urethane, and thioether determine the speed of drug release from the polymer *(56)*. For instance, amide bonding shows a slower release profile whereas ester bonding shows rapid release *(56)*. This release can be pH dependent where polymer membrane destabilization occurs by the protonation of the polymer at acidic pH that leads to water influx, swelling, and eventual breakdown of the membrane *(83,84)*. It can also be

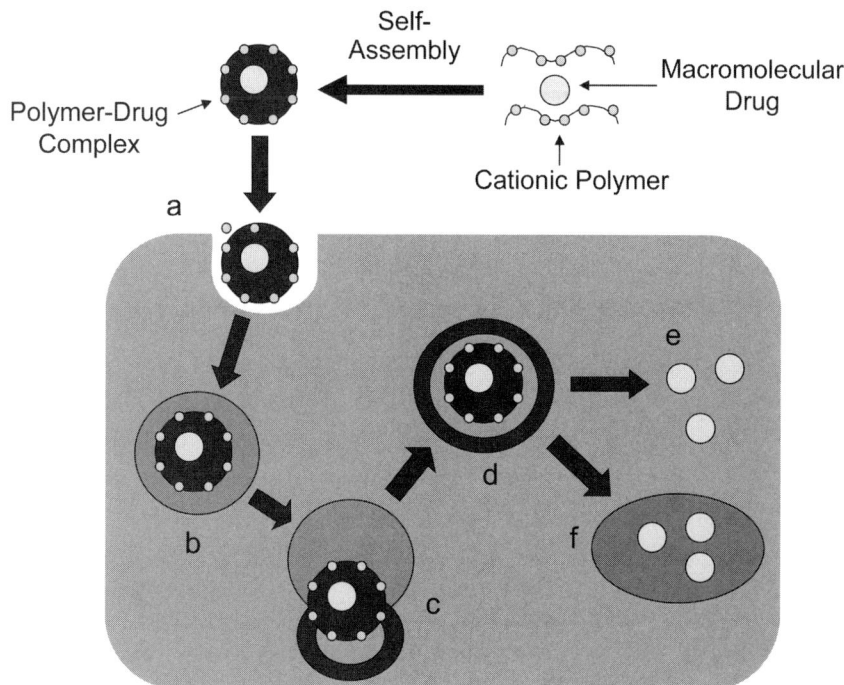

Fig. 5. Endocytosis-mediated uptake of macromolecular drugs facilitated by its attachment to delivery systems such as cationic polymers. (a) Endocytosis, (b) endosome, (c) endolysosome, (d) lysosome, (e) and (f) drug release into the cytosol and nucleus, respectively.

enzyme dependent where specific linkers which are generally biodegradable oligopeptides are used to attach the active drug moiety to the polymeric carrier that gets cleaved by specific intracellular enzymes. For example, the peptide linker Gly-Phe-Leu-Gly is preferably used to conjugate HPMA polymer and doxorubicin *(85)*.

The overall drug–polymer complex should be hydrophilic for proper water solubility. As most conventional drugs are hydrophobic in nature, polymeric micelles are widely used to confer hydrophilicity to an otherwise hydrophobic drug (Fig. 6). Micellar entrapment can be achieved by covalent or even noncovalent bonds including ionic, hydrogen bond, hydrophilic, or hydrophobic interactions, usually involving a block polymer or copolymer *(56)*. Micelle formation using hydrophilic moieties like PEG side chains is well documented. This hydrophilic backbone and the pendant PEG chains are in contact with water protecting the hydrophobic drug that forms the inner core of the micelle

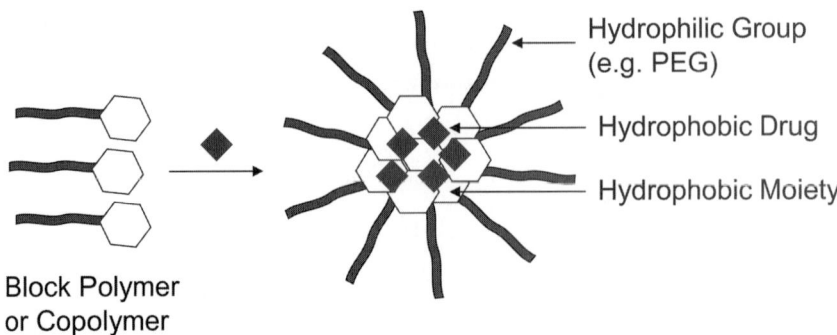

Fig. 6. Polymeric micelles facilitate the delivery of hydrophobic drugs by conferring hydrophilicity to them. Micellar entrapment can be achieved by covalent or noncovalent bonds usually involving a block polymer or copolymer. Micelle formation using hydrophilic moieties like PEG side chains protects hydrophobic drugs that form the inner core of the micelle.

from the water *(86)*. Conjugating peptides and proteins with polymers interferes with binding to proteases, thus increasing resistance against enzymatic degradation and leading to lowered immunogenicity *(42)*. For instance, conjugation of peptides and proteins with poly (styrene-co-maleic acid anhydride) (SMA) protects these macromolecules from enzymatic degradation and decreases the immunogenicity of the modified proteins. Furthermore, the conjugates bind to plasma albumin increasing the circulation time of proteins *(87,88)*. Currently, protein-based neocarzinostatin–SMA conjugate (SMANC) is approved in Japan for treating hepatoma *(89)*. PEG attached to therapeutic proteins has been shown to modify their activity by increasing their circulation life and stability *(90)*. PEG blocks the interactions between serum opsonins and therapeutic proteins as well as particulate drug carriers such as liposomes *(10)*, as discussed in the previous section. PEG-modified L-asparaginase (Oncospar®) and PEGylated interferon α-2b were approved in the United States for use in lymphoma and chronic hepatitis C treatment, respectively *(91,92)*. In addition, human IgG1 *(93)*, dextrans *(94)*, ganglioside-G_{M1} *(95)*, polysialic acid *(96)*, and other macromolecules have also been used to extend the circulatory life of proteins and liposomes.

Polyethylenimine (PEI) has emerged as a potential nonviral cationic polymer that can achieve nucleic acid transfection efficiency comparable to some viral vectors *(97)*. Conjugation of nucleic acids with PEI provides increased stability in extracellular environment, improved cellular uptake, and enhanced endosomal release *(98)*. PEI has been remarkably successful for gene delivery as evident from the recent report of the successful trial of PEI in human bladder cancer therapy *(99)*. Despite significant advances, more investigations regarding the systemic stability of PEI polyplexes and their interaction with the body are needed. This includes control over polyplex dissociation, aggregation, interaction with biomolecules, and activation of the complement system. Also, more detailed studies characterizing acute and long-term toxic response related with the use of PEI polymers are required.

Cationic polyamidoamine (PAMAM)-based dendrimers or highly branched polymers, Superfect® and Polyfect®, have been used for siRNA delivery. Superfect®-mediated siRNA delivery showed 50% inhibition of Erbin gene-product in rat pheochromocytome cells and Polyfect®-mediated siRNA delivery targeting P120RasGAP gene inhibited the target protein by greater than 90% in target cells *(100,101)*. Additionally, siPORT® inhibited the focal adhesion kinase (FAK) gene expression in human pancreatic cells by an exogenously delivered siRNA *(102)*. ONs encapsulated in biodegradable polymers can be implanted locally or can even be administered parenterally as a systemic formulation. Polylactides and copolymers of PLGA have been widely evaluated for the delivery of ONs conferring protection to the ONs from enzymatic degradation *(103,104)*. Nevertheless, macromolecular drugs formulated in advanced delivery systems may physically and chemically interact with the polymeric or encapsulating agent. Polymers are a potential class of delivery systems and should be investigated thoroughly for efficient macromolecular drug delivery.

3.3. Peptide Delivery

CPPs, also referred to as protein transduction domains (PTDs) *(105)*, can transport macromolecules across the cell membrane into the cytoplasm or nucleus. These peptides not only self-translocate into the cell but also help in the transport of various small and large molecules in a receptor, energy, and transporter-independent fashion and are thus called cell-penetrating peptides *(53,106–108)*. These peptides contain alternate domains of hydrophobic and charged residues that protect them from the harsh physiological environment. After endosomal internalization, the low pH protonates the acidic residues leading to the formation of an amphipathic helix that fuses and disrupts the endosomal membrane. In some cases, interaction of the peptide with plasma membrane phospholipids forms inverted vesicles or reverse micelles depositing the embedded drug molecules in the cytoplasm *(109)*. This mechanism is followed by the discovery of penetrating peptides that have been reported to successfully deliver tumor suppressor proteins and antisense ONs *(110)*. However, as the uptake of these peptides depends on binding to negatively charged phospholipids, they lack cell specificity, and the use of large molecules that cannot adjust to the micelles is also limited. CPPs comprise 7–30 amino acids and are further classified into homeodomain-derived peptides originating from transcription activating proteins, *Drosophila* antennapedia, and Tat-derived peptides originating from HIV-1 Tat protein.

The transport capabilities of Tat peptides for the delivery of macromolecules have been studied extensively. Initially, Tat peptide fused proteins such as Tat-p27Kip1 and Tat-p16INK4a were demonstrated to localize intracellularly followed by the report of in vivo delivery of Tat peptide and caspase-3 fusion protein that selectively induced cell death in HIV-infected cells by the activation of HIV protease *(111–113)*. Additionally, VP22 protein from the herpes simplex virus type-1 has been reported to deliver intact and functionally active proteins such as p53 *(54)*. Naturally occurring peptides including protamine, LL-37, HIV-tat, and VP-22 as well as synthetic peptides such as polylysines and polyarginines are known to form polyplexes with DNA *(114)*. Such peptide–DNA polyplexes efficiently enter cells through endocytosis *(115)* or by direct membrane transduction. The Tat peptide has also been demonstrated to be very useful in plasmid DNA delivery producing better gene expression as compared to polyarginine presumably because of an embedded NLS in its sequence that imparts importin-mediated passage through the nuclear pore *(1)*. Also, the attachment of PEI allows the plasmid DNA complex to escape the endosomal compartment, after which they are efficiently directed to the nucleus by the Tat peptide. It has also been demonstrated that NLS alone cannot achieve nuclear gene targeting. They usually require Tat peptide on their surface to deliver the gene possibly owing to the cell-penetrating capability of Tat peptides *(115)*.

The recombinant peptide myristoylphosphatidylglycerol (MPG) was the first CPP demonstrated to deliver siRNAs *(116)*. MPG is derived from the fusion peptide domain of HIV1 gp41 protein and SV40 NLS *(117)*. Recent reports suggest that the conjugation of CPPs transportan and penetratin to the 5′-end of one strand of the siRNA via a disulfide linkage increases its uptake *(27)*. However, the mechanism is not clear. SiRNA delivery with CPPs is still in its initial phases and more research is needed. Tat-protein and chimeric MPG peptide have also been employed to improve the uptake of ONs in a nonendocytic manner. A study demonstrated that the attachment of Tat peptide to β-galactosidase increased its distribution in all tissues including the brain of mice suggesting that these peptides may be used for the improved in vivo delivery of macromolecules such as ONs *(118)*. Both the Tat and Antp peptides have been shown to deliver antisense ONs to the cytoplasm as well as the nucleus, resulting in an increased pharmacological activity and cell specificity of the antisense ONs *(36)*.

CPP systems open new possibilities for the cytoplasmic and nuclear delivery of macromolecules, specifically fragile therapeutic moieties such as DNA and proteins. As the mechanism of action of these peptides is still unclear, the main challenge lies in understanding the different pathways that may be related to CPP-drug uptake. In the future, combination drug delivery systems may be designed to include the benefits of other macromolecular carriers and the unique advantages of CPPs facilitating superior intracellular delivery.

3.4. Nanodelivery

The field of nanoparticulate drug delivery is still in its infancy but it has great potential for therapeutic applications. One of the chief advantages of these systems is higher intracellular uptake as compared to other delivery systems. Nanoscale science and engineering has accelerated the development of novel drug delivery systems such as nanoscale needles, pumps, valves, and implantable devices that can improve the therapeutic effect of a macromolecular drug by targeted delivery *(119)*. Currently, nanostructured tissue scaffolds and nano/micro-arrays are being applied for reparative medicine, genetic, and other biological applications *(120,121)*. Self-assembly is being applied to create new nanoscale devices such as nanofiber peptides, protein scaffolds, and polymer networks *(122)*. In the case of polymer networks, functional monomers are preassembled with a target molecule and then the structure is locked with network formation *(123)*. Nanoscale polymer particles and spheres have been modulated to be used in various therapeutic systems *(124–126)*. It is known that environmentally responsive hydrogels that can swell or shrink in response to various pH or temperature conditions can be employed to control the release of macromolecules such as proteins and peptides. Based on this concept, nanoscale polymeric carriers, responsive to the pH of the environment, have been molecularly designed such that they release fragile therapeutic peptides and proteins directly in the upper small intestine protecting them in the acidic environment of the stomach *(127,128)*. Moreover, these nanoparticle carriers are designed such that they are stable for an extended period of time before being eliminated by the immune system *(129)*.

Reports also suggest the use of synthetic delivery systems, including polymeric nanoparticles for application in gene delivery *(130,131)*. DNA can be transfected with high efficiency and protected from lysosomal enzymes by encapsulating it in nanoparticles *(132)*. A novel nanodevice composed of ON DNA covalently attached to titanium dioxide nanoparticles has been reported to target, bind, and cleave DNA *(133)*. In vivo administration of DNA encapsulated within gelatin *(134)* or chitosan *(135)* nanospheres in mice via intramuscular injection showed enhanced transgene expression as compared to plasmids delivered either free or as cationic liposome–DNA complexes. Recently, it has been demonstrated that small micelles formed from polymers loaded with drugs could traverse rat cells *(136)*. Although these biocompatible nanocarriers could not reach the cell nucleus, they could access important drug targets such as mitochondria and golgi apparatus *(55)*. Biodegradable polyalkylcyanoacrylate (PACA) nanoparticles carrying ONs have been demonstrated to increase ex vivo serum stability and cellular uptake of ONs as well as in vivo plasma half-life of the nucleic acids targeted to the liver *(137)*. PACA nanoparticles are prepared by the polymerization of alkylcyanoacrylate monomers in acidic medium. The negatively charged PACA nanoparticles generally comprise a cationic copolymer or hydrophobic detergent in order to facilitate ON binding *(35)*. However, the in vivo application of this system may be limited because of the toxicity conferred by the hydrophobic cations and the production of formaldehyde on polymer degradation. Nanoscale devices have been fabricated using integrated circuit processing techniques and have demonstrated to allow strict regulation over the temporal control of drug release *(119)*. For instance, silicon microchips that provide controlled release of single or multiple chemical substances by electrochemical dissolution and resorbable polymeric microchip that can exhibit multipulse drug delivery have been developed and tested *(138,139)*.

On account of their small size, safe doses of nanoparticles have to be optimized that will provide manageable mass or volume for administration. The widespread application of nanoparticles and nanodevices for the delivery of macromolecules will require extensive research in order to improve the entrapment efficiencies and drug-loading techniques. Improved technologies and therapeutic methods will bring out the unlimited potential of nanoscale science and engineering leading to enhanced treatment.

4. Summary

Despite of the development of a large number of new therapeutic systems that improve drug efficacy and specificity, undesirable side effects still deter the potential therapeutic outcome of the drugs. Macromolecular drug delivery with altered pharmacokinetics and subcellular trafficking allows both tissue and subcellular specificity. Extensive clinical studies with various macromolecular drugs suggest that the problems associated with clinical trials of these drugs are basically similar to those involving conventional drugs. One of the most difficult features of macromolecular therapeutics is optimizing the delivery of high molecular weight drugs to their target sites. Most of the macromolecular drugs can elicit an immune response leading to increased toxicity and decreased therapeutic response. Despite all these limitations, interest and investment in the field of macromolecular drug delivery is growing because of the exquisite specificity that can be attained by using these molecules as therapeutic agents. An appropriate delivery system is the major determinant of the therapeutic efficacy of the drug. Besides the strategies that have been investigated so far, innovative approaches are unquestionably needed. Macromolecules as targetable drug carriers have been the focus of extensive research. Despite their shortcomings, advances in genomic and proteomic technologies continue to increase the macromolecular therapeutic candidates. Currently, about 350 macromolecules are undergoing clinical trials and approximately 84 have been approved for marketing in the United States. Research in this area will continue with thorough investigation of the macromolecular drugs and delivery systems that can successfully achieve site-specific delivery. The resulting technologies combined with the existing ones will dramatically revolutionize the field of medicine.

References

1. Nori, A. and Kopecek, J. (2005) Intracellular targeting of polymer-bound drugs for cancer chemotherapy. *Adv. Drug Deliv. Rev.* **57,** 609–636.
2. Langer, R. (1998) Drug delivery and targeting. *Nature* **392,** 5–10.
3. Takakura, Y. and Hashida, M. (1996) Macromolecular carrier systems for targeted drug delivery: pharmacokinetic considerations on biodistribution. *Pharm. Res.* **13,** 820–831.
4. Agrawal, S., Temsamani, J., Galbraith, W., and Tang, J. (1995) Pharmacokinetics of antisense oligonucleotides. *Clin. Pharmacokinet.* **28,** 7–16.
5. Srinivasan, S. K. and Iversen, P. (1995) Review of in vivo pharmacokinetics and toxicology of phosphorothioate oligonucleotides. *J. Clin. Lab. Anal.* **9,** 129–137.
6. Belting, M., Sandgren, S., and Wittrup, A. (2005) Nuclear delivery of macromolecules: barriers and carriers. *Adv. Drug Deliv. Rev.* **57,** 505–527.
7. Pouton, C. W. (1998) Nuclear import of polypeptides, polynucleotides and supramolecular complexes. *Adv. Drug Deliv. Rev.* **34,** 51–64.
8. Soyez, H., Schacht, E., and Vanderkerken, S. (1996) The crucial role of spacer groups in macromolecular prodrug design. *Adv. Drug Deliv. Rev.* **21,** 81–106.
9. de Duve, C., de Barsy, T., Poole, B., Trouet, A., Tulkens, P., and van Hoof, F. (1974) Commentary. Lysosomotropic agents. *Biochem. Pharmacol.* **23,** 2495–2531.
10. Cho, M. J. and Juliano, R. (1996) Macromolecular versus small-molecule therapeutics: drug discovery, development and clinical considerations. *Trends Biotechnol.* **14,** 153–158.
11. Olsnes, S. and Sandvig, K. (1988) How protein toxins enter and kill cells. *Cancer Treat. Res.* **37,** 39–73.
12. Luo, D. and Saltzman, W. M. (2002) Preface. *Somat. Cell Mol. Genet.* **27,** 1–3.
13. Dass, C. R. (2004) Oligonucleotide delivery to tumours using macromolecular carriers. *Biotechnol. Appl. Biochem.* **40,** 113–122.
14. Gansbacher, B. (2003) Report of a second serious adverse event in a clinical trial of gene therapy for X-linked severe combined immune deficiency (X-SCID). Position of the European Society of Gene Therapy (ESGT). *J. Gene Med.* **5,** 261–262.
15. Merdan, T., Kopecek, J., and Kissel, T. (2002) Prospects for cationic polymers in gene and oligonucleotide therapy against cancer. *Adv. Drug Deliv. Rev.* **54,** 715–758.
16. Duzgunes, N., De Ilarduya, C. T., Simoes, S., Zhdanov, R. I., Konopka, K., and Pedroso de Lima, M. C. (2003) Cationic liposomes for gene delivery: novel cationic lipids and enhancement by proteins and peptides. *Curr. Med. Chem.* **10,** 1213–1220.

17. Lee, M. and Kim, S. W. (2005) Polyethylene glycol-conjugated copolymers for plasmid DNA delivery. *Pharm. Res.* **22,** 1–10.
18. Pitard, B. (2002) Supramolecular assemblies of DNA delivery systems. *Somat. Cell Mol. Genet.* **27,** 5–15.
19. Gilmore, I. R., Fox, S. P., Hollins, A. J., Sohail, M., and Akhtar, S. (2004) The design and exogenous delivery of siRNA for post-transcriptional gene silencing. *J. Drug Target* **12,** 315–340.
20. Martinez, J. and Tuschl, T. (2004) RISC is a 5′ phosphomonoester-producing RNA endonuclease. *Genes Dev.* **18,** 975–980.
21. Lucentini, J. (2004) Silencing cancer — As therapeutic RNAi applications evolve, multidrug- resistant cancers come into the crosshairs. *Scientist* **18,** 14–15.
22. Pallarito, K. (2004) Fueling the fires of RNA interference is big business, but can companies deliver the goods? *Scientist* **18,** 18–19.
23. Chiu, Y. L. and Rana, T. M. (2002) RNAi in human cells: basic structural and functional features of small interfering RNA. *Mol. Cell.* **10,** 549–561.
24. Czauderna, F., Fechtner, M., Dames, S., et al. (2003) Structural variations and stabilising modifications of synthetic siRNAs in mammalian cells. *Nucleic Acids Res.* **31,** 2705–2716.
25. Holen, T., Amarzguioui, M., Wiiger, M. T., Babaie, E., and Prydz, H. (2002) Positional effects of short interfering RNAs targeting the human coagulation trigger tissue factor. *Nucleic Acids Res.* **30,** 1757–1766.
26. Parrish, S., Fleenor, J., Xu, S., Mello, C., and Fire, A. (2000) Functional anatomy of a dsRNA trigger: differential requirement for the two trigger strands in RNA interference. *Mol. Cell.* **6,** 1077–1087.
27. Muratovska, A. and Eccles, M. R. (2004) Conjugate for efficient delivery of short interfering RNA (siRNA) into mammalian cells. *FEBS Lett.* **558,** 63–68.
28. Beale, G., Hollins, A. J., Benboubetra, M., et al. (2003) Gene silencing nucleic acids designed by scanning arrays: anti-EGFR activity of siRNA, ribozyme and DNA enzymes targeting a single hybridization-accessible region using the same delivery system. *J. Drug Target* **11,** 449–456.
29. Lewis, D. L., Hagstrom, J. E., Loomis, A. G., Wolff, J. A., and Herweijer, H. (2002) Efficient delivery of siRNA for inhibition of gene expression in postnatal mice. *Nat. Genet.* **32,** 107–108.
30. McCaffrey, A. P., Meuse, L., Pham, T. T., Conklin, D. S., Hannon, G. J., and Kay, M. A. (2002) RNA interference in adult mice. *Nature* **418,** 38–39.
31. Makimura, H., Mizuno, T. M., Mastaitis, J. W., Agami, R., and Mobbs, C. V. (2002) Reducing hypothalamic AGRP by RNA interference increases metabolic rate and decreases body weight without influencing food intake. *BMC Neurosci.* **3,** 18.
32. Reich, S. J., Fosnot, J., Kuroki, A., et al. (2003) Small interfering RNA (siRNA) targeting VEGF effectively inhibits ocular neovascularization in a mouse model. *Mol. Vis.* **9,** 210–216.
33. Zhang, X., Shan, P., Jiang, D., et al. (2004) Small interfering RNA targeting heme oxygenase-1 enhances ischemia-reperfusion-induced lung apoptosis. *J. Biol. Chem.* **279,** 10,677–10,684.

34. Akhtar, S. and Agrawal, S. (1997) In vivo studies with antisense oligonucleotides. *Trends Pharmacol. Sci.***18**, 12–18.
35. Akhtar, S., Hughes, M. D., Khan, A. et al. (2000) The delivery of antisense therapeutics. *Adv. Drug Deliv. Rev.* **44**, 3–21.
36. Dokka, S. and Rojanasakul, Y. (2000) Novel non-endocytic delivery of antisense oligonucleotides. *Adv. Drug Deliv. Rev.* **44**, 35–49.
37. Wickstrom, E. (1986) Oligodeoxynucleotide stability in subcellular extracts and culture media. *J. Biochem. Biophys. Methods* **13**, 97–102.
38. Akhtar, S., Kole, R., and Juliano, R. L. (1991) Stability of antisense DNA oligodeoxynucleotide analogs in cellular extracts and sera. *Life Sci.* **49**, 1793–1801.
39. Hudson, A. J., Normand, N., Ackroyd, J., and Akhtar, S. (1999) Cellular delivery of hammerhead ribozymes conjugated to a transferrin receptor antibody. *Int. J. Pharm.* **182**, 49–58.
40. Leonetti, J. P., Mechti, N., Degols, G., Gagnor, C., and Lebleu, B. (1991) Intracellular distribution of microinjected antisense oligonucleotides. *Proc. Natl Acad. Sci. U. S. A.* **88**, 2702–2706.
41. Cryan, S. A. (2005) Carrier-based strategies for targeting protein and peptide drugs to the lungs. *AAPS J.* **7**, E20–E41.
42. Torchilin, V. P. and Lukyanov, A. N. (2003) Peptide and protein drug delivery to and into tumors: challenges and solutions. *Drug Discov. Today* **8**, 259–266.
43. Bai, J. P. and Chang, L. L. (1993) Comparison of site-dependent degradation of peptide drugs within the gut of rats and rabbits. *J. Pharm. Pharmacol.* **45**, 1085–1087.
44. Gombotz, W. R. and Pettit, D. K. (1995) Biodegradable polymers for protein and peptide drug delivery. *Bioconjug. Chem.* **6**, 332–351.
45. Lyczak, J. B. and Morrison, S. L. (1994) Biological and pharmacokinetic properties of a novel immunoglobulin-CD4 fusion protein. *Arch. Virol.* **139**, 189–196.
46. Zelphati, O., Wang, Y., Kitada, S., Reed, J. C., Felgner, P. L., and Corbeil, J. (2001) Intracellular delivery of proteins with a new lipid-mediated delivery system. *J. Biol. Chem.* **276**, 35,103–35,110.
47. Tinsley, J. H., Hawker, J., and Yuan, Y. (1998) Efficient protein transfection of cultured coronary venular endothelial cells. *Am. J. Physiol.* **275**, H1873–H1878.
48. Liu, X. H., Castelli, J. C., and Youle, R. J. (1999) Receptor-mediated uptake of an extracellular Bcl-x(L) fusion protein inhibits apoptosis. *Proc. Natl Acad. Sci. U. S. A.* **96**, 9563–9567.
49. Hawiger, J. (1999) Noninvasive intracellular delivery of functional peptides and proteins. *Curr. Opin. Chem. Biol.* **3**, 89–94.
50. Yoneda, Y., Semba, T., Kaneda, Y., et al. (1992) A long synthetic peptide containing a nuclear localization signal and its flanking sequences of SV40 T-antigen directs the transport of IgM into the nucleus efficiently. *Exp. Cell. Res.* **201**, 313–320.
51. Dworetzky, S. I., Lanford, R. E., and Feldherr, C. M. (1988) The effects of variations in the number and sequence of targeting signals on nuclear uptake. *J. Cell. Biol.* **107**, 1279–1287.

52. Derossi, D., Calvet, S., Trembleau, A., Brunissen, A., Chassaing, G., and Prochiantz, A. (1996) Cell internalization of the third helix of the Antennapedia homeodomain is receptor- independent. *J. Biol. Chem.* **271,** 18,188–18,193.

53. Vives, E., Brodin, P., and Lebleu, B. (1997) A truncated HIV-1 Tat protein basic domain rapidly translocates through the plasma membrane and accumulates in the cell nucleus. *J. Biol. Chem.* **272,** 16,010–16,017.

54. Phelan, A., Elliott, G., and O'Hare, P. (1998) Intercellular delivery of functional p53 by the herpesvirus protein VP22. *Nat. Biotechnol.* **16,** 440–443.

55. Orive, G., Hernandez, R. M., Rodriguez Gascon, A., Dominguez-Gil, A., and Pedraz, J. L. (2003) Drug delivery in biotechnology: present and future. *Curr. Opin. Biotechnol.* **14,** 659–664.

56. Greish, K., Fang, J., Inutsuka, T., Nagamitsu, A., and Maeda, H. (2003) Macromolecular therapeutics: advantages and prospects with special emphasis on solid tumour targeting. *Clin. Pharmacokinet.* **42,** 1089–1105.

57. Cao, A., Briane, D., Coudert, R., et al. (2000) Delivery and pathway in MCF7 cells of DNA vectorized by cationic liposomes derived from cholesterol. *Antisense Nucleic Acid Drug Dev.* **10,** 369–380.

58. Farhood, H., Bottega, R., Epand, R. M., and Huang, L. (1992) Effect of cationic cholesterol derivatives on gene transfer and protein kinase C activity. *Biochim. Biophys. Acta* **1111,** 239–246.

59. Mahato, R. I., Takakura, Y., and Hashida, M. (1997) Development of targeted delivery systems for nucleic acid drugs. *J. Drug Target* **4,** 337–357.

60. Leonetti, J. P., Machy, P., Degols, G., Lebleu, B., and Leserman, L. (1990) Antibody-targeted liposomes containing oligodeoxyribonucleotides complementary to viral RNA selectively inhibit viral replication. *Proc. Natl Acad. Sci. U. S. A.* **87,** 2448–2451.

61. Renneisen, K., Leserman, L., Matthes, E., Schroder, H. C., and Muller, W. E. (1990) Inhibition of expression of human immunodeficiency virus-1 in vitro by antibody-targeted liposomes containing antisense RNA to the env region. *J. Biol. Chem.* **265,** 16,337–16,342.

62. Senior, J. H. (1987) Fate and behavior of liposomes in vivo: a review of controlling factors. *Crit. Rev. Ther. Drug Carrier Syst.* **3,** 123–193.

63. Kamps, J. A., Morselt, H. W., and Scherphof, G. L. (1999) Uptake of liposomes containing phosphatidylserine by liver cells in vivo and by sinusoidal liver cells in primary culture: in vivo-in vitro differences. *Biochem. Biophys. Res. Commun.* **256,** 57–62.

64. Dass, C. R. and Burton, M. A. (1999) Lipoplexes and tumours. A review. *J. Pharm. Pharmacol.* **51,** 755–770.

65. Klibanov, A. L., Maruyama, K., Torchilin, V. P., and Huang, L. (1990) Amphipathic polyethyleneglycols effectively prolong the circulation time of liposomes. *FEBS Lett.* **268,** 235–237.

66. Gao, X. and Huang, L. (1991) A novel cationic liposome reagent for efficient transfection of mammalian cells. *Biochem. Biophys. Res. Commun.* **179,** 280–285.

67. Gao, X. and Huang, L. (1995) Cationic liposome-mediated gene transfer. *Gene Ther.* **2,** 710–722.

68. Dass, C. R. (2002) Biochemical and biophysical characteristics of lipoplexes pertinent to solid tumour gene therapy. *Int. J. Pharm.* **241,** 1–25.

69. Gabizon, A. A. (2001) Pegylated liposomal doxorubicin: metamorphosis of an old drug into a new form of chemotherapy. *Cancer Invest.* **19,** 424–436.

70. Cornelis, S., Vandenbranden, M., Ruysschaert, J. M., and Elouahabi, A. (2002) Role of intracellular cationic liposome-DNA complex dissociation in transfection mediated by cationic lipids. *DNA Cell. Biol.* **21,** 91–97.

71. Lechardeur, D., Sohn, K. J., Haardt, M., et al. (1999) Metabolic instability of plasmid DNA in the cytosol: a potential barrier to gene transfer. *Gene Ther.* **6,** 482–497.

72. Bohula, E. A., Salisbury, A. J., Sohail, M., et al. (2003) The efficacy of small interfering RNAs targeted to the type 1 insulin-like growth factor receptor (IGF1R) is influenced by secondary structure in the IGF1R transcript. *J. Biol. Chem.* **278,** 15,991–15,997.

73. Troussard, A. A., Mawji, N. M., Ong, C., Mui, A., St -Arnaud, R., and Dedhar, S. (2003) Conditional knock-out of integrin-linked kinase demonstrates an essential role in protein kinase B/Akt activation. *J. Biol. Chem.* **278,** 22,374–22,378.

74. Leu, Y. W., Rahmatpanah, F., Shi, H., et al. (2003) Double RNA interference of DNMT3b and DNMT1 enhances DNA demethylation and gene reactivation. *Cancer Res.* **63,** 6110–6115.

75. Potente, M., Fisslthaler, B., Busse, R., and Fleming, I. (2003) 11,12-Epoxy-eicosatrienoic acid-induced inhibition of FOXO factors promotes endothelial proliferation by down-regulating p27Kip1. *J. Biol. Chem.* **278,** 29,619–29,625.

76. Sorensen, D. R., Leirdal, M., and Sioud, M. (2003) Gene silencing by systemic delivery of synthetic siRNAs in adult mice. *J. Mol. Biol.* **327,** 761–766.

77. Farhood, H., Gao, X., Son, K., et al. (1994) Cationic liposomes for direct gene transfer in therapy of cancer and other diseases. *Ann. N.Y. Acad. Sci.* **716,** 23–34.

78. Felgner, P. L., Gadek, T. R., Holm, M., et al. (1987) Lipofection: a highly efficient, lipid-mediated DNA-transfection procedure. *Proc. Natl Acad. Sci. U. S. A.* **84,** 7413–7417.

79. Akhtar, S. (1998) Antisense technology: selection and delivery of optimally acting antisense oligonucleotides. *J. Drug Target* **5,** 225–234.

80. DeLong, R. K., Yoo, H., Alahari, S. K., et al. (1999) Novel cationic amphiphiles as delivery agents for antisense oligonucleotides. *Nucleic Acids Res.* **27,** 3334–3341.

81. Ringsdorf, H. (1975) Structure and properties of pharmacologically active polymers. *J. Polym. Sci. Polym. Symp.* **51,** 135–153.

82. Duncan, R. and Kopecek, J. (1984) Soluble synthetic polymers as potential drug carriers. *Adv. Polym. Sci.* **57,** 51–101.

83. Boussif, O., Lezoualc'h, F., Zanta, M. A., et al. (1995) A versatile vector for gene and oligonucleotide transfer into cells in culture and in vivo: polyethylenimine. *Proc. Natl Acad. Sci. U. S. A.* **92,** 7297–7301.

84. Haensler, J., Szoka, F. C., Jr. (1993) Polyamidoamine cascade polymers mediate efficient transfection of cells in culture. *Bioconjug. Chem.* **4,** 372–379.

85. Vasey, P. A., Kaye, S. B., Morrison, R., et al. (1999) Phase I clinical and pharmacokinetic study of PK1 [N-(2-hydroxypropyl)methacrylamide copolymer doxorubicin]: first member of a new class of chemotherapeutic agents-drug-polymer conjugates. Cancer Research Campaign Phase I/II Committee. *Clin. Cancer Res.* **5,** 83–94.

86. Kwon, G. S. and Kataoka, K. (1995) Block copolymer micelles as long-circulating drug vehicles. *Adv. Drug Deliv. Rev.* **16,** 295–309.

87. Maeda, H. (2001) SMANCS and polymer-conjugated macromolecular drugs: advantages in cancer chemotherapy. *Adv. Drug Deliv. Rev.* **46,** 169–185.

88. Mu, Y., Kamada, H., Kaneda, Y., et al. (1999) Bioconjugation of laminin peptide YIGSR with poly(styrene co-maleic acid) increases its antimetastatic effect on lung metastasis of B16-BL6 melanoma cells. *Biochem. Biophys. Res. Commun.* **255,** 75–79.

89. Maeda, H., Sawa, T., and Konno, T. (2001) Mechanism of tumor-targeted delivery of macromolecular drugs, including the EPR effect in solid tumor and clinical overview of the prototype polymeric drug SMANCS. *J. Control. Release* **74,** 47–61.

90. Abuchowski, A., Kazo, G. M., Verhoest, C. R., Jr., et al. (1984) Cancer therapy with chemically modified enzymes. I. Antitumor properties of polyethylene glycol-asparaginase conjugates. *Cancer Biochem. Biophys.* **7,** 175–178.

91. Ettinger, A. R. (1995) Pegaspargase (Oncaspar). *J. Pediatr. Oncol. Nurs.* **12,** 46–48.

92. Youngster, S., Wang, Y. S., Grace, M., Bausch, J., Bordens, R., and Wyss, D. F. (2002) Structure, biology, and therapeutic implications of pegylated interferon alpha-2b. *Curr. Pharm. Des.* **8**(24), 2139–2157.

93. Capon, D. J., Chamow, S. M., Mordenti, J., et al. (1989) Designing CD4 immunoadhesins for AIDS therapy. *Nature* **337,** 525–531.

94. Fujita, T., Nishikawa, M., Tamaki, C., Takakura, Y., Hashida, M., and Sezaki, H. (1992) Targeted delivery of human recombinant superoxide dismutase by chemical modification with mono- and polysaccharide derivatives. *J. Pharmacol. Exp. Ther.* **263,** 971–978.

95. Allen, T. M. and Chonn, A. (1987) Large unilamellar liposomes with low uptake into the reticuloendothelial system. *FEBS Lett.* **223,** 42–46.

96. Gregoriadis, G., McCormack, B., Wang, Z., and Lifely, R. (1993) Polysialic acids: potential in drug delivery. *FEBS Lett.* **315,** 271–276.

97. Abdallah, B., Hassan, A., Benoist, C., Goula, D., Behr, J. P., and Demeneix, B. A. (1996) A powerful nonviral vector for in vivo gene transfer into the adult mammalian brain: polyethylenimine. *Hum. Gene Ther.* **7,** 1947–1954.

98. Michael Neu, M., Fischer, D., and Kissel, T. (2005) Recent advances in rational gene transfer vector design based on poly(ethyleneimine) and its derivatives. *J. Gen. Med.* **8,** 992–1009.

99. Ohana, P., Gofrit, O., and Ayesh, S. (2004) Regulatory sequences of the H19 gene in DNA based therapy of bladder cancer. *Gene Ther. Mol. Biol.* **8,** 181–192.

100. Tsubouchi, A., Sakakura, J., Yagi, R., et al. (2002) Localized suppression of RhoA activity by Tyr31/118-phosphorylated paxillin in cell adhesion and migration. *J. Cell. Biol.* **159,** 673–683.

101. Huang, Y. Z., Zang, M., Xiong, W. C., Luo, Z., and Mei, L. (2003) Erbin suppresses the MAP kinase pathway. *J. Biol. Chem.* **278,** 1108–1114.
102. Duxbury, M. S., Ito, H., Benoit, E., Zinner, M. J., Ashley, S. W., and Whang, E. E. (2003) RNA interference targeting focal adhesion kinase enhances pancreatic adenocarcinoma gemcitabine chemosensitivity. *Biochem. Biophys. Res. Commun.* **311,** 786–792.
103. Lewis, K. J., Irwin, W. J., and Akhtar, S. (1998) Development of a sustained-release biodegradable polymer delivery system for site-specific delivery of oligonucleotides: characterization of P(LA-GA) copolymer microspheres in vitro. *J. Drug Target* **5,** 291–302.
104. Putney, S. D., Brown, J., Cucco, C., et al. (1999) Enhanced anti-tumor effects with microencapsulated c-myc antisense oligonucleotide. *Antisense Nucleic Acid Drug Dev.* **9,** 451–458.
105. Schwarze, S. R., Hruska, K. A., and Dowdy, S. F. (2000) Protein transduction: unrestricted delivery into all cells? *Trends Cell. Biol.* **10,** 290–295.
106. Derossi, D., Joliot, A. H., Chassaing, G., and Prochiantz, A. (1994) The third helix of the Antennapedia homeodomain translocates through biological membranes. *J. Biol. Chem.* **269,** 10,444–10,450.
107. Frankel, A. D. and Pabo, C. O. (1988) Cellular uptake of the tat protein from human immunodeficiency virus. *Cell* **55,** 1189–1193.
108. Lindgren, M., Hallbrink, M., Prochiantz, A., and Langel, U. (2000) Cell-penetrating peptides. *Trends Pharmacol. Sci.* **21,** 99–103.
109. Prochiantz, A. (2000) Messenger proteins: homeoproteins, TAT and others. *Curr. Opin. Cell. Biol.* **12,** 400–406.
110. Allinquant, B., Hantraye, P., Mailleux, P., Moya, K., Bouillot, C., and Prochiantz, A. (1995) Downregulation of amyloid precursor protein inhibits neurite outgrowth in vitro. *J. Cell. Biol.* **128,** 919–927.
111. Nagahara, H., Vocero-Akbani, A. M., Snyder, E. L., et al. (1998) Transduction of full-length TAT fusion proteins into mammalian cells: TAT-p27Kip1 induces cell migration. *Nat. Med.* **4,** 1449–1452.
112. Gius, D. R., Ezhevsky, S. A., Becker-Hapak, M., Nagahara, H., Wei, M. C., and Dowdy, S. F. (1999) Transduced p16INK4a peptides inhibit hypophosphorylation of the retinoblastoma protein and cell cycle progression prior to activation of Cdk2 complexes in late G1. *Cancer Res.* **59,** 2577–2580.
113. Vocero-Akbani, A. M., Heyden, N. V., Lissy, N. A., Ratner, L., and Dowdy, S. F. (1999) Killing HIV-infected cells by transduction with an HIV protease-activated caspase-3 protein. *Nat. Med.* **5,** 29–33.
114. Sandgren, S., Wittrup, A., Cheng, F., et al. (2004) The human antimicrobial peptide LL-37 transfers extracellular DNA plasmid to the nuclear compartment of mammalian cells via lipid rafts and proteoglycan-dependent endocytosis. *J. Biol. Chem.* **279,** 17,951–17,956.
115. Eguchi, A., Akuta, T., Okuyama, H., et al. (2001) Protein transduction domain of HIV-1 Tat protein promotes efficient delivery of DNA into mammalian cells. *J. Biol. Chem.* **276,** 26,204–26,210.

116. Simeoni, F., Morris, M. C., Heitz, F., and Divita, G. (2003) Insight into the mechanism of the peptide-based gene delivery system MPG: implications for delivery of siRNA into mammalian cells. *Nucleic Acids Res.* **31,** 2717–2724.

117. Morris, M. C., Vidal, P., Chaloin, L., Heitz, F., and Divita, G. (1997) A new peptide vector for efficient delivery of oligonucleotides into mammalian cells. *Nucleic Acids Res.* **25,** 2730–2736.

118. Schwarze, S. R., Ho, A., Vocero-Akbani, A., and Dowdy, S. F. (1999) In vivo protein transduction: delivery of a biologically active protein into the mouse. *Science* **285,** 1569–1572.

119. Hilt, J. Z. (2004) Nanotechnology and biomimetic methods in therapeutics: molecular scale control with some help from nature. *Adv. Drug Deliv. Rev.* **56,** 1533–1536.

120. Saltzman, W. M. and Olbricht, W. L. (2002) Building drug delivery into tissue engineering. *Nat. Rev. Drug Discov.* **1,** 177–186.

121. Gershon, D. (2002) Microarray technology: an array of opportunities. *Nature* **416,** 885–891.

122. Zhang, S. (2003) Fabrication of novel biomaterials through molecular self-assembly. *Nat. Biotechnol.* **21,** 1171–1178.

123. Byrne, M. E., Park, K., and Peppas, N. A. (2002) Molecular imprinting within hydrogels. *Adv. Drug Deliv. Rev.* **54,** 149–161.

124. Peppas, N. A., Bures, P., Leobandung, W., and Ichikawa, H. (2000) Hydrogels in pharmaceutical formulations. *Eur. J. Pharm. Biopharm.* **50,** 27–46.

125. Uhrich, K. E., Cannizzaro, S. M., Langer, R. S., and Shakesheff, K. M. (1999) Polymeric systems for controlled drug release. *Chem. Rev.* **99,** 3181–3198.

126. Chien, Y. W. and Lin, S. (2002) Optimisation of treatment by applying programmable rate-controlled drug delivery technology. *Clin. Pharmacokinet.* **41,** 1267–1299.

127. Morishita, M., Goto, T., Peppas, N. A., et al. (2004) Mucosal insulin delivery systems based on complexation polymer hydrogels: effect of particle size on insulin enteral absorption. *J. Control. Release* **97,** 115–124.

128. Donini, C., Robinson, D. N., Colombo, P., Giordano, F., and Peppas, N. A. (2002) Preparation of poly(methacrylic acid-g-poly[ethylene glycol]) nanospheres from methacrylic monomers for pharmaceutical applications. *Int. J. Pharm.* **245,** 83–91.

129. Gref, R., Minamitake, Y., Peracchia, M. T., Trubetskoy, V., Torchilin, V., and Langer, R. (1994) Biodegradable long-circulating polymeric nanospheres. *Science* **263,** 1600–1603.

130. Luo, D. and Saltzman, W. M. (2000) Synthetic DNA delivery systems. *Nat. Biotechnol.* **18,** 33–37.

131. Cohen, H., Levy, R. J., Gao, J., et al. (2000) Sustained delivery and expression of DNA encapsulated in polymeric nanoparticles. *Gene Ther.* **7,** 1896–1905.

132. Bonadio, J., Smiley, E., Patil, P., and Goldstein, S. (1999) Localized, direct plasmid gene delivery in vivo: prolonged therapy results in reproducible tissue regeneration. *Nat. Med.* **5,** 753–759.

133. Paunesku, T., Rajh, T., Wiederrecht, G., et al. (2003) Biology of TiO_2-oligonucleotide nanocomposites. *Nat. Mater.* **2,** 343–346.

134. Truong-Le, V. L., August, J. T., and Leong, K. W. (1998) Controlled gene delivery by DNA- gelatin nanospheres. *Hum. Gene Ther.* **9,** 1709–1717.

135. Leong, K. W., Mao, H. Q., Truong-Le, V. L., Roy, K., Walsh, S. M., and August, J. T. (1998) DNA-polycation nanospheres as non-viral gene delivery vehicles. *J. Control. Release* **53,** 183–193.

136. Savic, R., Luo, L., Eisenberg, A., and Maysinger, D. (2003) Micellar nanocontainers distribute to defined cytoplasmic organelles. *Science* **300,** 615–618.

137. Fattal, E., Vauthier, C., Aynie, I., et al. (1998) Biodegradable polyalkylcyano-acrylate nanoparticles for the delivery of oligonucleotides. *J. Control. Release* **53,** 137–143.

138. Santini, J. T., Jr., Cima, M. J., and Langer, R. (1999) A controlled-release microchip. *Nature* **397,** 335–338.

139. Grayson, J. (1967) Thermal conductivity of normal and infracted heart muscle. *Nature* **215,** 767–768.

15

Changes in Biologic Drug Approval at FDA

Richard E. Lowenthal and Robert G. Bell

Abstract

Over the next 5 yr at FDA, there is a great need for the agency to continue its efforts to evolve and collaborate with the pharmaceutical industry to promote rationale drug development. The agency is dynamic, and whereas plans are in place that may provide some concept of the direction of future changes for both drug and biologics reviews, local and world events often cause a reactionary response at the FDA that is often unpredictable. What is clear is that the agency will need strong leadership and should continue to make every effort to improve the regulatory review process and its ability to assess safety and efficacy of new medicinal products in a carefully managed and balanced process. Some of the safety concerns raised after the withdrawal of a number of commonly used medicinal products of the past 10 yr have shifted FDA's attention and resources away from critical programs intended to facilitate drug development and provide some regulatory relief to a struggling industry. To achieve improvements in both safety and efficacy assessments as well as review timelines and consistency in reviews within the agency, the FDA will have to further consolidate medicinal product reviews in alignment with therapeutic indications.

The FDA can provide both the biotechnology industry and the public great benefits in the future by revitalizing the effort to define *critical path* for medicinal product development and creating new methodologies to work with industry to accelerate product approval. The next 5 yr will be a critical period during which the FDA will need strong leadership to take managed risks to move forward a number of important initiatives intended to accelerate the drug development process. The leadership must also consolidate and focus internal resources within the agency to maintain consistent requirements for a therapeutic indication and continue to improve review processes.

Key Words: Food and drug administration; center for drug evaluation and research; center for biologics evaluation and research; critical path initiative; PDUFA III; medicinal product development; continual marketing applications; electronic common technical dossier.

From: *Biopharmaceutical Drug Design and Development*
Edited by: S. Wu-Pong and Y. Rojanasakul © Humana Press Inc., Totowa, NJ

1. Introduction

After a long period without a permanent Food and Drug Administration (FDA) Commissioner, Dr Lester Crawford was confirmed in July 2005. After the 5-mo confirmation process in Congress needed to gain consensus on the nomination, Dr Crawford resigned suddenly in September 2005 again leaving the FDA without a clear leadership at a time when the agency was facing significant challenges in its efforts to balance improvements in the drug development and approval process with ensuring safety of new medicinal products.

Acting quickly, the president appointed Dr Andrew von Eschenbach as acting FDA Commissioner on September 23, 2005 and it is widely anticipated that this will eventually become a permanent appointment. Dr von Eschenbach will face significant challenges to shift the agency's evaluation of safety and efficacy back to a more balanced approach that takes into consideration new safety risks as well as proven efficacy of a medicinal product. In the coming years, as FDA continues to evolve, the need to revitalize a number of significant initiatives becomes increasingly critical to the pharmaceutical industry and the agency.

Between 2002 and 2005, the FDA has become increasingly more preoccupied with safety events and a need to re-engineer the assessment and oversight of medicinal products to better assess safety during development and in the first few years after marketing a new therapy. At the same time, PDUFA III has been implemented placing significant pressure on the agency to accelerate medicinal product reviews and meet obligations to Congress defined in the law. Further distractions include bioterrorism efforts, international efforts (e.g., International Committee for Harmonization, ICH), and internal reorganizations to align therapeutic review criteria and consolidate resources.

What is critical in the next 5 yr for FDA are efforts to refocus on initiatives that will accelerate the drug development process and improve the life cycle management of medicinal products. Dr von Eschenbach has already placed emphasis on revitalizing the Critical Path Initiative, which has stagnated in 2004 and has seen little progress without clear leadership at the agency. A strong leader at FDA is needed to move the Critical Path Initiative forward and further evolve the concept into a collaborative and circular process that allows the agency to collaborate with industry partners early in the development cycle and continue cooperation through the life time of the product.

It will also be important for the leadership of FDA to clarify the role of the agency in evaluating the benefit–risk of a medicinal product and explain this process to both the Congress and the public. The need to take managed risks in the development of new medicinal products and defining those risks in light of potential therapeutic benefits will become increasingly important in the future to help avoid lengthy development timelines and excessive costs that delay or eliminate beneficial drugs from coming to market that could help patients in need for new therapies to treat or cure serious illnesses.

Over the long history of FDA, the agency has evolved and changed. In some cases, these changes have been in reaction to sudden and negative events, whereas other changes have been well conceived and managed. With a significant need for the agency to consolidate resources, improve processes, improve consistency and work with industry to facilitate good medicinal products coming to market, leadership with the foresight and ability to manage future perceptions and changes is critical. Pharmaceutical development remains one of the largest industries of United States and scientific advances that bring patients new medicines one of the greatest needs in society. With an industry seeking a clear pathway to develop cost-effective drugs, the need for FDA to participate and collaborate in the development process is becoming increasingly important. The next 5 yr will be an important and significant period in the history of FDA that will define the future of the pharmaceutical industry in the United States and worldwide.

2. History of Biologic Regulation at FDA *(1,2,3,4)*

Regulatory oversight of biologics by the US Federal Government has dramatically evolved over the 100 yr since the passing of the 1902 Biologics Control Act by Congress. The Act was in reaction to the death of 13 children after receiving diphtheria antitoxin contaminated with tetanus, which emphasized the need for regulatory oversight for the production of biologics. The Act also required that producers of vaccines be licensed annually for the manufacture and sale of vaccines, serum, and antitoxins. Prior to this time no regulatory oversight of biologics existed in the United States.

The Food and Drugs Act was passed by Congress in 1906, which outlawed adulterated and misbranded foods and drugs, but did not provide for additional specific regulation of biologics. As the efforts to regulate both drugs and biologics continued in the face of mounting public health events, the Public Health Service was created in 1912. Congress also made efforts to improve the government's ability to support and conduct medical research, which led to the formation of the National Institute of Health (NIH) in 1930 as a scientific and medical research organization within the government. The Division of Biologics Control was established in 1937 to support oversight of commercial biologics, but was limited in their ability to regulate products sold to the public.

It was well recognized by this time that the Pure Food and Drug Act had serious flaws and experienced numerous failures in efforts to enforce the Act. As the complexities of drug and biologics manufacturing increased, some medicinal products with associated risks fell outside of the legal and regulatory oversight of the Pure Food and Drug Act, and efforts to amend the Act were initiated within Congress. In 1937, a diethylene glycol-contaminated elixir of sulfanilamide resulted in the death of over 100 children. This event renewed efforts to amend the Act within Congress resulting in the Federal Food, Drug and Cosmetic Act that was signed into law in June 1938 creating the modern Food and Drug Administration (FDA) *(5)*. Between the 1902 and 1938 Acts, appropriate provisions in the law began to emerge that would facilitate regulation of biologics.

Throughout the 1940s, amendments to the Act brought further oversight of biologics that included the Insulin and Penicillin Amendments and requirements for prescriber labeling. The Penicillin Amendment was of particular importance as it required that all batches of drug products containing penicillin were required to undergo a formal batch certification by the FDA prior to use in human medicinal products. This certification requirement was expanded to include other antibiotics and was managed within FDA's Drug Division. The Public Health Service Act was passed in 1944 creating the Laboratory of Biologics Control to facilitate testing of biologics for the bureau.

In the 1950s significant breakthroughs in medicine were made with the development of the first vaccine for polio by Jonas Salk, arguably the greatest advance in medicine at the time. The polio vaccine had to be tested before licensing with injections given to over 1.8 million children prior to its licensure in 1950. Later emphasis on the manufacturing and controls was further pursued when a batch of vaccine produced by Cutter Laboratories led to more than 260 people contracting polio because of a failure to completely inactivate the live virus vaccine. Polio vaccine use was suspended in May 1955 until all manufacturing facilities could be inspected and procedures for production approved by the bureau.

In 1951, the Durham-Humphrey Amendment defined the types of drugs that would require a prescription from a medical professional to be sold in the United States. In 1962, the Kefauver-Harris Amendments to the Act were introduced, after thalidomide was found to cause birth defects, to establish additional regulatory authority to evaluate drug and biologic safety and to require that such products be tested in clinical trials. These amendments also required that companies provide "substantial evidence" of the effectiveness of a drug or biologic prior to approval of a labeled claim for the product. Enforcement was also enhanced by these amendments through the establishment of "Good Manufacturing Practices" (GMPs) and the requirement that commercial manufacturers be inspected every 2 yr and register annually with the FDA. The amendments passed in 1962, also extended the FD&C Act to blood products and blood banks. In 1972, the Bureau of Biologics was formally established within the FDA to review the safety, effectiveness, and labeling of all previously licensed biologics.

During the 1980s and 1990s there were a series of organizational changes within the FDA that ultimately transformed the Bureau of Biologics into the Center for Biologics Evaluation and Research (CBER) in 1988. At this time, there was a rapid evolution of biologics with the advent of biotechnology-derived products as well as therapeutic protein products. Protein products, such as cytokines and monoclonal antibodies, were being developed for a host of diseases and the number of product reviews increased rapidly at CBER. In 1993, CBER was reorganized after the introduction of the Prescription Drug User Fee Act (PDUFA) of 1992 to separate vaccines, blood, and therapeutic product regulation in an effort to improve efficiency within the center. Responsibilities to license manufacturing facilities and compliance activities were also separated within the center to better define roles and oversight.

In the 1990s the FDA also established international efforts to collaborate with the European Union and Japan in the development of a new set of standardized guidelines for drug development. The ICH was established to create global standards for drug and biologic research through a series of international guidelines. To date the ICH effort has made substantial progress in defining a consistent approach to drug development and approval in the three regions.

During these efforts to harmonize drug review internationally, the differences between approaches to the review of therapeutic products for treatment of diseases and separate cultures within the FDA's Center for Drug Evaluation and Research (CDER) and CBER became more pronounced and apparent. The separation of responsibilities between CDER and CBER for review of therapeutic products led to increasing discrepancies between the approval process and requirements within the FDA at the same time FDA was making efforts to harmonize such requirements internationally. Whereas the politics and considerations within the agency that led to the most recent significant reorganization are complex *(14)*, the ultimate outcome of the announcement of Dr Lester M. Crawford, Deputy Commissioner of FDA on September 6, 2002, that the review of all new pharmaceutical products (therapeutics) would be consolidated into CDER ultimately will prove to be a positive change for the industry *(6)*. The restructuring of FDA to fully align therapeutic divisions within CDER is still ongoing and substantial changes are expected over the next 5 yr. However, the concept of having a single office and division within the agency reviewing all therapeutic products intended to treat an indication is a significant advance for the agency and corrects the earlier FDA underestimation of the evolution of biologics into therapeutic uses.

3. Changing Paradigms at FDA (PDUFA III)

The 2002 reorganization of the agency and ongoing changes are a new paradigm in the approach to the review of therapeutic agents for treatment of a disease. The reorganization refocused CBER's efforts on the areas of vaccines and blood safety, but also allowed CBER to retain responsibilities for gene therapy, cell therapy, and tissue transplantation. The rationale for leaving some therapeutic products within CBER is based on the scientific expertise required for such products and the close working relationship of CBER with NIH, who also regulates such products as those used in gene therapy or somatic cell therapy. One can foresee future changes in the agency to further consolidate these therapeutic areas as the FDA evolves and the concept of a single gold standard for pharmaceutical development and approval for an indication continues to develop.

Currently, FDA is executing a 5-yr plan under the PDUFA III that extends from 2002 to 2007 *(7)*. The PDUFA III goals further extend improvements in the agency under PDUFA I and II with current goals of completing 90% of drug and biologic reviews within 6 mo for priority drug and biologic designations and 10 mo for standard applications. The current 5-yr plan outlines major initiatives for the agency and appropriation of user fee funds collected under the current PDUFA III legislation. The plan also outlines specific initiatives and staffing levels for CDER, CBER, and the Office of Regional Affairs (ORA) responsible for compliance activities.

3.1. The Center for Drug Evaluation and Research

Under PDUFA III, CDER will gain the majority of resources provided to the agency by product review user fees. Of the 1464 additional full-time employees (FTEs) above base funding levels provided under the User Fee Act by year 2007, a total of 1054 are anticipated to be hired within CDER for review of new therapeutic pharmaceutical products. The distribution of headcount reflects the expanded review responsibilities of CDER to include therapeutic biologic products.

The key initiatives of CDER during this period will include enhancements in staffing and training efforts to improve reviewers' developmental activities and train reviewers within CDER on Good Review Management Principles (GRMPs) and Continuous Marketing Application Pilot Programs. Training is also being enhanced to address improvements in counterterrorism efforts, current GMP initiatives, and pediatric labeling information.

The history of FDA as described earlier is a process of evolution and continual improvement. These evolutionary changes have often been a reaction to dramatic events that were not anticipated by the agency, resulting from serious adverse effects with pharmaceutical products in the research or market place and unanticipated, even fatal, side effects of drugs approved for use by the FDA. Through the 1990s there were a number of drugs removed from the market because of unanticipated side effects, such as QTc wave prolongation and most recently cardiovascular side effects of Cox-2 inhibitors that have emphasized the need for improved efforts in the area of risk management at the agency. In response, FDA has defined Efficient Risk Management as one of five strategic goals during the current 5-yr PDUFA III period. Within CDER careful evaluation of safety throughout the development process and review during the approval phase have been significant efforts. It is acknowledged by FDA that this is an evolving science and it is not possible to evaluate all considerations regarding safety during the development of a drug product. The objective within CDER is clearly to balance benefit and risk, leading to some minimum criteria for evaluation of safety over time. In the enhanced risk management initiative, CDER is focusing more attention and resources on surveillance of the safety of approved medicines with emphasis on the first 2 yr after launch for standard drug approvals and 3 yr for drugs with significant safety concerns at the time of approval. This amounts to a periapproval process where safety reporting will have to occur more frequently initially as the drug usage is expanded and once safety is established in practice, safety reporting requirements would be reduced to expedited and annual reporting requirements more typical of the prior decade. One can envision that this effort will continue to expand as scientific advances identify new risk factors and as the system is improved over the coming years. In the future, the ability to file safety data electronically and data mining technologies will dramatically improve the center's ability to manage risk of new therapeutic products after approval.

Another key strategic initiative emerging at the FDA and within CDER is the Continuous Marketing Application initiative. The initiative takes into account the entire life cycle of drug development leading to approval and creates a continuous process for drug review and approval. The concept, although simple, is complex to implement and represents a significant change in how CDER manages review of therapeutic products. The concept is that review of the product for approval is initiated with the Investigational Drug Product (IND) Application and continues throughout development of the product with a close partnership between the agency and the company. In theory, on completion of the final clinical trials, CDER would be in a position to rapidly review the final clinical data and integrated summaries. This would allow more rapid approval of the product without conducting a retrospective review of the entire scope of data on the medicinal product. Given the review is continuous and builds on itself, review of prior completed studies and other information would be complete even before the final submission for approval. Currently, the center is conducting two pilot programs in an effort to more fully evaluate both the benefits and the cost of this program, given the more intensive early review of data necessary on medicinal products that may not succeed during development.

Other objectives outlined in the center's 5-yr plan include efforts to improve first cycle review performance and training on GRMPs in an effort to increase the efficiency of first cycle reviews of new applications. Whereas the center presents plans to hire outside expert consultants to support several efforts, including the review of biologics and the first cycle review initiatives, the budget is very limited and unlikely to support significant efforts in consultant support.

3.2. The Center for Biologics Evaluations and Research

The CBER's 5-yr plan focuses on similar objectives as with CDER, but given the shift in responsibilities to CDER for many therapeutic biologics, CBER will have limited resources from PDUFA III with only 221 additional staff members of the 1054 available to the agency under the user fee bill assigned to CBER by 2007. This is a reduction from the 269 staff members allocated in 2003. Interestingly, CBER has a much greater emphasis on enhanced training and staff development in their 5-yr plan than in that of CDER. Some of these efforts are focused on training to help facilitate the drug development process and first cycle review management efforts. Risk management is also a primary objective of CBER over the 5-yr period defined in the plan, with emphasis on the periapproval surveillance responsibilities of the center. CBER will also follow initiatives outlined by the agency in developing systems for Continuous Marketing Applications by revising management review processes and tracking systems with the center to aid in tracking application review and coordination.

3.3. The Office of Regional Affairs

ORA, who has responsibility for field compliance activities, has no significant initiatives outlined in the FDA's 5-yr plan and maintains a constant level of staffing and budget from the user fees available through the PDUFA III years. The lack of changes result from fewer drug applications filings since 2000, and more preapproval inspections are being waived because of improved compliance status of companies based on routine biannual inspections.

4. Evolution of Safety Monitoring and Assessment at FDA

Great emphasis on drug safety has always been placed on drug reviews and in postmarketing surveillance at FDA. In 2004, the agency initiated efforts to further improve the processes for postmarketing surveillance and review of marketed drug safety (8). At the forefront of these efforts is the revitalized Office of Drug Safety (ODS) made up of a diverse staff of medical professionals from various disciplines. The office has been highly productive in preparing new guidelines on both pre- and postmarketing safety surveillance as well as developing guidance around risk management or Risk Minimization Action Plans.

It is anticipated that over the years these efforts will continue to mature and improve safety monitoring; however, some caution is also warranted that industry and the agency maintain a fair balance between the benefits and risks of any pharmaceutical product and avoid overreaction to the potential discovery of unanticipated side effects in postmarketing surveillance. Such findings in an uncontrolled environment must be carefully evaluated, first to determine if they are in fact true events and second to ensure that they are adequately assessed vs the benefit of the therapeutic to the patient population involved.

Some of the unique approaches to drug safety surveillance include new sources of data that include managed care databases, international sources of data, as well as data mining technologies that look for trends in large bodies of information.

The Adverse Event Reporting System (AERS) is the primary effort of ODS to collect surveillance data of significant events for evaluation. In 2004, a total of 426,109 adverse event reports were received by ODS of which over 263,000 events were entered into the AERS database. The enormous workload and emphasis in the community on reporting adverse drug experiences has increasingly demonstrated the need for a fully electronic system for spontaneous adverse event reporting. By 2010, it is anticipated that essentially all serious adverse drug experiences from companies will be reported through electronic transmission to ODS.

The emphasis on risk management programs (i.e., Risk Minimization Action Plans) is also anticipated to continue and increase through 2010 in an effort to manage the use of pharmaceutical products that have significant safety concerns. These plans are designed to manage the safe use of a medication, often by limiting use to target populations, through safety monitoring, limiting advertising, limiting distribution channels, and in some cases, requiring drug registries. These risk management plans are anticipated to become more prevalent in future drug review cycles and the current draft guidelines published in May 2004 will soon become final as emphasis on drug safety continues to be a primary focus of the agency.

Another primary initiative of ODS through the 2010 time frame is the Healthy People 2010 program that includes improvements in adverse event monitoring, information systems for the public, physician oversight of medication use, patient information through pharmacies, oral counseling for patients on medication use, and increasing effort to get public blood donations, while maintaining the safety of the blood supply.

ODS is also active with the ICH efforts with the initiation of M5 (Data Elements and Standards for Drug Dictionaries) in 2004, which will develop a new international guideline supporting all pre- and postapproval pharmacovigilance activities as well as communication of regulatory information. The ICH

E2B Implementation Working group and E2B(M) guideline for electronic submission of adverse drug experiences continues to be evaluated and revised with plans to modify this guideline in the coming years. ODS is also working with ICH on Pharmacovigilance Planning (E2E) to define a framework for the organization and analysis of preclinical and clinical safety data obtained from clinical trials and postmarketing surveillance.

ODS staff also participate in the Management Board of the Medical Dictionary Regulatory Activities (MedDRA) in an effort to continue to improve and develop the current gold standard in adverse event reporting definitions. Additional activities in the CIOMS Working Groups, with the World Heath Organization, collaboration with other health authorities, as well as pharmacovigilance videoconferences highlight the continued efforts to coordinate internationally the focus on safety review and evaluation of marketing pharmaceutical products.

Within ODS, the Division of Drug Risk Evaluation (DDRE) will play a key role in working with the therapeutic divisions to assess safety signals, improve product labeling, and compliance with both labeling and risk management plans. Industry and the public should anticipate that by 2010 the number of labeling changes to increase warnings or include black box warnings on drug products will steadily increase as surveillance activities and technologies for data mining continue to improve. Supporting this effort is the Division of Medication Errors and Technical Support (DMETS) which supports the technical efforts within ODS as well as evaluates medication dispensing errors and potential for name confusion prior to approval. Also within ODS, the Division of Surveillance, Research and Communication Support (DSRCS) provides access to safety data and risk information on medications to CDER, CBER, and ODS.

The great emphasis on improving drug safety and appropriate labeling of medications, both biologics and drugs, is anticipated to continue as a rapid pace through 2010. Companies will be hard pressed to keep pace with these efforts on the part of the agency and compliance will become an increasingly important consideration as the agency turns towards stronger measures to enforce guidelines for safety reporting and evaluation requirements in future years. One should also anticipate some consolidation with ODS to streamline the process and better utilize resources over time. One of the more significant considerations of the agency for the future is how to balance emerging safety considerations for approved medicinal products, with additional data on efficacy and benefit to the patient. The FDA will have to consider new methodologies in the ongoing assessment of the risk–benefit for a medicinal product during the product's life cycle, while avoiding the temptation to overreact to political and public pressures that, in the long run, may be counterproductive.

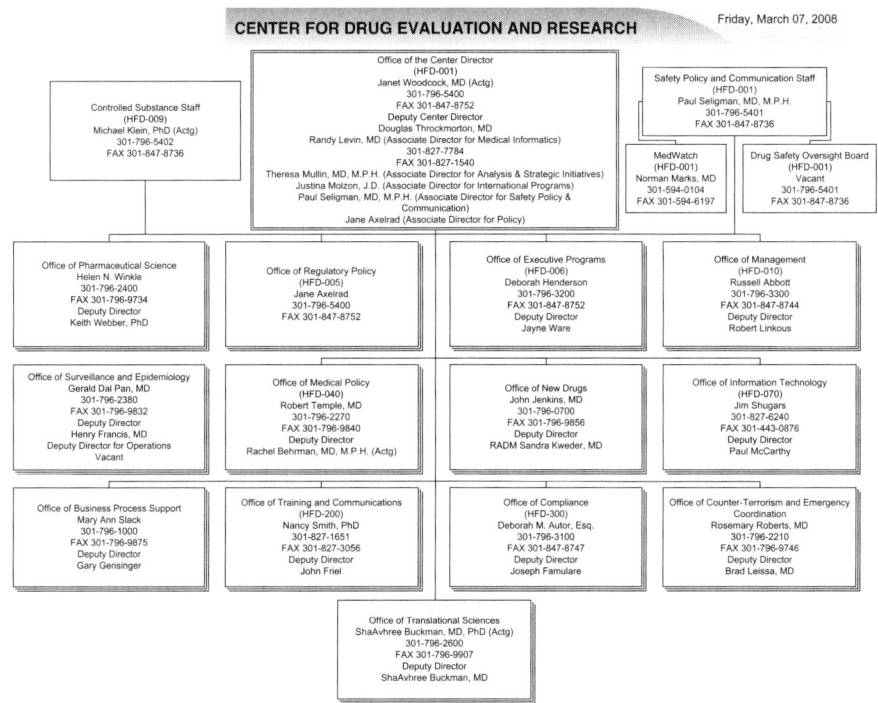

Fig. 1.

5. Review of Therapeutic Biologics

As CDER continues to incorporate the review of biologics into the therapeutic areas, there has been a period of transition. The current center structure is shown (Fig. 1) and outlines the review divisions and makeup of the center in the fourth quarter of 2005. After the announcement of the transfer of responsibility for many of the therapeutic biologics in 2002, the center created the Office of Drug Evaluation VI to incorporate the biologic product reviewer process into CDER, but this is anticipated to be an interim measure and eventually all review activities should be within the therapeutic divisions. The consolidation of these groups will also lead to more consistent and predictable medicinal product reviews. Although CBER in the past has had significant scientific expertise, the divergence of biologic product evaluations from those criteria used to evaluate drugs in the therapeutic divisions of CDER, as well as those criteria outlined by ICH, has led to the need for change at the agency. Although some in the biotechnology industry viewed these changes in a negative manner because of the fear that such changes would increase the burden on biologics to obtain approval, these changes are necessary given the different criteria for approval that developed during the evolution of CBER and CDER. One must consider that the requirements for approval of a therapeutic treatment for a particular indication should primarily be the same for both simple drug molecules and complex biologics. Thus the artificial separation that developed over the past 30 yr, which also lead to the divergence of requirements for approval, was justifiably corrected by the agency. Faced with a consistent approach to review both drugs and biologics, industry will eventually benefit by having a single standard.

Within CBER responsibilities remain the regulation of blood products, vaccines, cellular and gene therapies, live tissues, xenotransplantation, some devices, and allergenics *(9)*. Some of these technologies, such as gene therapies, in reality are focused on therapeutic indications. Whereas, currently, they remain within the responsibilities of CBER because of the scientific and technical expertise these products require, ultimately one can theorize that FDA will decide to move these products as well into the therapeutic divisions of CDER based on the same rationale that emphasized this need for other biologic products.

5.1. Critical Path Initiative

For the past 15 yr at FDA, the Critical Path Initiative has taken shape into a concept that continues to expand and is anticipated to play a critical role in the development of future biologic and drug products *(10–12)*. This initiative is especially interesting to the development of biologic products given the fast pace of scientific advances within the biotechnology industry. The Critical Path Initiative analyzes the various stages of drug development from basic research through final approval and launching of a new medicinal product. In the face of increases in pharmaceutical development costs and decreases in new medicinal products coming to market, the agency has continued to expand its efforts in defining the critical path and implementing appropriate reforms to facilitate drug development. The three dimensions outlined by FDA on the critical path include: assessing safety, demonstrating medical utility, and industrialization. Whereas the Critical Path Initiative has led to many meaningful changes, including the 1995 IND Guidelines, more emphasis on expanding these efforts to define the critical path and implement regulatory changes to facilitate medicinal product development is critical for the future of the industry and to patients in need of new therapies for serious diseases.

There is a significant need for the agency to take acceptable risks during drug development that may accelerate the process and, for wider acceptance, novel and improved scientific assessments of drug safety and efficacy be utilized to reduce unnecessary regulatory burdens. There is a clear need to continuously update guidance to keep current and consistent. It is common in industry that

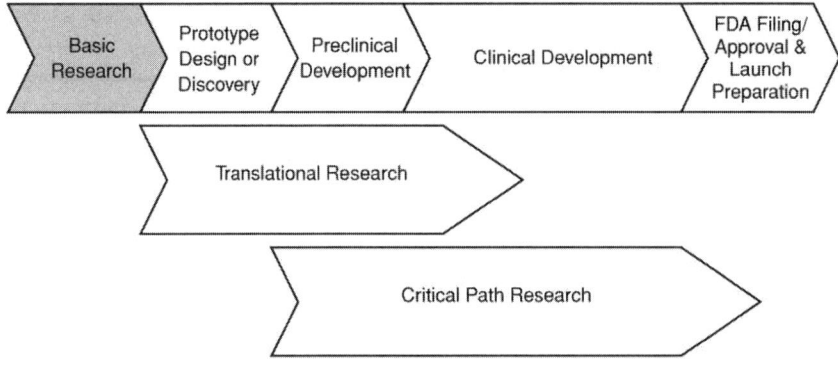

Fig. 2.

uncertainty over regulatory requirements can stagnate the development process or make funding of novel medicinal therapies difficult because of the lack of a clear regulatory pathway. The review practices at FDA will also need to be kept consistent to improve predictability of companies during the development process. There is also a significant need to have FDA more involved in collaboration with academia, industry, and other regulatory bodies to ensure consistent standards and promote scientific advances that may lead to new medicinal products. These efforts are of particular importance to the biotechnology industry given the unique, and sometimes atypical, technologies that are currently under development and which may not be fully addressed by current FDA guidance.

One of the burdens for the biotechnology field is the rigid nature of regulations and guidance currently existing at FDA. Whereas FDA maintains that guidance documents are not binding, reviewers are often stressed to deviate significantly from these guidelines during the drug development and review process. An essential part of the Critical Path Initiative needs to include improved flexibility of the agency to work with companies and accept rigorous science, even when novel.

Through 2010 it will be critical for the agency to continue to work with academia and industry to more fully define and implement the initial efforts of the Critical Path Initiative. FDA will need to expend more resources to understand the basic research phase of development and help define which scientific techniques may be useful to identify medicinal products with a higher potential for success. During the translational research phase of development the agency will have to more fully embrace the need to take acceptable risks to facilitate the drug development process. These early nonclinical and clinical evaluations are critical to moving products into full development and there is a need for FDA to find new ways to facilitate these efforts in collaboration with medicinal product developers. Of most significance is the effort on critical path research, which is directed toward improving the medicinal product development process by establishing new development tools and consistent guidance. FDA is expected to emphasize this later phase of development through 2010 to facilitate the introduction of new therapies and help manage the increasing cost of development because of new regulatory requirements.

6. The Future of Regulation for Blood Products

CBER maintains as one of its core responsibilities the regulation of the nation's blood supply to minimize the risk of infectious disease transmission and other risks to the public. This responsibility includes whole blood, plasma, blood-derived products, artificial blood products, as well as products used to prepare blood and cellular products. CBER works closely with the Centers for Disease Control and Prevention (CDC) as well as the NIH in developing new scientific approaches to the detection of infectious diseases in blood products and regulation to ensure a safe blood supply. Over the next 5 yr, the efforts necessary to protect the national blood supply from new or existing infectious agents and potentially terrorist acts will require a substantial effort on the part of the agency.

7. Recent Regulatory Changes for Vaccines

Vaccines and virus materials are also still regulated by CBER after the reorganization in 2002. Currently there are significant concerns with the ability of manufacturers to produce adequate supplies of safe vaccines to prevent seasonal illnesses as well as to fight bioterrorism. The center has taken the initiative to work with industry and within their resources to develop new technique for the production of vaccines in a cost-effective manner. The recent shortages of flu vaccine in 2004 as a result of GMP compliance problems at a single Chiron vaccine production plant emphasized the vulnerability of the vaccine supply in the United States. CBER is focused on new techniques to correlate vaccine gene changes with actual production from the specific infectious disease in patients to predict the effectiveness of a vaccine (as a surrogate market). These efforts along with recombinant vaccine production techniques are anticipated to accelerate the development of vaccines while reducing the cost of development by allowing manufacturers to demonstrate the effectiveness with fewer numbers of patients in clinical trials.

The focus on the safety of vaccines and vaccine products will continue to expand at CBER through the next 5 yr. CBER is focused on studying methods to detect side effects in live vaccines by measuring the gene changes with attenuated and virulent species of the vaccine. The Vaccine Adverse Event Reporting System (VAERS), which is jointly managed by CBER and CDC, will continue to be the primary postmarketing safety surveillance program responsible for monitoring adverse side effects potentially related to vaccines.

8. Regulation of Gene Therapy and Somatic Cell Therapy Products

8.1. Gene Therapy

The promise of gene therapy in the 1980s and 1990s has not been realized as many had prophesized, in part because of the lack of specific vector technologies that have led to increased concern about serious adverse effects from these therapies. Between 2000 and 2002 a number of gene therapy programs were discontinued owing to safety and only those viral vector-mediated gene therapies for the most severe diseases and plasmid-mediated gene therapies were allowed to continue in clinical development. Much of the effort of CBER in the area of gene therapy is done in conjunction with the NIH and the Recombinant Advisory Committee (RAC), which operates as an independent organization, managed by NIH, to review and approve gene therapy trials in the United States. The efforts of CBER and the NIH RAC are in some respects redundant and inefficient for the industry. The burdensome process of getting approval from FDA, an IRB, the NIH RAC, and a Biologics Safety Committee at the clinical trial site has created an unnecessary burden for industry. As the gene therapy industry recovers and new more targeted technologies are developed to facilitate gene transfer, future reform and consolidation of responsibilities between FDA and NIH is needed to make gene therapy research a manageable endeavor for industry.

Another consideration for the future is the internal FDA regulation of Gene Therapy. Although currently maintained in CBER after the 2002 reorganization, the Division of Gene Therapy is in fact primarily focused on products to treat therapeutic indications. Although the unique scientific expertise needed to properly review gene therapy products was the primary reason to maintain this function within CBER, the primary clinical focus of treating a disease with a therapeutic agent is still consistent with the rationale for moving other similar functions to CDER. Ultimately one must ask if the review of therapeutic treatments involving gene therapies should be managed within the CDER review divisions with either a consult from CBER, or perhaps a technology group within the Office of Pharmaceutical Sciences, to manage the scientific product manufacturing review. In the coming years, FDA will be challenged to rethink the regulation of gene therapy as new and more innovative methods of gene transfer are developed that facilitate future clinical development in this area of medical science.

8.2. Somatic Cell Therapy

Similar to the issues facing the regulation of gene therapies, cellular therapies derived from hematopoietic stem cells, myoblasts, and other cells differentiated beyond the embryonic state are being investigated not only as replacement therapies, but also to treat a wide range of illnesses. Such therapies show significant promise for cardiovascular diseases, a variety of autoimmune diseases, and neurological diseases. Again the rationale for FDA maintaining the review of cellular therapies within CBER is the science and research base necessary for review of these products. The dilemma of having duplication of the clinical therapeutic function and potential divergence of regulatory requirements still exists and FDA will have to consider options in the future to consolidate such functions.

Another emphasis that CBER places on their role in the cell therapy field is a series of research projects focused on advancing the science of evaluating such products. This work is done in conjunction with various groups within the NIH, who is the primary government agency for such research and development. One has to ask if such efforts on the part of FDA, a regulatory body, are in fact appropriate or value added to the public and the industry. Such considerations will have to be addressed and perhaps future reorganizations will realize more clearly the role of FDA as a regulatory body and NIH as a scientific research organization.

8.3. Role of NIH in the Review of Biologics

The FDA and NIH implemented a memorandum of understanding on September 19, 2005, in an effort to clarify their respective roles and collaborate in their efforts to regulate medicinal products and, in particular, biologics (13). The memorandum takes a step towards clarifying the role of FDA as the primary regulatory body that reviews and approves therapeutic agents, and NIH as the government organization that is responsible for biomedical research in the United States. The agreement outlines areas and efforts to collaborate to protect public health and define appropriate regulation for biologics and new biomedical technologies. The process for sharing nonpublic information was also clarified to facilitate the exchange of data between the two agencies. The agreement, however, falls short of addressing the duplication of efforts in regulating some types of biotechnologies, such as gene therapies and cellular therapies, which remain a burden on the industry for the foreseeable future.

9. Electronic Filings (eCTD)

The agency is rapidly moving towards all electronic filings with the implementation of the electronic Common Technical Dossier (eCTD) in 2003. The eCTD format is harmonized among the United States, European Union, and

Japan to facilitate the preparation of a marketing authorization application. The eCTD concept, format, and implementation were carried out through the ICH process and represent one of the most critical initiatives of ICH. Although the format does not define content, for the first time common format of marketing applications will be possible.

The eCTD is an electronic form of the CTD, which in the United States may be submitted without any paper copies of the application. The European Union still requires paper copies of applications for archives and the Rapporteurs, and module one for all EU countries. However, it is anticipated in the future that all regions will move to all electronic filings without the requirement for printing hundreds or thousands of volumes of data. The ability to file fully electronic filings alone will save at least 2 and 4 wk in the approval time of new chemical entities, purely because of the time saving from not printing and quality controlling volumes of the application. At present, the advantages in review time for the various regulatory agencies is not clear, but it is anticipated that the eCTD will also accelerate review cycles and allow for more specific comments from reviewers who will focus less on the administrative aspects of the review process and more on evaluation of the application content.

The greatest potential benefit of the eCTD is to facilitate the efforts of ICH to harmonize the actual content of a market authorization application for a new chemical entity and significant line extensions. Currently, the FDA will still require the full Integrated Safety and Efficacy Summaries as required in past New Drug Application (NDA) submissions. At present there are two primary strategies in the industry for the preparation of an eCTD. The first philosophy is similar to the historical approach of customizing the eCTD for each region, including specific data, reports, and summaries designed for each ICH region's preferences. The second approach is to make eCTD submissions as consistent as possible between the various ICH regions, with the exception of administrative information in Module 1 of the application.

As the ICH progresses, it can be argued that industry should make every effort to promote consistency and identify areas for improvement within the ICH process. The concept of customizing applications, while understandable in the short-term objective of a single submission, may in fact negatively impact any effort to identify areas of inconsistency and facilitate change through the ICH process. Filing applications which are as similar as possible across ICH regions, or identical if possible, would help identify areas of difference that remain between regions and differing interpretations of current ICH guidelines. These differences could be addressed during the review process for an individual application and also used by industry to identify and promote new discussions within ICH to modify existing guidelines, clarify interpretation of guidelines, and identify areas where new guidelines are needed to facilitate the harmonization process.

10. Into the Future

Over the next 5 yr at FDA, the agency will continue to evolve and the overall direction of any topic outlined in this chapter is difficult to assess with great accuracy. The agency is dynamic, and whereas plans are in place that may provide some concept of the direction of future changes for both drug and biologics reviews, local and world events often cause a reactionary response at the FDA that is often unpredictable. What is clear is that the agency will continue to make every effort to improve the regulatory review process and its ability to assess safety and efficacy of new medicinal products. Some of the safety concerns raised after the withdrawal of a number of commonly used medicinal products in the past 10 yr have shifted FDA's approach to pharmaceutical product reviews and requirements for approval. To achieve improvements in both safety and efficacy assessments as well as review timelines and consistency in reviews within the agency, the FDA will have to further consolidate medicinal product reviews aligned with therapeutic indications. As such, the biotechnology industry should expect continued consolidation of therapeutic drug reviews under a single center within the agency given the need for consistency of clinical assessments for a single indication.

It can be anticipated that future reviews for each therapeutic indication with all medicinal products, including somatic cell and gene therapies, will be contained within a single therapeutic division. Specialized technology reviews can be managed through the Division of Pharmaceutics or through other internal consults within the agency. Whereas the agency has currently established a separate division for some biologic reviews (e.g., ODE VI), the future should bring continued consolidation and centralization of medicinal product reviews within the therapeutic divisions.

Throughout the history of the FDA, safety events have been the most influential driver for change and improvement in the review process for medicinal products. During the 1990s, a number of drugs were discovered to cause QTc wave prolongation and were withdrawn off the market owing to a reassessment of the risk–benefit by the agency and companies involved. As the science around cardiovascular safety evolved, new requirements for approval have been established to reduce the likelihood of unknown serious cardiovascular side effects. In the past couple of years, the field of pain management has been dramatically impacted with the findings that the Cox-2 mechanism may also lead to serious adverse cardiovascular side effect with prolonged use. Such events are inevitable and the agency will continue to assess these safety risks and adjust the approval requirements based on science as well as perception of risk–benefit.

Where the FDA can provide both the industry and the public the greatest benefit in the future evolution is a continued effort on the Critical Path Initiative and creating new methodologies to work with industry to facilitate development.

The past achievements of the agency in this area have provided the pharmaceutical development industry with some relief and have already facilitated drug development. Some of the benefits have been taken back with increased safety requirements, but ultimately the agency shall have to carefully consider appropriate risk during development and collaborative methodologies to facilitate the pharmaceutical development process to bring good drugs to the public more quickly than in the past two decades. Such efforts will benefit the general public, one of the United States' most critical industries, as well as the agency.

References

1. McDonald, S. US Department of Health, History of Biologics. Food and Drug Administration website, September 2002.
2. Center for Biologics Evaluation and Research Centennial – Commemorating 100 Years of Biologics Review, Food and Drug Administration website (www.FDA.gov/CBER/Inside/Centscireg.htm) September 25, 2002.
3. A Brief History of the Center for Drug Evaluation and Research, Food and Drug Administration website (www.FDA.gov/CDER/About/History/Histex.htm).
4. Milestones in Food and Drug Law History, Food and Drug Administration website (www.FDA.gov/opacom/loachgrounders/miles.html).
5. 1938 Food, Drug and Cosmetic Act.
6. "FDA to Consolidate Review Responsibilities for New Pharma Products", Food and Drug Administration Press Release September 6, 2002.
7. Food and Drug Administration PDUFA III Five-Year Plan, July 2003.
8. Food and Drug Administration Center for Drug Evaluation and Research Office of Drug Safety Annual Report FY 2004, Food and Drug Administration website (www.FDA.gov/CDER/Office/ODS/AnnRep2004/default.htm).
9. Center for Biologics Evaluation and Research FY 2004 Annual Report, Food and Drug Administration website (www.FDA.gov/CBER/Inside/Annrpt.htm).
10. "FDA Critical Path", California Healthcare Institute White Paper 2005.
11. News Along the Pike, Critical Path report calls for modernization tools, July 15, 2004.
12. Center for Drug Evaluation and Research CDER 2004 Report to the Nation:Improving Public Health Through Human Drugs (www.fda.gov/cder/reports/rtn/2004/rtn2004.htm).
13. Memorandum of Understanding Between the Food and Drug Administration Center for Biologics Research and Evaluation (CBER) and the National Institute of Health (NIH)/National Institute of Neurological Disorders and Stroke (NINDS) Fed. Register, Vol 70, No 180/Monday September 19, 2005.
14. Hawthorne, F. (2005) *Inside the FDA*. John Wiley and Sons.

16

Follow-On Protein Products: The Age of Biogenerics?

Robert G. Bell

Abstract

Follow-on protein products (FOPP), also known as follow-on biologics, biogenerics, and biosimilars, are available outside of western markets; however, no FOPP have yet been approved in the US or European pharmaceutical market. This is because of a lack of registration process for the demonstration of pharmaceutical and therapeutic/clinical equivalence. The production of FOPP is a significant opportunity for patients, managed health care providers, and manufacturers with bioprocessing capabilities, assuming the follow-on manufacturer can produce a FOPP at a lower cost without compromising the quality, safety, and efficacy of the product. This may be difficult because there are significant barriers to entry associated with biological production which may result in a smaller price differential between brand and generic products than that seen in regular generics. Presently, there is sparse regulatory guidance from the FDA concerning the development of FOPP, although guidance is expected shortly for well-characterized biologics such as insulin and growth hormone. In addition, legislative changes will be required. Legal issues associated with incorporation of reference label information and therapeutic equivalence designation have to be addressed. A manufacturer needs to demonstrate that the biologic produced is safe and effective as the reference product in the intended patient population. Relevant guidances exist today from the FDA and EMEA pertaining to production, scaleup, chemistry, manufacturing controls, and comparability of human biological products. FOPP manufacturers must demonstrate the identity, purity, strength, quality, potency, stability, and safety of the biologic. Appropriate testing of the chemical, physical, biological, pharmacokinetic, and immunologic properties of the FOPP and postapproval pharmacovigilance would ensure the therapeutic equivalence and comparability of the follow-on biologic to a reference biologic.

Key Words: Biologics; biogenerics; biosimilars; comparability; follow-on protein products; immunogenicity; pharmaceutical equivalence; therapeutic equivalence.

1. Introduction

Many of the patents associated with blockbuster biotechnology-derived protein products have expired, making them theoretically eligible for multisource,

From: *Biopharmaceutical Drug Design and Development*
Edited by: S. Wu-Pong and Y. Rojanasakul © Humana Press Inc., Totowa, NJ

follow-on, or "biogeneric" competition (*see* Table 1) *(1–15)*. This includes recombinant products such as human insulin, human growth hormone, and interferon. Follow-on protein products (FOPP) that are therapeutically equivalent offer a significant opportunity for patients, managed health care providers, and manufacturers with bioprocessing capabilities. This assumes, however, that the follow-on manufacturer can produce a FOPP at a lower cost without compromising the quality, safety, and efficacy of the product. This may be difficult because there are significant barriers to entry associated with biological development and production which is costlier and will result in a smaller price differential between brand and generic products than that is typically seen with small molecules *(2–15)*. The European Agency for the Evaluation of Medicinal Products (EMEA) has issued guidances for biosimilar protein products (adopted or released for consultation), including specific guidances for insulin, growth hormone, GCSF, and erythropoietin *(16–28)*. Recently, the EMEA's Committee on Medicinal Products for Human Use recommended approval for a generic version of a growth hormone; however, the EMEA has yet to approve any FOPP. Australia approved a FOPP for human growth hormone in 2005. Presently, there is sparse regulatory guidance from the FDA concerning the development of FOPP *(29,30)*, although guidance is expected shortly for immunogenicity assessment of FOPP and well-characterized biologics such as insulin and growth hormone. The US registration process for FOPP is not defined and presently, there is no abbreviated pathway for the registration of FOPP as there is with generic drugs. At this time, preclinical and clinical data would be required for the approval of a FOPP utilizing either the new drug (NDA) or biologic license application (BLA) registration route. In addition, legislative changes will be required. Legal and intellectual property issues regarding the incorporation of reference product information and therapeutic equivalence designation must be addressed. A FOPP manufacturer needs to demonstrate that the biologic produced is safe and effective as the reference product in the intended patient population. This should also include immunological assessments of the product. It may be possible to conduct abbreviated comparative safety and efficacy assessments of well-characterized proteins such as insulin and growth hormone; however, as the complexity of the biologic increases, additional safety, clinical and quality assurances of the FOPP will be required. Follow-on biologic manufacturers must demonstrate the identity, purity, strength, quality, potency, stability, and safety of the FOPP. In addition, post-approval pharmacovigilance safety programs should be initiated. Appropriate testing of the chemical, physical, biological, pharmacokinetic, pharmacodynamic, and immunologic properties of "generically" produced biotechnology-derived drug products would ensure the therapeutic equivalence and comparability of a FOPP to a reference biologic. This chapter will provide an overview of nomenclature, registration process, pharmaceutical and therapeutic equivalence approaches for a FOPP.

Table 1
Biotechnology-Derived Products and Patent Expiry

Brand name	Generic name	Indication	Manufacturer	US patent expiration
Rebetron	Ribavirin and Interferon α-2b	Hepatitis C	Schering Plough	2001
Ceredase	Alglucerase	Gaucher disease	Genzyme	2001
Cerezyme	Imiglucerase	Gaucher disease	Genzyme	2001
Intron A	Interferon α-2b	Malignant melanoma	Schering Plough	2002
Humulin	Human insulin	Diabetes	Lilly	2002
Humatrope	Human growth hormone	Growth hormone deficiency	Lilly	2003
Nutropin/ Nutropin AQ	Human growth hormone	Growth hormone deficiency	Genentech	2003
Avonex	Interferon β-1a	Multiple sclerosis	Biogen	2003
Epogen/Procrit	Erythropoietin α	Anemia	Amgen, Johnson & Johnson and Sankyo	2004
Geref	Sermorelin	Growth hormone deficiency	Serono	2004
Synagis	Palivizumab	Respiratory syncytial virus	Abbott	2004
Activase	Alteplase	Myocardial infarction, stroke, pulmonary embolism	Genentech, Boehringer Ingelheim, Mitsubishi, and Kyowa Hakko Kogyo	2005
Novolin	Human insulin	Diabetes	Novo Nordisk	2005
Protropin	Somatrem	Growth hormone deficiency	Genentech	2005
Neupogen	Filgrastim	Neutropenia	Amgen and Roche	2006
ReoPro	Abciximab	Ischemic complications	Lilly	2019

2. Terminology and Definitions

Pharmaceutical biotechnology is a general term that describes the use of biology and engineering to produce products from genetically modified organisms. Typical products include peptides, monoclonal antibodies, cytokines, hormones, clotting factors, vaccines, and cell-based therapies. In general, a biologic product is any virus, serum, toxin, antitoxin, vaccine, blood, blood component or derivative, allergenic product, or analogous product applicable to the prevention, treatment, or cure of diseases or injuries *(31,32)*. Biotechnology-derived protein products are described as "proteins and polypeptides, their derivatives and products produced from recombinant or non-recombinant cell-culture expression systems and can be highly purified and characterized using appropriate analytical techniques" *(33)* and well-characterized proteins are proteins where the natural molecular heterogeneity, impurity profile, and potency are defined with a high degree of confidence *(34)*.

A drug is defined as a substance recognized by an official pharmacopoeia or formulary and is intended for use in the diagnosis, cure, mitigation, treatment, or prevention of disease *(35)*. Biologic products are included within this definition and are generally covered by the same laws and regulations, but differences typically exist regarding their manufacturing processes.

Drugs in most cases are small molecules that are chemically synthesized, highly purified, and well characterized. Biotechnology-derived pharmaceutical products are typically macromolecules derived from living sources that can result in complex heterogeneous mixtures that are difficult to characterize and widely diverse in their form and function.

The terminology of how to describe a FOPP has evolved over the years. In general, "generic" biotechnology-derived products are biologics developed by manufacturers other than the original manufacturer utilizing similar or different manufacturing processes, and the product must be bioequivalent, clinically comparable, and pharmaceutically equivalent to the original reference product. A "follow-on" protein product is a biological product which is intended to be a similar version or a duplicate of an already approved or licensed protein product *(36)*. Other similar terms include "biogeneric, biosimiliar, second generation protein product, therapeutically equivalent biologic, and subsequent entry proteins" (*see* Table 2). The term "second generation protein" is a product, similar to an already approved or licensed product, but which has been deliberately modified to change one or more of the product's characteristics *(37)*. The European regulators use the term "biosimiliar" or similar biological medicinal product when referring to a biologic that purports to be similar to a reference biological medicinal product manufactured by an innovative biopharmaceutical company *(16)*. Biosimilar products may or may not be intended to be molecular copies of the innovator's product. They do, however, depend on the same mechanism of action and are intended to be used for the same therapeutic indication. The terminology "generic medicinal products" and "similar biological medicinal products" is because of the opinion that the current analytical methodology is not sufficient at discriminating subtle differences in the structure and conformation of complex proteins or adequately able to address their impurity profiles *(16)*. This requires manufacturers to produce biologics utilizing validated

Table 2
Follow-On Biologics: Definitions

Term	Definitions
Biologic Product	"Any virus, therapeutic serum, toxin, antitoxin, vaccine, blood, blood component or derivative, allergenic product, or analogous product…applicable to the prevention, treatment, or cure of disease or injuries of man" In regard to regulations, biologics are also drugs.
Drug Products	"Articles recognized in official compendia; intended for the use in the diagnosis, cure, mitigation, treatment and prevention of disease…"
Biotechnology-Derived Drug Product	"Proteins and polypeptides…produced from recombinant or non-recombinant cell-culture expression systems…can be highly purified and characterized…".
Well-Characterized Proteins	The natural molecular heterogeneity, impurity profile, and potency are defined with a high degree of confidence.
Biogeneric	Therapeutically Equivalent Biotechnology-Derived Products that are pharmaceutically and therapeutically equivalent (substitutable) to a reference product or products. Virus, therapeutic serum, toxin, antitoxin, vaccine, blood, blood component or derivative, allergenic product, or analogous product, or arsphenamine, applicable to the prevention, treatment, or cure, of diseases or injuries of man w/regard to regulations, biologics are also drugs.
Biosimilar	Second and subsequent versions of biologics that are inde pendently developed and approved after a pioneer has developed an original version. Biosimilar products may or may not be intended to be molecular copies of the innovator's product. They do, however, depend on the same mechanism of action and are intended to be used for thesame therapeutic indication.
Follow-On Protein Product	A protein product which is intended to be a similar version or duplicate of an already approved or licensed protein product.
Second Generation Protein Product	A product similar to an already approved or licensed product, but which has been deliberately modified to change one or more of the product's characteristics (e.g., to provide more favorable pharmacokinetic parameters or to decrease immunogenicity).

manufacturing processes and quality systems that will assure a product with a predictable safety, efficacy, and quality profile. It may not be possible for the FOPP manufacturer to have access to the production experience of the reference manufacturer and may have to generate additional data, including preclinical and clinical data, in order to demonstrate the safety and efficacy of a biological product claimed to be similar to marketed biological medicine.

The definition of FOPP indicates the product will be therapeutically equivalent, pharmaceutically equivalent (similar rather than identical), produce the equivalent efficacy and safety of the reference protein product. Issues of therapeutic substitution and switching during the patient's pharmacotherapy have to be addressed. The sponsor of the FOPP needs to demonstrate the substitutability of the product by conducting comparability studies with the reference protein product. It is desired that a therapeutic rating code, similar to the existing Orange book rating (AB rating), be similarly applied to FOPP.

3. Regulations and Registration Process

Typically in the United States, drugs are developed utilizing an Investigational Drug Application (IND), and a NDA is submitted for approval under Section 505 of the Food, Drug & Cosmetic Act (FD&CA). After the patents and exclusivity expire for drugs approved under the FD&CA, they become eligible for generic competition through Drug Price Competition and Patent Term Res lent to a reference-listed drug product (brand product). The registration process for a generic drug is "abbreviated" because the generic products generally are not required to include preclinical (animal) and clinical (human) data to establish safety and efficacy. It is required that the generic submission demonstrates that their product is bioequivalent and pharmaceutically equivalent to the reference drug product listed in the publication, *Approved Drug Products with Therapeutic Equivalence Evaluations* (the Orange Book) *(38)*.

Biologics are also developed utilizing an IND, but the BLA is submitted for license under Section 351 of the Public Health Service Act (PHSA) and in specific sections of FD&CA. However, there are no provisions or process in the PHSA that would allow the submission of an abbreviated "generic" biologic to the FDA. Biologics tend to be large, complex, heterogeneous molecules, for which the manufacturing process has a significant impact on the quality and structure of the final product. In addition, the FOPP manufacturer would have to demonstrate their product is as safe and effective as the reference protein product. Investigations of the immunogenicity of the products should be examined through comparability studies to demonstrate that any undetected differences would not impact the safety or efficacy of the biological product.

In addition to the scientific and legal issues, there is uncertainty regarding the appropriate registration process for FOPP. Brand companies suggest that a process for approving generic biologics permitting reliance on proprietary data in approving subsequent applications could be an unconstitutional taking of private property under the Fifth Amendment *(39)*. The brand companies assert that proprietary data and other information submitted in support of any application for agency approval constitute trade secrets, which, coupled with FDA's long-standing practice of nondisclosure, creates a reasonable investment-backed expectation that agency use of the data in approving a generic version would constitute a taking. A review of Takings Clause case law, with particular attention to its application to regulated industries, demonstrates that the FDA's proposed change in approving biologics will not constitute a taking *(40)*. However, it is not certain how much of the reference protein product label can be incorporated into the FOPP-approved label.

It appears that legislative changes to Hatch–Waxman or PHSA will be required to provide a defined regulatory registration pathway. Biological guidances and monographs from the FDA and USP will also be needed for the approval of FOPP. Section 505(b)(2) of FD&CA has been suggested as a regulatory submission mechanism for FOPP. This submission type allows a sponsor to submit an NDA that relies on safety and efficacy studies not conducted by that applicant and supplemented, if necessary, by their own clinical data. A regulatory submission under FD&CA 505(b)(2) can be viewed as a hybrid approval process that contains more data than an ANDA, but less data than an NDA. The FDA ruled in 2002 that this type of NDA could be used for biologics originally approved as drugs (including insulin and human growth hormone). However, legal challenges have been put forth regarding the use and incorporation of reference product protein safety and efficacy data by the FOPP manufacturer because it would violate trade secrets and proprietary confidential information of the brand companies *(40–42)*. These legal and legislative issues, combined with the present lack of registration process and product-specific scientific comparability guidance, will continue to delay the introduction of FOPP in the United States. At this time, a follow-on manufacturer of a FOPP would have to pursue a FD&CA 505(b)(2) with prior agreement from the FDA or submit a full BLA to seek approval of a similar biologic. It is also not assured that therapeutic substitutability would be granted by these submission routes.

4. Pharmaceutical Equivalence and Product Comparability of Biologics

Drug products are considered pharmaceutical equivalents if they contain the same active ingredient(s), are of the same dosage form, route of administration, and are identical in strength or concentration. Pharmaceutically equivalent drug products are formulated to contain the same amount of active ingredient in the same dosage form and to meet the same compendial or other applicable standards (i.e., strength, quality, purity, and identity), but they may differ in characteristics such as shape, scoring configuration, release mechanisms, packaging, excipients (including colors, flavors, preservatives), expiration time, and, within certain limits, labeling *(38,43)*.

Typically, biologics can range from simple to complicated, and tend to be more structurally complex than chemically synthesized drugs. Production of a biotechnology-derived substance typically uses living organisms or their products to make or modify the active substances. Biologics typically tend to have relatively large molecular weights (e.g. >5000 Da) with greater structural complexity compared with small chemically derived molecules. Biologics can be heterogeneous and can have multiple patterns of crosslinking, glycosylation, and molecular species present. As with chemically synthesized drugs, biological quality is associated with valid production processes and testing methods, and applying quality systems and process analytical technologies to the protein production assures the quality of the biological product.

Applying strict pharmaceutical equivalence standards to many biologics may not be possible. For instance, it will be difficult to produce an identical pharmaceutically equivalent thrombolytic protein, having an approximate molecular weight of 300,000, with six heterogeneous domains, multiple glycosylation patterns, and disulfide bridges, from two different manufacturers. Compounds of this magnitude will be difficult to analytically characterize, let alone demonstrate the two proteins are identical. Pharmaceutical equivalence can be demonstrated with well-characterized proteins such as insulin, growth hormone, calcitonin, glucagon, IGF, FSH; however, as the complexity of the protein increases, the present ability to analytically compare the compounds decreases. In situations where the proteins are not well characterized, the concept of "sameness" is applied to protein comparability. Sameness for a macromolecule is defined as a "…drug that contains the same principal molecular structural features (but not necessarily all of the same structural features) and is intended for the same use as a previously approved drug…" *(44)*. Furthermore, two protein products would be considered the same "…if the only differences in structure between them were due to post-translational events or infidelity of translation or transcription or were minor differences in amino acid sequence…" *(45)*. For complex biologics, pharmaceutical equivalence between a brand biologic and a FOPP would demonstrate "sameness" and may not be identical to the reference biologic. The FOPP should share the same critical quality attributes such as primary, secondary, tertiary, and quaternary structure, posttranslational modifications, potency, amount/concentration, dosage form, and route of administration.

To demonstrate pharmaceutical equivalence, analytical comparability between the products must be shown. Guidance for comparability testing is available from the EMEA, FDA, ICH, and pharmacopeias (USP, EP, etc.) *(16,29,33,46)*. Comparability does not necessitate that two products have identical attributes. Comparability should demonstrate that the protein products are highly similar and, based on existing safety and structure activity knowledge, any differences do not predictably affect safety and efficacy of the product. The production process for active biologic ingredient must be validated and the product- and process-related substances, impurities, and contaminants should be qualified and characterized. Appropriate reference standards must be established, preferably with pharmacopoeia designations. This may be difficult to establish because there are multiple reference products. Physicochemical characterization should include the ability to characterize clinically relevant aspects of protein products, such as process/product impurities and aggregation. Comparability testing would include structural characterization, physicochemical properties, biological activity, purity, and impurities. There are a myriad of traditional and modern analytical/ bioanalytical techniques that can be applied to the analysis of biologics that enables comprehensive comparative characterization of protein products (Table 3). Peptide mapping with high-resolution mass spectroscopy identifies the covalent structure and peptide backbone in most proteins. Other techniques such as CD, HPLC, SEC, IEC, SDS-PAGE, IEF, FTIR, fluorescence, NMR, SEC, light scattering, ultracentrifugation, ELISA, surface plasmon resonance are sensitive methods for comparing higher-order fingerprint structure.

The United States Pharmacopoeia *(46)* has general chapters and monographs for many biological products. The test procedures contained in the biological monographs discern the quality attributes of a biologic and can be used as a demonstration of comparability between FOPPs. The analytical methods used to compare biologics are sophisticated, but may not discern all subtle differences associated with very large molecules. Well-characterized proteins, such as insulin and growth hormone, contain 51 and 191 amino acids, respectively (Table 4), and can be characterized with a high degree of certainty. Modern analytical techniques, such as MALDI-TOF, hold promise of accurately comparing biological compounds of 100,000 Da. However, as the biological compound gets larger, the analytical techniques lose their ability to discern minor changes between the products. It is not known if these differences result in a clinical effect or immunogenic response. Presently, there is no penultimate analytical technique that can prove pharmaceutical equivalence of complex biologics; thus, complimentary analytical techniques are required.

Optimally, a FOPP will share the same critical quality attributes of the reference protein. Each FOPP should be addressed on a case-by-case basis of the products' history, safety, structure/activity relationship, and complexity. A compliment of analytical and bioanalytical techniques will be required to demonstrate comparability. Appropriate reference products, standards, specifications, monographs, and quality assurances should be developed for the actives, final product, excipients, and adjuvants. The USP is active in providing information and

Table 3
Follow-On Protein Products: Techniques for Physico-Bio-Chemical Comparability

A. Gross structural Comparison
  • Mass
  • Hydrodynamic radius
  • Sedimentation rate
  • Diffusion rate
B. Optical Properties
  • Absorbance
  • Light scattering
  • Fluorescence
  • Optical rotation
C. Electrical Properties
  • Charge
  • Isoelectric point
  • Electrophoretic mobility
  • Ion exchange
D. Magnetic Resonance
  • 1D, 2D, 3D, 4D
E. 3D Crystal Structure
  • X-ray diffraction
  • Electron microscopy
    ○ Electron protein tomography
    ○ Cryoelectron microscopy
F. Thermal Properties
  • DSC
  • TGA
  • Titration calorimetry
G. Surface Properties
  • Immunological
  • Chromatographic
  • H/D exchange
H. Native/Denatured States
I. Fragmentation
  • Gas phase (MS)
  • Chemical
  • Enzymatic
J. Temporal Changes
  • Fluorescence
  • NMR
  • MS
K. Multiple Principles
  • LC/MS/MS

(*Continued*)

Table 3 (***Continued***)

- Peptide mapping/LC/MS/MS
- MALDI-TOF
- LC/light scattering

L. Other
  - Cell culture bioassays
  - Ligand binding
  - In vitro analysis
  - Immunoassays
  - Animal-based assays

Table 4
Biotechnology-Derived Product Complexity

Product	Approximate molecular weight	# Amino acids
Human insulin	5,808	51
Erythropoietin	30,400	165
hGH	22,125	191
TPA	70,000	527
Factor VIII	280,000	2351

guidance for biotechnology-derived products, recently publishing the monograph for growth hormone (Table 5) *(47)*. Additional guidance from the FDA and USP will be required for each biologic product that claims to be FOPP.

5. Biological Manufacture

The manufacturing processes for biologics typically derive from living systems which can be varied and complex. Biological productss can be isolated from naturally derived sources or manufactured using bacteria, yeast, fungi, insect, plant, chimerics, mammalian cell cultures, transgenics, and other systems. Bioprocessing of proteins involves the integration and scaleup of upstream and downstream processing, process monitoring, optimization, and control. The final protein product can be influenced at any single step in the production process and the mantra of "product equals process" arises from this fact. The manufacture of proteins requires strict control of the starting raw materials, genetic engineering, expression systems, optimization of growth conditions, batch culture design, purification, protein analyses, formulation, analytical testing, stability, aseptic filling, packaging, and the validation of these processes. Product quality assurance is provided by careful monitoring of critical process variables and the understanding of how these process variables affect the final product (quality by

Table 5
USP 29 Monograph Tests for Somatropin USP

API	Finished product
• Packaging and storage	Packaging/storage
• Labeling	Labeling
• USP reference standards <11>	USP reference standards <11>
• Identification	Identification
◦ A (HPLC <621>)	◦ HPLC <621>
◦ B (peptide mapping <1047>)	Bioidentity
• Bacterial endotoxins <85>	Bacterial endotoxins <85>
• Microbial limits	Sterility <71>
• Water *Method Ic* <921>	Chromatographic purity <621>
• Chromatographic purity <621>	Limit high molecular proteins
• Limit of high molecular proteins	Assay—HPLC <621>
• Content of protein <851>	
• Bioidentity—API or FP	
◦ Rat weight gain test	
• Assay—HPLC <621>	

design). Minor changes of the critical process variables can result in changes in the protein's secondary, tertiary, and quaternary structures and affect other higher-order posttranslational modifications such as glycosylation, acetylation, phosphorylation, and other amino acid modifications that can profoundly affect activity and toxicology.

Owing to patent and intellectual property protection, FOPP manufacturers must develop noninfringing manufacturing processes that result in protein products with similar, if not identical, structure and equivalent therapeutic behavior to the original reference protein product. In addition, the FOPP manufacturer must demonstrate similar profiles of quality, safety, and efficacy between a new or modified process producing the protein product to a reference protein product. The quality of the FOPP manufacture should be demonstrated by process validation which includes comparisons of several production batches using the same manufacturing process. In addition, the FOPP manufacturer must minimize the immumogenic potential of the protein. It may be possible to identify surrogate markers by which immunogenicity of the protein product can be determined during development or at postapproval marketing. Process-related impurities, such as host proteins, medium components, downstream reagents, and product-related impurities (truncated and modified forms, aggregates, etc.) should be minimized to predetermined specifications. If this specification cannot be met, the impurities should be qualified through appropriate toxicology assessments.

As with small molecules, the biological manufacturing process is intimately associated with the quality of the biological product. In general, biological manufacture can be divided into two main processes—upstream and downstream processing. Upstream activities produce the protein of interest, usually by cell culture or fermentation. Upstream considerations include integrity and quality of the process, cell banks, expression systems, cultivation, media, process/product purity, impurities, and contaminants. Downstream processing refers to the separation and purification of the bulk bioproduct into a form suitable for its end use. This usually includes the recovery of the product from the media, purification, sterilization, and final formulation. Typically, downstream processing techniques include filtration, centrifugation, precipitation, numerous chromatographic separations, and sterilization by aseptic processing, terminal filtration, or lyophilization. The impurity profile and related substances are process and formulation dependent. The complexity associated with the manufacturing of biotechnology-derived products by the many available biotechnology processes may result in different impurity profiles, which if not controlled could lead to therapeutic inequivalence or immunogenic responses between similar biologics. Proving pharmaceutical equivalence (sameness) between multisource biologics may be possible; however, the analytical/bioanalytical procedures may be limited in their ability to detect heterogeneity, glycosylation, and conformational changes associated with complex biologics. When the biologic comparability is limited by the analytical methods, further in vitro/in vivo studies are warranted.

The technology and expertise are available in the pharmaceutical industry to produce "generic" biologics that have the same quality attributes and therapeutic effect as a reference biologic. In many cases, the compound will be identical, but as the complexity of the protein increases, the products may exhibit a high degree of similarity, but not identical, to the reference protein. In addition, because of the living process that produces biologics, minor changes in process parameters can have significant effects on the quality of the biologic and impurity profiles. The impurities, both process and product, should be identified and kept at a minimum level with the appropriate limits and specifications. The uniqueness of the impurity profiles between similar biologics may lead to different immunogenic responses in the clinical setting.

Producing a consistent product with predefined quality by design specifications requires validation and documentation of the integrity of the cell banks, expression systems, cultivation, media, process/product purity, impurities, contaminants, stability, and quality attributes. Other techniques, such as transgenics in plants or animals, can be used as biofactories for the production of commercial products. However, regulatory authority tend to consider that the "product equals process", and although the different manufacturing methods may produce the same molecule, they may be considered new biologics.

Formulation and container closure components have a greater impact on the safety of a biologic than with a small molecule formulation. Changes in formulation and container closure have been implicated as the cause of adverse events and death with erythropoietin products *(48)*. Quality, process validation, and process analytical technologies should be integrated into the development of a FOPP. A thorough knowledge of manufacturing process and impact of any production changes on the final product must be understood by the FOPP manufacturer and relationships of the changes to product quality immunogenic potential should be defined. The use of surrogate markers should be developed and examined during development, scaleup, and changes to product formulation *(49)*.

6. Therapeutic Bioequivalence

In addition to demonstrating pharmaceutical equivalence, a sponsor of a generic product has to demonstrate bioequivalence. Bioequivalence describes pharmaceutical equivalent or pharmaceutical alternative products that display comparable bioavailability when studied under similar experimental conditions. Products are considered bioequivalent when the rate (C_{max}) and extent (AUC) of absorption of the test drugs do not show a significant difference from the rate and extent of absorption of the reference drug when administered at the same molar dose of the therapeutic ingredient under similar experimental conditions in either a single dose or multiple doses to healthy volunteers *(38)*. In other situations, bioequivalence may sometimes be demonstrated through comparative clinical trials or pharmacodynamic studies. Traditional bioequivalence studies assume that two chemically identical products that demonstrate similar pharmacokinetic profiles will also have identical clinical effects. Presently, it is argued that pharmacokinetic bioequivalence is not sufficient for ensuring the safety and efficacy of FOPP *(50)*. It has been recommended that until analytical methods are able to better characterize the complexity of biopharmaceuticals and establish predictive comparability, full clinical trial data are proposed to demonstrate the quality, safety, and effectiveness of FOPP. In addition, biological products are more complex and heterogeneous than conventional small molecules, and differences in the molecular characteristics between biologic products in the same therapeutic class have the potential to influence or alter their biological activity and immunogenic potential. Complex proteins can also often have multiple sites of action, and actual or surrogate markers of efficacy are often not clearly established. In addition, the pharmacokinetic/pharmacodynamic relationship of the biologic may not be fully understood. An accurate prediction of the clinical and immunologic properties of a biologic may not be possible with current analytical techniques or single-dose bioequivalence studies. Therapeutic comparability of a FOPP should be evaluated on a case-by-case basis that examines the history, safety, analytical complexity, structure activity, and quality aspects of the biological product *(49)*.

The EMEA has issued guidelines (adopted or released for consultation) for the development of insulin, growth hormone, GCSF, and erythropoietin *(21,23,24,27)*. The guidelines suggest that before going into clinical development, comparative nonclinical studies should be performed to detect differences in the response to the similar biological medicinal product and the reference product. Pharmacodynamic studies (e.g., insulin in vitro bioassays for affinity, insulin- and IGF-1-receptor binding assays) with the similar biological product and the reference product should be performed. The assays used to demonstrate equivalence should have the appropriate sensitivity. Comparative studies of pharmacodynamic effects are not anticipated to be sensitive enough to detect any nonequivalence that is not identified by in vitro assays and are normally not required as part of the comparability exercise. Data from at least one repeat dose toxicity study in a relevant species (e.g., rat) should be provided and the study duration should be at least 4 wk with special emphasis on the determination of immune responses. Data on local tolerance in at least one species should be provided, and local tolerance at the injection site reactions (usually subcutaneous dermal reactions) testing should be performed as part of the repeat dose toxicity study.

The pharmacokinetic properties of the similar biological product and the reference product should be determined in a single-dose crossover study using the appropriate administration (e.g., subcutaneous). Comprehensive comparative pharmacokinetic data should be provided on the time–concentration profile (including C_{max}, T_{max}, AUC, and half-life). Studies should be performed preferably in the patient population (e.g., type1 diabetes, etc.) and factors contributing to pharmacokinetic variability (e.g., insulin dose and site of injection/thickness of subcutaneous fat) should be taken into account.

For insulin, the EMEA guidance *(27)* states the clinical activity of an insulin preparation is determined by its time–effect profile of hypoglycemic response, which incorporates components of pharmacodynamics and pharmacokinetics. A double-blind, crossover hyperinsulinemic euglycemic clamp model is considered suitable for this characterization, and data on comparability regarding glucose infusion rate and serum-free insulin concentrations should be made available. The study population and study duration should be described and justified. If equivalence is concluded from the pharmacokinetic and pharmacodynamic, there is no anticipated need for efficacy studies on intermediary or clinical variables.

The EMEA guidelines suggest the issue of immunogenicity can only be settled through clinical trials of sufficient duration (i.e., for insulin, at least 12 mo using subcutaneous administration). The comparative phase of this study should be at least 6 mo in duration and the primary outcome measure should be the frequency of antibodies to the test and reference product. If any concern is raised through nonclinical and short-term clinical studies, additional evaluation of local tolerability may be needed prior to approval. Additionally, the sponsor should present pharmacovigilance—risk management program for the postapproval surveillance of the biosimilar product which takes into account risks identified during product development and potential risks, especially as regards immunogenicity.

7. Immunogenicity

A major issue surrounding the approval and substitution of FOPP is immunogenicity. An antibody response to a therapeutic protein can compromise the biologic activity and lead to an altered safety or efficacy profile. Many of the commercially approved therapeutic proteins have an identical or similar amino acid sequence of endogenous proteins and may not be immunogenic if they are recognized as such through immune tolerance mechanisms. However, many of the biologics currently approved for therapeutic use are known to have immunogenic responses (50,51). Differences in protein structure such as sequence variation and glycosylation patterns, contaminants, impurities, aggregation, formulation, adjuvants, processing differences, container closure, administration route, dose, length of treatment, and patient characteristics may result in the generation of an antibody response. In most cases of antibody formation to a therapeutic protein, the immunogenic response from these products does not cause a negative clinical effect (50–53). If an immune response is elicited, the antibodies may be neutralizing, which can negate the effect of the therapy, or crossreact with the endogenous protein, depleting the effect of the protein (50). This has been observed in chronic kidney disease patients with anemia, where crossreactive neutralizing antibodies were found to inhibit the activity of both the administered recombinant erythropoietin and the patient's own endogenous erythropoietin, resulting in pure red-cell aplasia (PRCA) (54,55). Though the exact immunological mechanism responsible for the PRCA associated with erythropoietin is not clear, it appears to be related to changes in product formulation, possibly because of the replacement of human serum albumin with polysorbate 80 and glycine, releasing leachates from the stoppers that resulted in protein aggregation (50).

The EMEA guidelines on comparability of biosimilars assert that preclinical data may be insufficient to demonstrate immunogenic safety (i.e., nonimmunogenicity) of some biosimilar products (21,23,24,27). For these products, immunological safety of a biosimilar product can only be demonstrated in cohorts of patients enrolled in clinical trials and from postmarketing surveillance, performed at predetermined intervals (e.g., ≥ 1 yr) after the approval of the product. Immunological safety studies also rely on the availability of highly sensitive, validated assays for measuring antibodies (18). There is also a need to standardize the immunogenicity assays and reference standards to allow comparisons of results obtained from different testing laboratories. Receptor binding assays and relevant animal models should be considered to identify differences in toxicology and immunology to that of the reference product. Depending on the safety history and complexity of the protein, population exposure, both pre- and postmarketing pharmacovigilance may be required. The EMEA guidelines suggest demonstrating comparability of a biosimilar to a reference protein product by nonclinical studies, receptor binding assays, rodent models, animal toxicology studies (4-wk repeat dose (erythropoietin 12 wks) and local tolerance), clinical pharmacokinetic, pharmacodynamic, safety, efficacy (except for insulin), immunogenicity (6–12 mo), local reactions (insulin) and pharmacovigilance evaluations *(21,23,24,27)*. It is thought that the future FDA guidelines will reflect the EMEA biosimilar guidances.

The induction of antibody formation in animals is not predictive of a potential for antibody formation in patients. Patients may develop serum antibodies against humanized proteins, and frequently the therapeutic response persists in their presence. The occurrence of severe anaphylactic responses to recombinant proteins is rare *(56)*.

Because of the complexity of biological substances, each compound seeking comparability to a reference protein should be examined on a case-by-case basis. The product's history, safety, and indications within the patient population should be reviewed. The microheterogeneity and variations to the production process and its effect on biological activity should be determined. The use of accepted preclinical models and surrogates for activity and immunological behavior should be examined. If the comparability criteria to a reference protein cannot be fully established, the FOPP will be viewed as a new (or slightly similar) molecular entity and complete investigations of preclinical and clinical assessments with the protein may be required for the biological license application.

8. Conclusion

Whether branded manufacturing lot to lot consistency or biologics produced from different manufacturers, the intrinsic heterogeneity of biologics has the potential to affect therapeutic efficacy and patient safety. Unlike conventional small molecule pharmaceutical drugs, complex therapeutic proteins cannot be completely characterized by physiochemical methods, bioassays, or modern analytical techniques. Additionally, the relationship between physiochemical similarities and differences to biological properties such as immunogenicity is often unclear. Whereas minor difference in the microheterogeneity of biologics is normal and usually not clinically significant, variability has been shown to produce differences in pharmacological properties that affect the safety and efficacy of the protein product. The development of FOPP is more complicated

Biologic Comparability

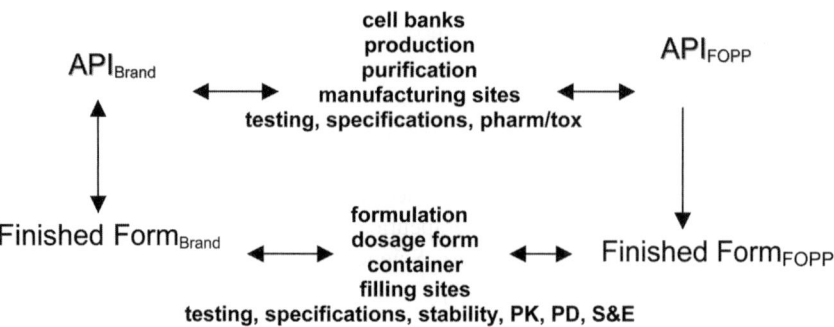

Fig. 1. Biologic comparability for FOPP.

than a conventional small molecule generic product (Fig. 1). The demonstration of biologic comparability is likely to require more than pharmaceutical equivalence and pharmacokinetic bioequivalence studies. Because of the issues of immunogenicity that can be observed with protein products, the safety of the biological compound is a major concern, especially in immunocompromised patients. Appropriate models and standardized testing for immunogenicity assessments for FOPP, as well as issues with substitution and therapeutic switching of protein products, need to be addressed. This is further complicated by the differences associated with the manufacture of protein products, the lack of established reference standards, testing monographs, and quality assurances for the active protein ingredients and finished protein products. Postapproval assurances for pharmacovigilance programs should be established.

Owing to the complexity of biological substances, each compound seeking comparability to a reference protein should be examined on a case-by-case basis, for the product's history, safety, and indications within the patient population. The microheterogeneity and variations to the production process and its effect on biological activity should be determined by the manufacturer utilizing accepted preclinical models and surrogates for activity and immunological behavior. If the comparability criteria to a reference protein cannot be fully established, the FOPP will be viewed as a new (or slightly similar) molecular entity and complete investigations of preclinical and clinical assessments with the protein may be required for the BLA.

However, there is a need for affordable, high-quality biological products. The technical competency and quality required to produce FOPP from multisource biotechnology manufacturers is readily available in the pharmaceutical community. Regulatory guidance and legislative change is required to make FOPP a reality.

References

1. Herrera, S. (2004) Biogenerics Standoff. *Nat. Biotechnol.* **22,** 1343–1346.
2. Bell, R. G. (2005) *Therapeutic Equivalence of Follow On Biologics—Current Issues and Challenges.* AAPS, Nashville, TN, November 7.
3. Bell, R. G. (2005) Proving Therapeutic Equivalence of Follow On Biologics—How, When and Where? AAPS National Biotechnology Conference, San Francisco, June 8.
4. Bell, R. G. (2005) Immunologic Comparability of Biologics. AAPS National Biotechnology Conference, San Francisco, June 6.
5. Bell, R. G. (2004) Establishing Therapeutically Equivalent Biotechnology Derived Products. IBC's 3rd Annual Business, Science and Partnering Strategies for Global Follow-On Biologics, Reston, VA, December 6.
6. Bell, R. G. (2004) Analytical Validation Requirements for Biotechnology Products. AAPS National Biotechnology Conference, Boston, MA, May 17.
7. Bell, R. G. (2004) Current Issues and Challenges for Establishing Therapeutically Equivalent Biotechnology Derived Products. AAPS National Biotechnology Conference, Boston, MA, May 17.
8. Bell, R. G. (2004) Challenges for Establishing Therapeutically Equivalent Biological Drug Products. Virginia Commonwealth University, College of Pharmacy, Richmond, VA, April 14.
9. Bell, R. G. (2003) Biogenerics—Understanding the Regulatory and Scientific Hurdles Impacting Development. 3rd Annual Generic Drug Conference, Washington, DC, November 19.
10. Bell, R. G. (2003) Current Issues and Challenges for Establishing Therapeutically Equivalent Biologics. AAPS, Salt Lake City, UT, October 27.
11. Bell, R. G. (2002) The Reality of Generic Biologics. 2nd Annual Generic Drug Conference, Washington, DC, November 22.
12. Bell, R. G. (2002) The Age of Biogenerics: Therapeutically Equivalent Biotechnology-Derived Drug Products. Chapter 6 in "Generic Drugs and Brand Lifecycle Strategies", pp. 46–58, Brigitta Tadmor (eds.), AdvanceTech Monitor Publishing, Woburn, MA, May.
13. Bell, R. G. (2001) Analysis and Pharmaceutical Quality of Biotechnology Derived Products: From Development to Production. AAPS, Denver, Colorado, October 22.
14. Bell, R. G. (1998) Therapeutic Biotechnology-Derived Products: Potential Pathways for Generic Development. AAPS, San Francisco, CA, November 18.
15. Bell, R. G. (1998) Therapeutic Biotechnology-Derived Products: Potential Pathways for Generic Development. ACS Biotechnology Conference, Boston, MA, August 24.
16. CHMP/437/04 Guideline on Similar Biological Medicinal Products (CHMP adopted September).
17. CPMP/3097/02 Note for Guidance on Comparability of Medicinal Products containing Biotechnology-derived Proteins as Drug Substance—Non Clinical and Clinical Issues (CPMP adopted December 2003).
18. CPMP/BWP/3207/00 Rev.1 Guideline on Comparability of Medicinal Products containing Biotechnology-derived Proteins as Active Substance—Quality Issues (CPMP adopted December 2003).

19. CHMP/146710/2004 Similar Biological Medicinal Products containing Recombinant Human Insulin—Annex to the Guideline for the Development of Similar Biological Medicinal Products containing Biotechnology Derived Proteins as Active Substance (non) Clinical Issues (Released for consultation by CHMP November 2004).

20. CHMP/146701/2004 Similar Biological Medicinal Products containing Recombinant Granulocyte-Colony Stimulation Factor—Annex to the Guideline for the Development of Similar Biological Medicinal Products containing Biotechnology Derived Proteins as Active Substance (non) Clinical Issues (Released for consultation by CHMP November 2004).

21. CHMP/146489/2004 Similar Biological Medicinal Products containing Recombinant Human Growth Hormone—Annex to the Guideline for the Development of Similar Biological Medicinal Products containing Biotechnology Derived Proteins as Active Substance (non) Clinical Issues (Released for consultation by CHMP November 2004).

22. CHMP/146664/2004 Similar Biological Medicinal Products containing Recombinant Human Erythropoietin—Annex to the Guideline for the Development of Similar Biological Medicinal Products containing Biotechnology Derived Proteins as Active Substance (non) Clinical Issues (Released for consultation by CHMP November 2004).

23. EMEA/CHMP/94526/05 Annex Guideline on Similar Biological Medicinal Products containing Biotechnology-Derived Proteins as Active Substance: Non-Clinical and Clinical Issues—Guidance on Biosimilar Medicinal Products containing Recombinant Erythropoietins (CHMP released for consultation June 2005).

24. EMEA/CHMP/31329/05 Annex Guideline on Similar Biological Medicinal Products containing Biotechnology-Derived Proteins as Active Substance: Non-Clinical and Clinical Issues—Guidance in Biosimilar Medicinal Products containing Recombinant Granulocyte-Colony Stimulating Factor (CHMP released for consultation June 2005).

25. EMEA/CHMP/42832/05 Guideline on Similar Biological Medicinal Products containing Biotechnology-Derived Proteins as Active Substance: Non-Clinical and Clinical Issues (CHMP released for consultation May 2005).

26. EMEA/CHMP/94528/05 Annex Guideline on Similar Biological Medicinal Products containing Biotechnology-Derived Proteins as Active Substance: Non-Clinical and Clinical Issues—Guideline on Similar Medicinal Products containing Somatropin (CHMP released for consultation May 2005).

27. EMEA/CHMP/32775/05 Annex Guideline on Similar Biological Medicinal Products containing Biotechnology-Derived Proteins as Active Substance: Non-Clinical and Clinical Issues—Guideline on Similar Medicinal Products containing Recombinant Human Insulin (CHMP released for consultation May 2005).

28. EMEA/CHMP/49348/05 Guideline on Similar Biological Medicinal Products containing Biotechnology-Derived Proteins as Active Substance: Quality Issues (CHMP released for consultation March 2005).

29. FDA Guidance Concerning Demonstration of Comparability of Human Biological Products, Including Therapeutic Biotechnology-derived Products. Center for Biologics Evaluation and Research (CBER) and Center for Drug Evaluation and Research (CDER). April, 1996.
30. FDA (2003) *Guidance for Comparability Protocols—Protein Drug Products and Biological Products—Chemistry, Manufacturing, and Controls Information.* September 3.
31. Code of Federal Regulations. Title 21, **7,** 600.3 (h), April 1, 2005.
32. Food, Drug and Cosmetic Act, Section 351 Public Health Service Act. Title 42, USC 262 (a).
33. ICH Guidance for Industry Q6B Specifications: Test Procedures and Acceptance Criteria for Biotechnological/Biological Products. August 1999.
34. Federal Register Notice (1996) in which the concept is defined, a well-characterized biologic is "a chemical entity whose identity, purity, impurities, potency, and quantity can be determined and controlled" **61,** 2733–2739.
35. Food, Drug and Cosmetic Act, Section 21 U.S.C. 321 (g)(1).
36. Webber, K. (2005) *Therapeutic Equivalence of Follow On Biologics—AN FDA Perspective.* AAPS National Biotechnology Conference, San Francisco, June 8.
37. Federal Register: August 16, 2004, **69**(157)**,** 50,386–350,388.
38. Approved Drug Products with Therapeutic Equivalence Evaluations, 25th Edition. U.S. Department of Health and Human Services, FDA/CDER, Office of Pharmaceutical Sciences, Office of Generic Drugs. 2005.
39. Arman, H. Nadershahi, A, and Reisman, J. (2003) BioProcess International, October, 26–31.
40. Wasson, A. (2005) Taking biologics for granted? Takings, trade secrets and off-patent biological patents. *Duke Law Technol. Rev.,* 4.
41. Genentech, Inc. (2004) Standard of "Similarity" or "Sameness" of Biotechnology-Derived Products Citizen's Petition 2004P-0171, April 8.
42. Pfizer Inc. (2004) Deny Approval of NDA 21-426 for Ominitrop 5.8 mg Somatropin (rDNA origin) for Injection, Lyophilized Powder and Diluent with Preservative Citizen's Petition 2004P-0231, May 14.
43. Code of Federal Regulations, Title 21, Section 320.1(c).
44. Code of Federal Regulations, Title 21, Section 316.3(b)(13)(ii).
45. Code of Federal Regulations, Title 21, Section 316.3(b)(13)(ii)(A).
46. United States Pharmacopeia 29. Sections <1041, 1045, 1046, 1047, 1049, 1050>. United States Pharmacopeial Convention, Rockville, MD, 2006.
47. United States Pharmacopeia 29, Somatropin Monograph. United States Pharmacopeial Convention, Rockville, MD, 2006.
48. van Regenmortel, M., Boven, K, and Bader, F. (2005) Immunogenicity of biopharmaceuticals: an example from erythropoietin. *BioPharm Int.* **18**(8)**,** 36–52.
49. Schellekens, H. (2005) Follow-on biologics: challenges of the "next generation". *Nephrol. Dial. Transplant* **20**(Suppl 4)**,** iv31–iv36.
50. BIO Follow On Citizen's Petition 2003 03P-176, April 23, 2003.

51. Porter, S. (2001) Human immune response to recombinant human proteins. *J. Pharm. Sci.* **90,** 1–11.
52. Schernthaner, G. (1993) Immunogenicity and allergic potential of animal and human insulins. *Diabetes Care* **16,** 155–165.
53. Laricchia-Robbio, L., Moscato, S., Genua, A., et al. (1997) Naturally occurring and therapy-induced antibodies to human colony-stimulating factor (G-CSF) in human serum. *J. Cell Physiol.* **173,** 219–226.
54. Casadevall, N., Dupuy, E., Molho-Sabatier, P., et al. (1996) Autoantibodies against erythropoietin in a patient with pure red-cell aplasia. *NEJM* **334,** 630–633.
55. Casadevall, N., Nataf, J., Viron, B., et al. (2002) Pure red-cell aplasia and antierythropoietin antibodies in patients treated with recombinant erythropoietin. *NEJM* **346,s** 469–475.
56. CH Guidance for Industry S6: Preclinical Safety Evaluation of Biotechnology-Derived Pharmaceuticals. July 1997.s

Index

Printed in the United States of America